ENZYME TECHNOLOGY

ENZYME TECHNOLOGY

ANUSHA BHASKAR
Principal
Dhanalakshmi Srinivasan College of
Arts and Science for Women
Perambalur

V.G. VIDHYA
Lecturer
Department of Biotechnology
Faculty of Science and Humanities
SRM University, Kattankulathur
Chennai

MJP PUBLISHERS

MJP PUBLISHERS
47, Nallathambi Street
Triplicane
Chennai 600 005

To

All the students of Biological Sciences

Preface

Without enzymes there can be no life. Enzymes are proteins employed by Mother Nature to catalyse the chemical reactions necessary to sustain life in plants and animals. Although enzymes are formed only in living cells, many can be separated from the cells and can continue to function *in vitro*. This unique ability of enzymes to perform their specific chemical transformations in isolation has led to an ever-increasing use of enzymes in industrial processes, this being collectively termed enzyme technology.

More than 3000 enzymes catalysing a wide array of reactions are known to exist. The disintegration of foodstuff into amino acids, sugars, and lipids is normally accomplished within three to six hours, depending on the amount and type of food. In the absence of enzymes, hydrolysis of the food stuff would take more than 30 years. With the development in the science of biochemistry, a complete understanding of the wide range of enzymes present in living cells and their mode of action has come to light.

A textbook on *Enzyme Technology* is a need felt by teachers and students of life sciences in India. This book is a result of the efforts taken to fulfil the need of the day. The book provides a comprehensive history of enzymes with the contribution of the pioneering scientists in this field. The book thoroughly discusses the mechanisms and kinetics, production, recovery and characterization of enzymes. Alongside the application of immobilized enzymes, bioreactor design and reaction engineering are also dealt with. The book also includes the latest field of bioinformatics and its application in enzyme technology, biosensors, ribozymes and artificial enzymes. We have tried to focus more on the current trends in enzyme technology.

This book would not have been possible without the help of Mr. J.C Pillai, Director, and Mr. C. Sajeesh Kumar, Managing Editor, MJP Publishers. We thank them for their encouragement, support and patience they have shown during the development of this book. We are also thankful to Parvath Radha, Ramalakshmi, Gnanasoundari, Yamuna Devi, Lissy John, Magesh—the editorial team members of MJP Publishers—for their efforts to produce this book.

Our thanks are due to the members of the management of Dhanalakshmi Srinivasan Educational Institutions, Perambalur, and SRM University, Chennai.

We also wish to acknowledge our family members for their support and patience while preparing the manuscript, and for putting up with us when we missed some important moments for the family.

Suggestions for improvement may be sent to anushaparthiban@gmail.com or vidhyavg@gmail.com

Anusha Bhaskar

V.G. Vidhya

Contents

1. **Historical Perspectives of Enzymes** **1**

2. **Enzyme Nomenclature** **9**

Introduction 9
General Principles 11
The First General Principle 11
The Second General Principle 12
The Third General Principle 14
Rules for Classification and Nomenclature 15
General Rules for Systematic Names and Guidelines for Common Names 15
Rules and Guidelines for Particular Classes of Enzymes 21
Classification 27
EC 1 Oxidoreductases 28
EC 2 Transferase 39
EC 3 Hydrolase 44
EC 4 Lyases 52
EC 5 Isomerases 55
EC 6 Ligases 57
Review Questions *59*

3. **Characteristics of Enzymes** **61**

General Characteristics of Enzymes 61
Chemical Nature of Enzymes 63
Properties of Enzymes 63
Enzyme Efficiency 63
Enzyme Specificity 65

Modes of Enzyme Action 69
Lock-and-Key Theory 69
Induced Fit Hypothesis 70
Factors Affecting Enzyme Activity 71
Effect of pH 71
Effect of Temperature 73
Effect of Enzyme Concentration 75
Effect of Substrate Concentration 80
Effect of Inhibitors 81
Review Questions *84*

4. Isoenzymes and Multienzyme Complexes 87

Isoenzymes 87
Nomenclature of Isozymes or Isoenzymes 89
Lactate Dehydrogenase 91
Hexokinase 95
Isozymes and Allozymes as Molecular Markers 97
Multienzyme Complex System 97
Pyruvate Dehydrogenase 97
Fatty Acid Synthase 100
Tryptophan Synthetase 101
Review Questions *103*

5. Coenzymes 105

Significance of Nucleotide Structure in Coenzymes 106
Mode of Reduction 107
Properties and Distribution 110
Function of Coenzymes 110
Role of Coenzymes in Clinical Enzymology 111
Coenzyme I (NAD^+) 112
Coenzyme II ($NADP^+$) 113
Coenzyme A 115
Acetyl-CoA 118
Acyl-CoA 119
Coenzyme M 120
Coenzyme B 121

Coenzyme B_{12} 121
Coenzyme F_{420} 123
Adenosine Triphosphate 123
Flavin Adenine Dinucleotide 125
Flavin Mononucleotide 126
Lipoate 126
Phosphopyridoxal 128
Coenzyme Q 129
Tetrahydrofolic Acid 130
Glutathione 131
Ascorbate 132
Thiamine Pyrophosphate 133
Pyrophosphothiamine 134
Uridine Phosphate 135
Phosphomutase Coenzymes 135
Glutathione Coenzymes 136
Cytochromes 137
Relationship between Coenzymes and Vitamins 139
Review Questions *142*

6. Active Site and its Determination 143

Determination of Amino Acids at the Active Site 145
Chemical Modification of
Amino Acids in the Active Site 146
Photo-oxidation 147
Modification by Substrate Analogue 147
Alkylation 151
Iodination, Nitration and Acylation 155
Enzyme Modification by Site-directed Mutagenesis 157
Review Questions *158*

7. Enzyme Catalysis 159

Mechanism of Enzyme Action 159
Catalytic Mechanisms 161
Acid–Base Catalysis 161
Mechanism of Keto-enol Tautomerization 162

	Covalent Catalysis	165
	Metal Ion Catalysis	167
	Metalloenzymes	170
	Electrostatic Catalysis	174
	Catalysis through Proximity and Orientation Effects	175
	Preferential Binding of the Transition State Complex	177
	Lysozyme	180
	Chymotrypsin	184
	Ribonuclease	186
	DNA Polymerase	186
	DNA Polymerase I	189
	DNA Polymerase II	190
	DNA Polymerase III	190
	Review Questions	*120*
8.	**Control of Enzyme Activity**	**191**
	Control of Activities of Single Enzymes	192
	Control of Activity by Changes in the Covalent Structures of Enzymes	192
	Control of Activity by Ligand-induced Conformational Changes in Enzymes	200
	Feedback Inhibition	203
	Review Questions	*206*
9.	**Kinetics of Multisubstrate–Enzyme Reactions**	**209**
	Assumption of Equilibrium	211
	Assumption of Steady State	211
	Derivation of Michaelis–Menten equation	213
	The Double-Reciprocal Plot	217
	The Haldane Relationship	218
	The Eadie–Hofstee and Hanes Plot	221
	Pre-steady-state Kinetics	223
	Steady-state Kinetics	223
	k_{cat}/K_m—Measure of Catalytic Efficiency	224
	Inhibition Kinetics	224
	Competitive Inhibition	224

Uncompetitive Inhibition 229
Noncompetitive Inhibition 232
Reactions Involving Two Substrates 235
Types of Bi-Bi Reactions 236
Enzymes Catalysing Two Reactions Simultaneously 240
Two Enzymes Acting on One Substrate 241
Review Questions *242*

10. Allosteric Modification of Enzymes 243

The Monod–Wyman–Changuex (MWC) Model 246
Cooperative Binding 247
The Koshland–Nemethy–Filmer (KNF) model 248
Allosteric Activation and Inhibition 249
Aspartate Transcarbamoylase 250
Pharmacological Applications of Allosteric Modulation 254
Allosteric Sites as Drug Targets 254
Review Questions *255*

11. Enzyme Isolation, Fractionation and Purification 257

Need for Enzyme Purification 257
General Methods in Enzyme Isolation 259
Estimation of Enzyme Activity 259
Unit Concentration of Enzyme 260
Units of Enzyme Activity 260
Specific Activity 261
Catalytic Activity 261
Katal 261
Quantities Derived from Catalytic Activity 262
Enzyme Assay 263
Spectrophotometric Methods 263
Spectrofluorimetric Method 263
Luminescence Method 264
Manometric Method 264
The Thunberg Method 265

Electrode Method 266
Polarimetric Method 267
Chromatographic Methods 268
Chemical Estimations 268
Protein Estimation 269
Kjeldahl Method 269
Colorimetric Methods 270
Folin–Ciocalteau Method 270
Spectrophotometric Measurements 270
Source of Enzymes 271
Extraction 271
Fractionation Methods 273
Fractional Precipitation by Change of pH 274
Fractional Denaturation by Heating 275
Fractional Precipitation with Organic Solvents 275
Fractional Precipitation by Salts 276
Fractional Adsorption 278
Chromatography 279
Electrophoresis 283
Concentration 285
Crystallization 286
Other Methods 288
Criteria of Purity of Enzymes 288
Review Questions *291*

12. Enzyme Immobilization 293

Historical Aspects of Enzyme Immobilization 293
Need of Immobilizing Enzyme 294
Immobilized Enzyme Characteristics 294
Enzymic Action 294
Stability 295
Kinetic Properties 295
Aspects of Immobilization Procedure 296
Enzyme Support 296
Support Activation and Enzyme Attachment 297

Methods of Immobilization 298
Carrier-binding Method 299
Carrier Disability Under Application Conditions 300
Covalent Binding Method 301
Ionic Binding Method 304
Physical Adsorption Method 305
Biospecific Binding Method 306
Cross-linking Method 307
Entrapment Method 307
Polyacrylamide Gel Method 308
Alginate Gel Method 309
κ-Carrageenan Gel Method 310
Synthetic Resin Prepolymer Method 311
Photo-crosslinkable Resin Prepolymer Method 312
Urethane Prepolymer Method 312
Beneficial Features of Immobilized Enzymes 313
Applications of Immobilized Enzymes 313
Production of Fructose Syrup 315
Use of Immobilized Invertase 315
Use of Lactose in Dairy Industry 315
Use of Immobilized Enzymes in Pharmaceutical Industry 316
Use of Glucose Oxidase in Enzyme Electrodes 317
Production of Amino Acids 317
Detergent Industry 318
Immobilization of Glucoamylase 318
Immobilized Enzymes in Organic Solvents 319
Immobilization of Enzymes for the Fabrication of Biosensors 319
Immobilized Multienzymes and Enzyme Cell Co-Immobilization 320
Immobilized Enzymes for the Regeneration of Cofactors 321
Medical/Clinical Applications 322
Applications in Food Processing 323

Immobilization of Raffinase 324
Review Questions *324*

13. Immobilization Bioreactors **325**

Key Issues in Bioreactor Design and Operation 325
Batch Bioreactions 326
Continuous Bioreactions 327
Types of Continuous Cultivation 329
Semi-continuous Bioreactions 329
Fed-batch Bioreactions 330
Application of Enzyme Immobilization in Industrial Bioreactors 331
Large-scale Immobilized Cell Bioreactors 331
Stirred Tank Reactors 332
Continuous Stirred Tank Reactor 333
Airlift Bioreactor 334
Packed Bed Reactor 337
Fluidized Bed Reactor 338
Recycle Reactor 340
Review Questions *341*

14. Ribozymes **343**

Classes of Ribozymes **344**
Hammerhead Ribozymes 345
Hairpin Ribozymes 348
Introns 350
Group I Introns 350
Group II Introns 353
RNase P 354
Hepatitis Delta Virus Ribozyme 356
The Varkud Satellite (VS) Ribozyme 358
Viroids 358
Artificial Ribozymes 360
Leadzyme 361
Ligase Ribozymes 361
Allosteric Ribozymes 362

The "10-23" RNA-Cleaving DNA Enzyme 362
Kinetics of Ribozymes 364
Mechanism of Ribozyme-mediated RNA Cleavage 365
Involvement of Metal Ions 365
Internal Transesterification 366
External Transesterification 367
Applications 368
Ribozymes as Therapeutic Agents 368
Methods to Deliver Ribozymes 369
Advantages of Ribozyme in Clinical Application 371
New Therapeutic Approach 372
Clinical Trials with Ribozymes 373
Ribozymes—Molecular Scissors for Investigating Gene Function 375
RiboaptDB—A Comprehensive Database of Ribozymes and Aptamers 376
Review Questions *383*

15. Abzymes **385**

Kinds of Abzymes 386
Antibodies Directed Against Transition State Analogues—the Substrate-based Approach 386
The Anti-idiotypic Approach—from Enzymes to Abzymes 388
Naturally Occurring Abzymes 389
Metallo- and Haemo-Abzymes 390
Strategy for Abzyme Production 391
Haldane Pauling Hypothesis of Transition State Stabilization as a Primary Effector of Catalysis 391
Designing Abzyme to Catalyse Aldol Reaction 391
Product of these Studies 392
Industrial and Medical Applications 392
Antigen-directed Abzyme Prodrug Therapy 393
Catalytic Antibodies—Targeting Cancer with Prodrugs 395
HIV Treatment 397
Review Questions *398*

16. Enzyme Engineering **399**
Principles of Enzyme Engineering 401
Strategies for Enzyme Engineering 402
Enzyme Engineering Tools 403
Application of Genetic Engineering Techniques to Enzyme Technology 405
Protein Engineering 406
Artificial Enzymes 407
Applications of Enzyme Engineering 408
To Degrade Explosives 408
To Treat Lysosomal Storage Disease 409
Review Questions *409*

17. Biosensor Technology **411**
Generations of Biosensors 412
Components of Biosensors 413
Use of Enzymes in Analysis 416
Types of Biosensors 418
Electrochemical Affinity Biosensors 418
Biocatalysis-based Biosensors 418
Bioaffinity-based Biosensors 420
Microorganism-based Biosensors 421
Potentiometric Biosensors 423
Amperometric Biosensors 427
Calorimetric Biosensors 430
Conductimetric Biosensors 434
Optical Biosensors 435
Immunosensors 436
Mass-detecting Immunosensors 437
Types of Immunosensors 439
Clinical Applications of Immunosensors 446
Principles of Detection 447
Optical Biosensors 447
Surface Plasmon Resonance Sensors 447
Electrochemical Biosensors 448

The Potentiometric Biosensor 448
Piezoelectric Sensors 449
Beneficial Features of a biosensor 449
Applications of biosensors 450
Review Questions *451*

18. Industrial Enzymes and Their Applications **453**

Naturally Occurring Enzymes 453
Primary Food Processing 454
Food Production 455
Enzymes in Fruit and Vegetable Juice Technology 456
Enzymes in Baking Technology 458
Bread Making 459
Cheese Making 460
Enzymes for Animal Feed 460
Enzymes in Brewing Industry 461
Beer-Making 462
Enzymes in Detergents 464
Enzymes in Washing Powders 464
Enzymes in Textile Industry 465
Enzymes in Leather Bating 467
Enzymes in Pulp and Paper Industry 468
Enzymes in Personal-Care Products 469
Enzymes in Fine Chemical Production 470
Chirally Pure Amino Acids and Aspartame 471
Rare Sugars 472
Semi-synthetic Penicillins 473
Lipase-based Reactions 473
Asymmetric Synthesis 473
Enzymatic Oligosaccharide Synthesis 474
High-Fructose Corn Syrup 474
Medical Applications of Enzymes 476
Cancer Treatment 477
Diabetes 477
Enzyme Deficiency Diseases 478

Heart Attacks 478
Drug Manufacture 478
Immobilized Enzymes 479
Enzymes in Biofuels 480
Enzymes in Recombinant DNA Technology 480
Protein Engineering 480

19. Bioinformatics in Enzymology 483

Catalytic Site Atlas 483
Sequence-Structure Plot 486
Histogram of Matches 486
Integrated Relational Enzyme Database IntEnz 487
REBASE 489
Enzyme Structures Database 491
BioCyc Open Chemical Database 494
Enzyme Nomenclature Database 497
Kinase Pathway Database 499
A Protein Phosphatase Information Resource 502
The Protein Kinase Resource 504
Predicted and Consensus Interaction Sites in Enzymes 507
A Database of Protein Catalytic Domains 508
Thermodynamics of Enzyme-catalyzed Reactions Database 509
ERGO 511
Data Types in ERGO 513
Enzyme Nomenclature Database 516
CAZy (Carbohydrate-Active Enzymes) 517
The Peptidase Database 519
SABIO-Reaction Kinetics Database 522
BRENDA—The Enzyme Database 523
Proteolytic Enzymes Database 524
BRENDA, AMENDA and FRENDA—The Enzyme Information System 526
ORENZA—a Web Resource for ORphan ENZyme Activities 528
MACiE—A Database of Enzyme Reaction Mechanisms 531

LIGAND—Chemical Database for Enzyme Reactions 533
The MetaCyc—Database of Metabolic Pathways and Enzymes 536
The BioCyc Database 537
A Bayesian Method for Identifying Missing Enzymes 539
EzCatDB—the Enzyme Catalytic-mechanism Database 541
Hierarchical Classification of Catalytic Mechanisms 542
The ENZYME Data Bank 544

Glossary *547*
Reference *601*
Index *607*

1

HISTORICAL PERSPECTIVES OF ENZYMES

Christian D.A. Hansen

The concept of enzyme machinery was first formulated in 1874, when the Danish chemist, **Christian Hansen,** produced the first specimen of rennet by extracting the dried stomach of calves, with saline solution. Apparently, this was the first enzyme preparation of relatively high purity used for industrial purposes. For thousands of years, man has been harnessing the power of enzymes.

Even though the action of enzymes was recognized and enzymes have been used throughout history, it was only quite recently that their importance has been realized. Enzymatic processes, particularly fermentation, were the focus of numerous studies in the 19th century and many valuable discoveries in this field were made.

Anselme Payen

Schwann, Theodor

A particularly important experiment was the isolation of the enzyme complex from malt by **Payen and Persoz** in 1833. This extract, like malt itself, converts gelatinized starch into sugars primarily maltose, and was termed "diastase". Development progressed during the following decades, particularly in the field of fermentation where the achievements by **Schwann**, **Leibig, Pasteur** and **Kuhne** were of the greatest importance.

Justus Von Leibig

Louis Pasteur

Eduard Buchner

A famous scientific dispute between Leibig and Pasteur concerning the fermentation process caused much heated debate. Leibig claimed that fermentation resulted from chemical process while Pasteur argued that fermentation did not occur unless viable organisms were present. The dispute was finally settled in 1897 when the **Buchner** brothers demonstrated that cell-free yeast extract could convert glucose into ethanol and carbon dioxide just like viable yeast cells.

Wilhelm Kuhne

In 1876, **William Kuhne** proposed that the term "enzyme" be used to denote phenomena previously known as "unorganized ferments," that is, ferments isolated from the viable organisms in which

James B. Sumner

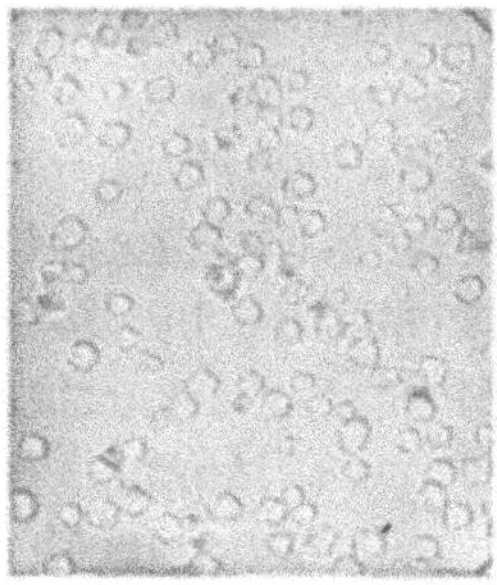

Urease crystals

they were formed. The word itself means "in yeast" and is derived from the Greek *en* meaning 'in' and *zyme* meaning "yeast" or "leaven". Research has rapidly accelerated throughout the 20th century from the year 1926 when **Sumner** isolated and crystallized urease, and proved that it was a form of protein. This enzyme is being used even in today's research.

Jokichi Takamine

Today, enzymes are in common usage in industry, food preparation, environmental management and medicine. Enzymes are still being discovered in nature and their individual properties are still largely a mystery. It is not an exaggeration to say that there is more still not known about enzymes than is known and that enzymes hold promise for many exciting scientific and medical breakthroughs. In the Far East, there prevailed an age-old tradition where the mould fungi called *koji* were used in the production of certain foodstuff and flavour additives. This formed the basis for the development of a fermentation process for the industrial production of fungal amylase, by the Japanese scientist, **Takamine.** The process included the culture of *Aspergillus oryzae* on moist rice or wheat bran. The product was called "**Takadiastase,**" and it is still used as a digestive aid.

At about the same time as Takamine was developing his novel fermentation techniques, another field was being opened up for the use of enzymes—the designing of textiles. Bacterial amylase derived from *Bacillus subtilis* was used for desizing for the first time by **Boidin** and **Effront** as early as 1917. Investigations carried out by the German chemist and industrialist **Magrate Otto Rohm** before World War I was of great importance for the further development of the industrial use of enzymes. Among other things he studied was the so-called bating process, a step in the preparation of hides and skins prior to tanning. Parallel to his studies of the problems involved in tanning, Rohm investigated other processes where enzymes would prove even more valuable.

Rohm actually developed the first method for washing protein-stained cloth in detergents containing enzymes and manufactured the first detergent containing enzymes. A breakthrough in detergents was made in 1959, when a Swiss chemist, **Dr. Jaag**, developed a new product called **Bio 40** containing a bacterial protease instead of trypsin. A very important field in which enzymes have proved to be of great value over the last 15–20 years is the starch industry. The real turning point was reached early in the 1960s when the enzyme, **glucoamylase** that could completely break down starch into glucose was launched for the first time.

Within a few years, almost all glucose production was reorganized and enzyme hydrolysis was used instead of acid hydrolysis because of the benefits such as greater yield, higher degree of purity and easier crystallization.

Years of research in biochemistry and biotechnology have boosted knowledge of enzymes for industries as well as research. Many new techniques have been established to modify enzymes or increase their yields. New techniques for purification of enzymes are constantly developing and so are new applications of enzymes in medicine, research and industries being discovered. Having shown that enzymes could function outside a living cell, the next step was to determine their biochemical nature.

John H. Northrop

Many early workers noted that enzymatic activity was associated with proteins, but several scientists (such as Nobel laureate Richard Willstatter) argued that proteins were merely carriers for the true enzymes and that proteins per se were incapable of catalysis. However, in 1926, **James B.Sumner** showed that the enzyme urease was a pure protein and crystallized it. Sumner did likewise for the enzyme catalase in 1937. The conclusion that pure proteins can be enzymes was definitively proved by **Northrop** and **Stanley**, who worked on the digestive enzymes pepsin, trypsin and chymotrypsin. These scientists were awarded the 1946 Nobel Prize in chemistry.

Wendell Stanley

David Chilton Phillips

This discovery that enzymes could be crystallized eventually allowed their structures to be solved by X-ray crystallography. This was first done for the lysozyme, an enzyme found in tears, saliva and egg whites, that digests the cell wall of some bacteria. The structure was solved by a group led by **David Chilton Phillips** and published in 1965. This high-resolution structure of lysozyme marked the beginning of the field of structural biology and the effort to understand how enzymes work at an atomic level of detail. Enzymes are the "true workers" in and out of our cells. As **Dr. Richard Gerber** states, "the enzymes catalyse specific reactions of chemicals either to create structure through molecular assemblies or to provide the electrochemical fire to run the cellular engines and ultimately keep the entire system working."

In *Today's Health* published in September 1960 by the American Medical Association, **Dr. Ratcliff** states that "many researchers believe that the aging process is the result of the slowing down and disorganization of enzyme activity. Might it eventually be possible to restore youthful patterns of activity by supplying those enzymes that are deficient?" According to **Dr. David Greenberg**, "enzymes are becoming increasingly important in medical research because an ever-growing number of diseases can be traced to some defect in the enzyme process." In 1941, Butler referred to the branch of enzyme kinetics as molecular kinetics.

John Alfred Valentine Butler

The first person to carry out kinetic studies with a pure enzyme, trypsin, was the British physical chemist **John Alfred Valentine Butler**. Laidler and Bunting in 1973, stated that "when an enzyme and a substrate are brought together, the steady state is usually established within a few milliseconds" and this had contributed to the understanding of enzyme action.

As a matter of experimental convenience, most investigations into enzyme kinetics have been concerned with the steady state to find out what is occurring while the steady state is being established—during what is called the transient phase of the reaction—and this requires special high-speed techniques. Two problems have to be overcome.

F.J.W. Roughton

The first is to bring the enzyme and substrate together rapidly (as otherwise the reaction may be over before they are properly mixed). The second is to make measurements within short periods of time. The first problem may sometimes be overcome by the use of flow methods, in which solutions are forced together very rapidly. Suitable techniques were developed, in particular by **Roughton** and his co-workers in 1963.

An important variation of their method was the stopped-flow method introduced by **Britton Chance** in 1940 and later developed further by Gibson and Milnes in 1964. Sometimes the individual steps are too fast for their rates to be measured by flow methods, and then relaxation methods have to be used.

Britton Chance

Leonor Michaelis

Maud Leonora Menten

During more recent years much further work has been done using high-speed techniques and many of the kinetic details of enzyme reactions have been worked out. The results have led to the realization that few such reactions appear to conform to the **Michaelis and Menten** equation.

2

ENZYME NOMENCLATURE

INTRODUCTION

Naming things is essential for people to understand, no matter what language or field of interest is involved. This is as true for enzymes, genes and chemicals as it is for birds, food, flowers, etc. In early days, the naming of enzymes was not systematic and trivial names were given that meant little or were ambiguous. Many different enzymes were given the same name and conversely, several names were given to the same enzyme, leading to much confusion.

In general, the suffix "ase" was added to the name of the substrate as in the case of urease (the first enzyme to be crystallized) or else the name gave some indication of the reaction catalysed, e.g. glucose oxidase. Although such arcane names may help to keep the subject area esoteric, the rapid increase in the number of enzymes being discovered necessitated the development of some method for naming them systematically. In the 1950s two groups of enzymologists set about addressing this problem.

The first system was instigated by **Otto Hofmann-Ostenhof** (1953), who classified enzymes using a system based on the number of molecules involved in the reaction. He proposed the following three general classes:

1. Hydrolases, transferases and oxidoreductases
2. Lyases and synthases
3. Racemases

Malcolm Dixon and Edwin Webb, who were compiling a list of all known enzymes for their influential book *Enzymes*, noted that despite their relatively large number of enzymes, the number of types of reactions involved was quite small. This included the synthetases, stereoisomers and enzymes that added groups to double bonds. This was the beginning of the current enzyme classification and nomenclature system, in which enzymes were divided into groups and subgroups according to the nature of the reaction catalysed.

In 1956, after being approached by Dixon and Hoffmann-Ostenhoff, the President of the International Union of Biochemistry, Marcel Florkin, established an International Commission on Enzymes to tackle the problems of classification and nomenclature. The commission's initial terms of reference were: "to consider the classification and nomenclature of enzymes and coenzymes, their units of activity and standard methods of assay together with the symbols used in the description of enzyme kinetics."

In 1958, they proposed an interim report, which was approved and finalized in 1961 and published in the second edition of *Enzymes* (Dixon and Webb, 1964) as well as in a separate book covering enzyme nomenclature and the units and symbols used in enzyme kinetics (*Enzyme Nomenclature*, 1965). In this, they extended the earlier system of Dixon and Webb (1958) by classifying enzymes into six categories.

The International Union of Biochemistry and Molecular Biology (IUBMB)

The mission of the IUBMB is to foster and support the growth and advancement of biochemistry and molecular biology as the foundation from which the biomolecular sciences (extending from cell biology to neurobiology, cancer biology, environmental biosciences, biotechnology and many others) derive their basic ideas and techniques in the service of mankind. This is done throughout the world with particular concern for areas where biochemistry is less well-developed, by promoting international cooperation and high standards in research, discussion, application and publication, and through international standardization of methods, nomenclature and symbols, in biochemistry and molecular biology. The IUBMB promotes the norms, values, standards and ethics of science and the free and unhampered movement of scientists of all nations interested in participation in activities related to biochemistry and molecular biology.

The International Union of Biochemistry and Molecular Biology has developed a nomenclature of enzymes, the EC numbers; each enzyme is described by a sequence of four numbers preceded by "EC." The first number broadly classifies the enzyme based on its mechanism and the second and third number represent the subset and sub-subset respectively.

GENERAL PRINCIPLES

The First General Principle

The first general principle of these recommendations is that names purporting to be names of enzymes, especially those ending in -ase, should be used only for single enzymes, i.e., single catalytic entities. They should not be applied to systems containing more than one enzyme. When it is desired to name such a system on the basis of the overall reaction catalysed by it, the word "system" should be included in the name. For example, the system catalysing the oxidation of succinate by molecular oxygen, consisting of succinate dehydrogenase, cytochrome oxidase and several intermediate carriers should not be named succinate oxidase but it may be called the succinate oxidase system, the similar 2-oxoglutarate dehydrogenase system, and the fatty acid synthase system.

In this context it is appropriate to express disapproval of a loose and misleading practice that is found in the biological literature. It consists in designation of a natural substance, responsible for a physiological or biophysical phenomenon that cannot be described in terms of a definite chemical reaction, by the name of the phenomenon in conjugation with the suffix -ase, which implies an individual enzyme. Some examples of such phenomenase nomenclature, which should be discouraged even if there are reasons to support that the particular agent may have enzymic properties, are: permease, translocase, reparase, joinase, replicase, codase, etc.

The Second General Principle

The second general principle is that enzymes are principally classified and named according to the reaction they catalyse. The chemical reaction catalysed is the specific property that distinguishes one enzyme from another, and it is logical to use it as the basis for the classification and naming of enzymes.

Several alternative bases for classification and naming had been considered, e.g. chemical nature of the enzymes (whether it is a flavoprotein, a haemoprotein, a pyridoxal phosphate protein, a copper protein and so on), or chemical nature of the substrate (nucleotides, carbohydrates, proteins, etc.). The former cannot serve as a general basis as only a minority of enzymes have such identifiable prosthetic groups. The chemical nature of the enzymes has, however, been used exceptionally in certain cases where classification based on specificity is difficult, for example, with the peptidases.

The second basis for classification is hardly practicable, owing to the great variety of substances acted upon and because it is not sufficiently informative unless the type of reaction is also given. It is the overall reaction, as expressed by the formal equation that should be taken as the basis. Thus, the mechanism of the reaction and the formation of intermediate complexes of the reactants with the enzyme is not taken into account, but only the observed chemical change produced by the complete enzyme reaction is considered. For example, in those cases in which the enzyme contains a prosthetic group that serves to catalyse transfer from a donor to an acceptor (e.g. flavin, biotin) the name of the prosthetic group is not normally included in the name of the enzyme. Nevertheless, where alternative names are possible, the mechanism may be taken into account in choosing between them.

A consequence of the adoption of the chemical reaction as the basis for naming enzymes is that a systematic name cannot be given to an enzyme until it is known what chemical reaction it catalyses. This applies, for example, to a few enzymes that have so far not been shown to catalyse any chemical reaction, but only isotopic exchanges; the isotopic exchange gives some idea of one step in the overall chemical reaction, but the reaction as a whole remains unknown.

A second consequence of this concept is that a certain name designates not a single enzyme protein but a group of proteins with the same catalytic property. Enzymes from different sources (various bacterial, plant or animal species) are classified as one entry. The same applies to isoenzymes, however, there are exceptions to this general rule. Some are justified because the mechanism of reaction or the substrate specificity is so different as to warrant different entries in the enzyme list. This applies, for example, to the two cholinesterases (EC 3.1.1.7 and EC 3.1.1.8), the two citrate hydrolyases (EC 4.2.1.3 and EC 4.2.1.4) and the two amine oxidases (EC 1.4.3.4 and EC 1.4.3.6). Others are mainly historical, e.g. acid and alkaline phosphatase (EC 3.1.3.1 and EC 3.1.3.2).

The Third General Principle

The third general principle adopted is that the enzymes are divided into groups on the basis of the type of the reaction catalysed, and this together with the name(s) of the substrate(s) provides a basis for naming individual enzymes. It is also the basis for classification and code numbers. Special problems arise in the classification and naming of enzymes catalysing complicated transformations that can be resolved by simplifying the reactions into several sequential or coupled intermediary reactions of different types, all catalysed by a single enzyme (not an enzyme system).

Some of the steps may be spontaneous non-catalytic reactions, while one or more intermediate steps depend on catalysis by the enzyme. Wherever the nature and sequence of intermediary reactions is known or can be presumed with confidence, classification and naming of the enzyme should be based on the first enzyme that catalysed the step which is essential to the subsequent transformations.

To classify an enzyme according to the type of reaction catalysed, it is occasionally necessary to choose between alternative ways of regarding a given reaction. One important extension of this principle is the question of the direction in which the reaction is written for the purposes of classification. To simplify the classification, the direction chosen should be the same for all enzymes in a given class, even if this direction has not been demonstrated for all. Thus the systematic names on which the classification and code numbers are based may be derived from a written reaction, even though only the reverse of this has been actually demonstrated experimentally.

RULES FOR CLASSIFICATION AND NOMENCLATURE

General Rules for Systematic Names and Guidelines for Common Names

RULE 1

Common names Generally accepted trivial names of substrates may be used in enzyme names. The prefix D- should be omitted for all D-sugars and L- for individual amino acids, unless ambiguity would be caused. In general, it is not necessary to indicate positions of substituents in common names, unless it is necessary to prevent two different enzymes having the same name. The prefix keto is no longer used for derivatives of sugars in which –CHOH has been replaced by –CO; they are named throughout as dehydro-sugars.

Systematic names To produce usable systematic names, accepted trivial names of substrates forming part of the enzyme names should be used. Where no accepted and convenient trivial names exist, the official IUPAC rules of nomenclature should be applied to the substrate name. The 1,2,3 system of locating substituents should be used instead of the α, β, γ system, although group names such

Types of Enzyme Classification

Enzyme classification	Function	Example
Oxidoreductases	Enzymes that do oxidation or reduction of substrate (BH_2) $BH_2 + A \square \ B' + AH_2$	Lactate dehydrogenase
Transferases	Transfer of a group-B from a donor D to an acceptor substrate A $D\text{-}B + A\text{-}H \square \ D\text{-}H + A\text{-}B$	NMP kinase
Hydrolases	Hydrolysis reactions removing group B from substrate A $A\text{-}B + H_2O \square \ A\text{-}H + B\text{-}OH$	Chymotrypsin
Lyases	Elimination reactions that split one molecule in two without an additional acceptor $A\text{-}B \square \ A' + B'$	Fumarase
Isomerases	Transfer of a group internally within a molecule $R\text{-}A\text{-}B \square \ A'\text{-}B'\text{-}R$	Triose phosphate isomerase
Ligases and synthetases	Enzymes in which their favoured direction is addition of joining (Lyase in reverse). $A\text{-}OH + BH \xrightarrow[ATP \ \to \ ADP + P_i]{} A\text{-}B$	Aminoacyl-tRNA synthetase

as β-aspartyl, γ-glutamyl and also β-alanine and γ-lactone are permissible; α , β should normally be used instead for indicating configuration, as in a D-glucose. For nucleotide groups, adenylyl (not adenyl), etc. should be the form used. The name oxo acids (not keto acids) may be used as a class name, and for individual compounds in which $-CH_2$ has been replaced by –CO, oxo should be used.

Rule 2

Where the substrate is normally in the form of an anion, its name should end in -ate rather than -ic, e.g. lactate dehydrogenase, not lactic dehydrogenase or lactic acid dehydrogenase.

Rule 3

Commonly used abbreviations for substrates, e.g. ATP, may be used in names of enzymes, but the use of new abbreviations (not listed in recommendations of the IUPAC-IUB Commission on Biochemical Nomenclature) should be discouraged. Chemical formulae should not normally be used instead of names of substrates. Abbreviations for names of enzymes, e.g. GDH, should not be used.

Rule 4

Names of substrates composed of two nouns, such as glucose phosphate, which are normally written with a space, should be hyphenated when they form part of the enzyme names and thus become adjectives, e.g. glucose 6-phosphate 1-dehydrogenase (EC 1.1.1.49).

Rule 5

The use of enzyme names as descriptions such as condensing enzyme, acetate-activating enzyme, pH 5 enzyme should be discontinued as soon as the catalysed reaction is known. The word "activating" should not be used in the sense of converting the substrate into a substance that reacts further, all enzymes act by activating their substrates, and the use of the word in this sense may lead to confusion.

RULE 6

Common names If it can be avoided, a common name should not be based on a substance that is not a true substrate, e.g. enzyme EC 4.2.1.17 should not be called "crotonase", since it does not act on crotonate.

RULE 7

Common names Where names in common use give some identification of the reaction and is not incorrect or ambiguous, its continued use is recommended. In other cases a common name is based on the same general principles as the systematic name but with a minimum of detail, to produce a name short enough for convenient use. A few names of proteolytic enzymes ending in -in are retained; all other enzyme names should end in -ase.

Systematic names Systematic names consist of two parts. The first contains the name of the substrate or, in the case of a bimolecular reaction, of the two substrates separated by a colon. The second part, ending in -ase indicates the nature of the reaction.

RULE 8

A number of generic words including a type of reaction may be used in either common or systematic names: oxidoreductases, oxygenase, transferase (with a prefix indicating the nature of group transferred), hydrolase, lyase, racemase, epimerase, isomerase, mutase, ligase.

Rule 9

Common names A number of additional generic words indicating reaction types are used in common names, but not in the systematic nomenclature, e.g. dehydrogenase, reductase, oxidase, peroxidase, kinase, tautomerase, deaminase dehydratase, etc.

Rule 10

Where additional information is needed to make the reaction clear, a phrase indicating the reaction or a product should be added in parentheses after the second part of the name, e.g. (ADP-forming), (dimerizing), (CoA-acylating).

Rule 11

Common names The direct attachment of -ase to the name of the substrate will indicate that the enzyme brings about hydrolysis.

Systematic names The suffix -ase should never be attached directly to the name of the substrate.

Rule 12

Common names The name "dehydrase" which was at one time used for both dehydrogenating and dehydrating enzymes, should not be used. Dehydrogenase will be used for the former and dehydratase for the latter.

Rule 13

Common names Where possible, common names should normally be based on a reaction direction that has been demonstrated, e.g. dehydrogenase or reductase, decarboxylase or carboxylase.

Systematic names In the case of reversible reactions, the direction chosen for naming should be the same for all the enzymes in a given class, even if this direction has not been demonstrated for all. Thus systematic names may be based on a written reaction, even though only the reverse of this has been actually demonstrated experimentally.

Rule 14

Systematic names When the overall reaction includes two different changes, e.g. an oxidative demethylation, the classification and systematic name should be based, whenever possible, on the one (or the first one) catalysed by the enzyme; the other function(s) should be indicated by adding a suitable participle in parentheses, as in the case of sarcosine: oxygen oxidoreductase (demethylating) (EC 1.5.3.1); D-aspartate: oxygen oxidoreductase (deaminating) (EC 1.4.3.1); L-serine hydrolyase (adding indoleglycerol phosphate) (EC 4.2.1.20).

Other examples of such additions are (decarboxylating), (cyclizing), (acceptor-acylating), (isomerizing).

Rule 15

When an enzyme catalyses more than one type of reaction, the name should normally refer to one reaction only. Each case must be considered on its merits, and the choice must be, to some extent arbitrary. Other important activities of the enzyme may be indicated in the list under "reaction" or "comments."

Similarly, when any enzyme acts on more than one substrate (or pair of substrates), the name should normally refer only to one substrate (or pair of substrates), although in certain cases it may be possible to use a term that covers a whole group of substrates, or an alternative substrate may be given in parentheses.

Rule 16

A group of enzymes with closely similar specificities should normally be described by a single entry. However, when the specificity of two enzymes catalysing the same reactions is sufficiently different (the degree of difference being a matter of arbitrary choice) two separate entries may be made, e.g. EC 1.2.1.4 and EC 1.2.1.7. Separate entries are also appropriate for enzymes having similar catalytic functions, but known to differ basically with regard to reaction mechanism or to the nature of the catalytic groups, e.g. amine oxidase (flavin-containing) (EC 1.4.3.4) and amine oxidase (copper-containing) (EC 1.4.3.6).

Rules and Guidelines for Particular Classes of Enzymes

CLASS I

Rule 17

Common names The terms **"dehydrogenase or reductase"** will be used much as hitherto. The latter term is appropriate when hydrogen transfer from the substance mentioned as donor in the systematic name is not readily demonstrated. **Transhydrogenase** may be retained for a few well-established cases. **Oxidases** is used only for cases where O_2 molecule (or part of it) is directly incorporated into the substrate. **Peroxidase** is used for enzymes using H_2O_2 as acceptor. **Catalase** must be regarded as exceptional. Where no ambiguity is caused, the second reactant is not usually named but where required to prevent ambiguity, it may be given in parentheses, e.g. EC 1.1.1.1 **alcohol dehydrogenase** and EC 1.1.1.2, **alcohol dehydrogenase (NADP)**.

Rule 18

Systematic names For oxidoreductases using NAD or NADP the coenzyme should always be named as the acceptor except for the special case of section 1.6 (enzymes whose normal physiological function is regarded as re-oxidation of the reduced coenzyme). The enzyme which can use either coenzyme, should be indicated by writing NAD(P).

Rule 19

Where the true acceptor is unknown and the oxidoreductase has only been shown to react with artificial acceptors, the word "acceptor" should be written in parentheses, as in the case of EC 1.3.99.1, Succinate: (acceptor) oxidoreductase.

Rule 20

Common names Oxidoreductases that bring about the incorporation of molecular oxygen into one donor or into either or both of a pair of donors are named **oxygenase.** If only one atom of oxygen is incorporated, the term "monooxygenase" is used; if both atoms of O_2 are incorporated, the term "dioxygenase" is used.

Systematic names Oxidoreductases bringing about the incorporation of oxygen into one of the paired donors should be named on the pattern donor, donor: oxygen oxidoreductase (hydroxylating).

CLASS 2

Rule 21

Common names Only one specific substrate or reaction product is generally indicated in the common names, together with the group donated or accepted. The form transaminases may be replaced if desired by the corresponding form aminotransferase.

A number of special words are used to indicate reaction types, e.g. "Kinase" to indicate a phosphate transfer from ATP to the named substrate (not phosphokinase), "diphosphokinase" for a similar transfer of diphosphate.

Systematic names Enzymes catalysing group-transfer reactions should be named transferase and the names formed on the pattern donor: acceptor group-transferred-transferase, e.g. ATP: acetate phosphotransferase (EC 2.7.2.1). e.g. ATP: D-fructose 1-phosphotransferase (EC 2.7.1.3). The spelling "transphorase" should not be used. In the case of the phosphotransferases, ATP should always be named as the donor. In the case of the transaminases involving 2-oxoglutarate, ATP should always be named as the acceptor.

RULE 22

Systematic names The prefix denoting the group transferred should, as far as possible, be non-committal with respect to the mechanism of the transfer, e.g. "phospho" rather than "phosphate".

CLASS 3

RULE 23

Common names The direct addition of *-ase* to the name of the substrate generally denotes a hydrolase. Where this is difficult, for e.g. EC 3.1.2.1, the word hydrolase may be used. Enzymes should not normally be given separate names merely on the basis of optimal conditions for activity. The acid and alkaline phosphatases (EC 3.1.3.1–2) should be regarded as special cases and not as examples to be followed. The common name lysozyme is also exceptional.

Systematic names Hydrolysing enzymes should be systematically named on the pattern "substrate hydrolase".

Where the enzyme is specific for the removal of a particular group, the group may be named as a prefix, adenine aminohydrolase (EC 3.5.4.4). In a number of cases, this group can also be transferred by the enzyme to other molecules, and the hydrolysis itself might be regarded as a transfer of the group to water.

CLASS 4

RULE 24

Common names The old names decarboxylase, aldolase, etc. are retained; and dehydrolase (not "dehydrase") is used for the hydro-lyases. "Synthetase" should not be used for any enzymes in this class. The term "synthase" may bc used for enzymes in this class (or any other class) when it is desired to emphasize the synthetic aspect of the reaction.

Systematic names Enzymes removing groups from substrates non-hydrolytically, leaving double bonds (or adding groups to double bonds) should be called as **lyases** in the systematic nomenclature. Prefixes such as hydro-, ammonia- should be used to denote the type of reaction, e.g. (S)-malate hydro-lyase (EC 4.2.1.2). Decarboxylases should be regarded as carboxy-lyases. A hyphen should be written before to avoid confusion with hydrolases, carboxylases.

RULE 25

Common names Where the equilibrium warrants it, or where the enzyme has long been named after a particular substrate, the reverse reaction may be taken as the basis of the name, using hydratase, carboxylase, e.g. fumarate hydratase for EC 4.2.1.2 (in preference to "fumarase" which suggests an enzyme hydrolysing fumarate).

Systematic names The complete molecule, not either of the parts into which it is separated, should be named as the substrate.

The part indicated as a prefix to -lyase is the more characteristic and usually, but not always, the smaller of the two reaction products. This may either be the removed (saturated) fragment of the substrate molecule, as in ammonia, hydro-, thiol-lyases, etc., or the remaining unsaturated fragment as in the case of carboxyl, aldehyde, or oxo-acid-lyases.

RULE 26

Various subclasses of the lyases include a number of strictly specific or group-specific pyridoxal 5-phosphate enzymes that catalyse **elimination** reactions of β- or γ- substituted α-amino acids. Some closely related pyridoxal 5-phosphate-containing enzymes, e.g. tryptophan synthase (EC 4.2.1.20) and cystathionine β-synthase (EC 4.2.1.22) catalyse replacement reactions in which α-, β- or γ- substituent is replaced by a second reactant without creating a double bond. Formally, these enzymes appear to be transferases rather than lyases. However, there is evidence that in these cases the elimination of the β- or γ- substituent and the formation of an unsaturated intermediate is the first step in this reaction. Thus, applying rule 14, these enzymes are correctly classified as lyases.

CLASS 5

RULE 27

In this class, the common names are, in general, similar to the systematic names which indicate the basis of classification.

RULE 28

Isomerase will be used as a general name for enzymes in this class. The types of isomerization will be indicated in systematic names by prefixes, e.g. maleate *cis-trans*-isomerase (EC 5.2.1.1), phenylpyruvate keto-enol-isomerase (EC 5.3.2.1), 3-oxosteriod D^5–D^4-isomerase (EC 5.3.3.1). Enzymes catalysing an aldose–ketose interconversion will be known as aldose-ketose isomerases, e.g. L-arabinose aldose–ketose isomerases (EC 5.3.1.4). When the isomerization consists of an intermolecular transfer of a group, the enzyme is named a mutase, e.g. phosphomutase (EC 5.4.1.1). When it consists of an intramolecular lyase-type reaction, it is systematically named a lyase (decyclizing), e.g. EC 5.5.1.1.

Rule 29

Isomerases catalysing inversions at asymmetric centres should be termed racemases or epimerases, according to whether the substrate contains one, or more than one centre of asymmetry. For example, compare EC 5.1.1.5 with EC 5.1.1.7. A numerical prefix to the word "epimerase" should be used to show the position of the inversion.

CLASS 6

Rule 30

Common names Common names for enzymes of this class were previously of the type XY synthase. However, as this use has not always been understood and synthetase has been confused with synthase (Rule 24), it is now recommended that as far as possible the common names should be similar in form to the systematic names.

Systematic names The class of enzymes catalysing the linking together of two molecules, coupled with the breaking of a diphosphate link in ATP, or other nucleoside triphosphate should be known as ligases. These enzymes were often previously known as synthetases, however, this terminology differs from all other systematic enzyme names as it is based on the product and not on the substrate. For these reasons, a new systematic class name was necessary.

Rule 31

Common names The common names should be formed on the pattern X–Y ligase, where X–Y is the substance formed by linking X and Y. In certain cases, where a trivial name is commonly used for XY, a name of the type XY synthase may be recommended (e.g. EC 6.3.2.11, carnosine synthase).

Systematic names The systematic names should be formed on the pattern X : Y ligase (ADP-forming), where X and Y are the two molecules to be joined together. The phrase shown in reactions is $X + Y + ATP \rightarrow X\text{–}Y + ADP + P_i$.

Rule 32

Common names In the special case where glutamine acts as an ammonia-donor, this is indicated in parentheses (glutamine-hydrolysing) along with ligase name.

Systematic names In this case, the name amido-ligase should be used in the systematic nomenclature.

CLASSIFICATION

The top-level classifications are:

EC 1 **Oxidoreductases** catalyse oxidation/reduction reactions.

EC 2 **Transferases** transfer a functional group (e.g. a methyl or phosphate group).

EC 3 **Hydrolases** catalyse the hydrolysis of various bonds.

EC 4 **Lyases** cleave various bonds by means other than hydrolysis and oxidation.

EC 5 **Isomerases** catalyse isomerization changes within a single molecule.

EC 6 **Ligases** join two molecules with covalent bonds.

Enzyme Identification Number

Each enzyme is given a systematic name and a unique 4-digit identification number for identification by the Enzyme Commission (E.C.) of IUBMB (since 1964).

Example

Lactate dehydrogenase (Lactate: NAD^+ oxidoreductase)

$$\text{Lactate} + NAD^+ \rightleftharpoons \text{Pyruvate} + NADH + H^+$$

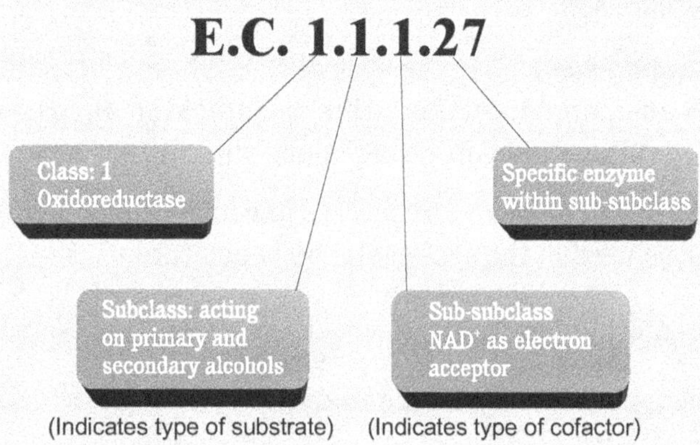

EC 1 Oxidoreductases

All enzymes catalysing oxido-reduction belong to this class. The substrate oxidized is regarded as hydrogen or electron donor. The classification is based on "donor: acceptor oxidoreductase". The accepted name is "dehydrogenase", wherever this is possible; as an alternative, "acceptor reductase" can be used. "Oxidase" is used only where O_2 is an acceptor. Classification is difficult in some cases, because of the lack of specificity towards the acceptor.

EC* 1.1 *Acting on the CH–OH group of donors This subclass contains all dehydrogenases acting on primary alcohols, secondary alcohols and hemi-acetals. They are further classified according to the acceptor which can be NAD^+ or $NADP^+$ (subclass EC 1.1.1), cytochrome (EC 1.1.2), oxygen (EC 1.1.3), a disulphide (EC 1.1.4), quinone (EC 1.1.5) or another acceptor (EC 1.1.99).

EC* 1.2 *Acting on the aldehyde or oxo group of donors This subclass contains all enzymes oxidizing aldehydes to the corresponding acids; when this acid is concomitantly phosphorylated or acetylated, this is indicated in parentheses. Oxo groups may be oxidized either with addition of water and cleavage of a carbon–carbon bond or, in the case of ring compounds, by addition of the elements of water and dehydrogenation. This subclass is subdivided according to the acceptor, which may be $NAD(P)^+$ (EC 1.2.1), a cytochrome (EC 1.2.2), oxygen (EC 1.2.3), a disulphide (EC 1.2.4), an iron–sulphur protein (EC 1.2.7) or some other acceptor (EC 1.2.99).

EC* 1.3 *Acting on the CH–CH group of donors In this subclass are listed all enzymes introducing a double bond into the substrate by direct dehydrogenation at a carbon–carbon single bond. Sub-subclasses are based on the acceptor; EC 1.3.1 with $NAD(P)^+$, EC 1.3.2 with cytochrome, EC 1.3.3 with oxygen, EC 1.3.5 with a quinone or related compound, EC 1.3.7 with an iron-sulphur protein and EC 1.3.99 with other acceptors.

EC* 1.4 *Acting on the CH–NH$_2$ group of donors These are the amino-acid dehydrogenases and the amine oxidases. In most cases, the imine formed is hydrolysed to give an oxo-group and NH_3. This is indicated as (deaminating). Sub-subclasses are formed according to acceptor: EC 1.4.1 with NAD(P)$^+$, EC 1.4.2 with a cytochrome, EC 1.4.3 with oxygen, EC 1.4.4 with a disulphide, EC 1.4.7 with an iron–sulphur protein as acceptor, and EC 1.4.99 with other acceptors.

EC* 1.5 *Acting on CH–NH group of donors Enzymes dehydrogenating secondary amines, introducing a C=N double bond as the primary reaction are listed in this subclass. In some cases, the C=N is later hydrolysed. Sub-subclasses are EC 1.5.1 with NAD(P)$^+$, EC 1.5.3 with oxygen as acceptor, EC 1.5.4 with a disulphide as acceptor, EC 1.5.5 with a quinone as acceptor, EC 1.5.57 with an iron–sulphur protein as acceptor, EC 1.5.8 with a flavin as acceptor and EC 1.5.99 with other acceptors.

EC* 1.6 *Acting on NADH or NADPH In general, enzymes using NADH or NADPH to reduce a substrate are classified according to the reverse reaction, in which NAD$^+$ or NADP$^+$ is formally regarded as acceptor. In subclass EC 1.6, only those enzymes are listed in which some other redox carrier is the acceptor. This can be either NAD(P)$^+$, in sub-subclass EC 1.6.1; a haem protein in EC 1.6.2; oxygen in EC 1.6.3; a disulphide in EC 1.6.4 ; a quinone in EC 1.6.5; a nitrogenous group in EC 1.6.6; a flavin in EC 1.6.8; some other acceptor in EC 1.6.99.

EC* 1.7 *Acting on other nitrogenous compounds as donors A small group of enzymes oxidizing diverse nitrogenous substrates with a cytochrome (EC 1.7.2), oxygen (EC 1.7.3), an iron–sulphur protein (EC 1.7.7), or with other acceptors (EC 1.7.99).

EC* 1.8 *Acting on a sulphur group of donors A small subclass acting either on inorganic substrates or organic thiols. Sub-subclasses again depend on the acceptor: EC 1.8.1 with NAD^+ or $NADP^+$; EC 1.8.2 with a cytochrome; EC 1.8.3 with oxygen; EC 1.8.4 with a disulphide; EC 1.8.5 with a quinone; EC 1.8.7 with an iron–sulphur protein; EC 1.8.98 with other known acceptors; and EC 1.8.99 with any other acceptor.

EC* 1.9 *Acting on a haem group of donors These are the cytochrome oxidases and nitrate reductases. Sub-subclasses include EC 1.9.3 with oxygen, EC 1.9.6 with a nitrogenous compound as acceptor and EC 1.9.99 with other acceptors.

EC* 1.10 *Acting on diphenols and related substances as donors These enzymes catalyse the oxidation of diphenols or ascorbate. There are four sub-subclasses: EC 1.10.1 with NAD^+ or $NADP^+$; EC 1.10.2 with cytochromes; EC 1.10.3 with oxygen as acceptor; and EC 1.10.99 with other acceptors. Some enzymes catalysing the oxidation of phenols are oxygenases (EC 1.14.18).

EC* 1.11 *Acting on peroxide as acceptor This has a single sub-subclass EC 1.11.1 that contains the peroxidases.

EC* 1.12 *Acting on hydrogen as donors Hydrogenases using iron–sulphur compounds as donors for the reduction of H^+ to H_2 are listed in subclass EC 1.18. Other hydrogenases have been left in sub-subclasses: EC 1.12.1 with NAD^+ or $NADP^+$ as acceptor, EC 1.12.2 with cytochromes, EC 1.6.5 with quinone, EC 1.12.7 with iron–sulphur protein, and EC 1.12.99 with other known acceptors.

EC* 1.13 *Acting on single donors with incorporation of molecular oxygen (oxygenases) The enzymes in this subclass differ from all those in earlier subclasses in that oxygen is actually incorporated from O_2 into the substance oxidized.

They differ from those in EC 1.14 in that a second hydrogen donor is not required. Sub-subclasses are EC 1.13.11, when two atoms of oxygen are incorporated; EC 1.13.12, when only one oxygen atom is used, and EC 1.13.99, for other cases. This classification replaces an earlier version. Accepted names in this subclass are of the form "monooxygenase" and "dioxygenase".

EC 1.14 ***Acting on paired donors with incorporation of molecular oxygen*** The enzymes in this subclass all act on two hydrogen donors, and oxygen from O_2 is incorporated into one or both of them. In sub-subclass EC 1.14.11, 2-oxoglutarate is one donor and one atom of oxygen goes into each donor; in EC 1.14.12, NADH or NADPH is one donor, and two atoms of oxygen go into the other donor; in EC 1.14.13, NADH or NADPH is again one donor, but only one atom of oxygen goes into the other donor; in sub-subclasses EC 1.14.14–1.14.18, one atom of oxygen is incorporated into one donor, the other donor being respectively a reduced flavin or flavoprotein, a reduced iron–sulphur protein, a reduced pteridine, ascorbate, or some other compound.

Sub-subclass EC 1.14.19 differs from others in subclass EC 1.14 in that hydrogen atoms removed from the two donors are combined with molecular oxygen to form two molecules of water. Sub-subclass EC 1.14.20 has 2-oxoglutarate as one donor, and the other is dehydrogenated. Sub-subclass EC 1.14.21 has NADH or NADPH as one donor, and the other is dehydrogenated. Sub-subclass EC 1.14.99 is for cases where information about the second donor is incomplete.

EC 1.15 ***Acting on superoxide radicals as acceptor*** Sub-subclass EC 1.15.1 contains a single enzyme that brings about the dismutation of superoxide radicals.

EC* 1.16 *Oxidizing metal ions Metal ions act as donor, being oxidized to a higher valency state. Two sub-subclasses are known: EC 1.16.1 with NAD^+ or $NADP^+$ as acceptor, EC 1.16.3 with oxygen as acceptor, and EC 1.16.8 with flavin as acceptor.

EC* 1.17 *Acting on CH or CH_2 groups The $-CH_2$ group of donors is oxidized by these enzymes to –CHOH (or CH to CHO); in the reverse direction, they are involved in the formation of deoxysugars. Five sub-subclasses are known: EC 1.17.1 with NAD^+ or $NADP^+$ as acceptor; EC 1.17.3 with oxygen as acceptor; EC 1.17.4 with a disulphide as acceptor; EC 1.17.5 with a quinone or similar compound as acceptor; and EC 1.17.99 with other acceptors.

EC* 1.18 *Acting on iron–sulphur proteins as donors Three sub-subclasses are known: EC 1.18.1 with NAD^+ or $NADP^+$ as acceptor; EC 1.18.6 with dinitrogen as acceptor; and EC 1.18.99 with H^+ as acceptor.

EC* 1.19 *Acting on reduced flavodoxin as donor The only sub-subclass is EC 1.19.6 with dinitrogen as acceptor.

EC* 1.20 *Acting on phosphorus or arsenic in donors Four sub-subclasses are known: EC 1.20.1 with NAD^+ or $NADP^+$ as acceptor; EC 1.20.4 with disulphide as acceptor; EC 1.20.98 with other, known acceptors; and EC 1.20.99 with other acceptors.

EC* 1.21 *Acting on X–H and Y–H to form an X–Y bond Three sub-subclasses are known: EC 1.21.3 with oxygen as acceptor; EC 1.21.4 with a disulphide as acceptor; and EC 1.20.99 with other acceptors.

EC* 1.97 *Other oxidoreductases This subclass is reserved for oxidoreductases not included in the previous categories.

The Enzyme Commission numbers and the classes of oxidoreductases are summarized in Table 2.1.

Table 2.1 EC 1 Oxidoreductases

Enzyme Commission Number	Classes
EC 1.1	**Acting on the CH–OH group of donors**
EC 1.1.1	With NAD^+ or $NADP^+$ as acceptor
EC 1.1.2	With a cytochrome as acceptor
EC 1.1.3	With oxygen as acceptor
EC 1.1.4	With a disulphide as acceptor
EC 1.1.5	With a quinone or similar compound as acceptor
EC 1.1.99	With other acceptors
EC 1.2	**Acting on the aldehyde or oxo group of donors**
EC 1.2.1	With NAD^+ or $NADP^+$ as acceptor
EC 1.2.2	With a cytochrome as acceptor
EC 1.2.3	With oxygen as acceptor
EC 1.2.4	With a disulphide as acceptor
EC 1.2.7	With an iron–sulphur protein as acceptor
EC 1.2.99	With other acceptors
EC 1.3	**Acting on the CH–OH group of donors**
EC 1.3.1	With NAD^+ or $NADP^+$ as acceptor
EC 1.3.2	With a cytochrome as acceptor
EC 1.3.3	With oxygen as acceptor
EC 1.3.5	With a quinone or related compound as acceptor
EC 1.3.7	With an iron–sulphur protein as acceptor
EC 1.3.99	With other acceptors
EC 1.4	**Acting on the CH–NH_2 group of donors**
EC 1.4.1	With NAD^+ or $NADP^+$ as acceptor
EC 1.4.2	With a cytochrome as acceptor
EC 1.4.3	With oxygen as acceptor
EC 1.4.4	With a disulphide as acceptor
EC 1.4.7	With an iron–sulphur protein as acceptor
EC 1.4.99	With other acceptors

(*Contd.*)

Table 2.1 (Continued)

Enzyme Commission Number	Classes
EC 1.5	**Acting on the CH–NH group of donors**
EC 1.5.1	With NAD^+ or $NADP^+$ as acceptor
EC 1.5.3	With oxygen as acceptor
EC 1.5.4	With a disulphide as acceptor
EC 1.5.5	With a quinone or similar compound as acceptor
EC 1.5.8	With a flavin as acceptor
EC 1.5.99	With other acceptors
EC 1.6	**Acting on NADH or NADPH**
EC 1.6.1	With NAD^+ or $NADP^+$ as acceptor
EC 1.6.2	With a haem protein as acceptor
EC 1.6.3	With oxygen as acceptor
EC 1.6.4	With a disulphide as acceptor
EC 1.6.5	With a quinone or similar compound as acceptor or
EC 1.6.6	With a nitrogenous group as acceptor
EC 1.6.7	With an iron–sulphur protein as acceptor
EC 1.6.8	With a flavin as acceptor
EC 1.6.99	With other acceptors
EC 1.7	**Acting on other nitrogenous compounds as donors**
EC 1.7.1	With NAD^+ or $NADP^+$ as acceptor
EC 1.7.2	With a cytochrome as acceptor
EC 1.7.3	With oxygen as acceptor
EC 1.7.7	With an iron–sulphur protein as acceptor
EC 1.7.99	With other acceptors

(*Contd.*)

Table 2.1 (Continued)

Enzyme Commission Number	Classes
EC 1.8	**Acting on a sulphur group of donors**
EC 1.8.1	With NAD^+ or $NADP^+$ as acceptor
EC 1.8.2	With a cytochrome as acceptor
EC 1.8.3	With oxygen as acceptor
EC 1.8.4	With a disulphide as acceptor
EC 1.8.5	With a quinone or similar compound as acceptor
EC 1.8.7	With an iron–sulphur protein as acceptor
EC 1.8.98	With other known acceptors
EC 1.8.99	With other acceptors
EC 1.9	**Acting on a haem group of donors**
EC 1.9.3	With oxygen as acceptor
EC 1.9.6	With a nitrogenous group as acceptor
EC 1.9.99	With other acceptors
EC 1.10	**Acting on diphenols and related substances as donors**
EC 1.10.1	With NAD^+ or $NADP^+$ as acceptor
EC 1.10.2	With a cytochrome as acceptor
EC 1.10.3	With oxygen as acceptor
EC 1.10.99	With other acceptors
EC 1.11	**Acting on a peroxide as acceptor**
EC 1.11.1	Peroxidases
EC 1.12	**Acting on hydrogen as donor**
EC 1.12.1	With NAD^+ or $NADP^+$ as acceptor
EC 1.12.2	With a cytochrome as acceptor

(*Contd.*)

Table 2.1 (Continued)

Enzyme Commission Number	Classes
EC 1.12.5	With a quinone or similar compound as acceptor
EC 1.12.7	With an iron–sulphur protein as acceptor
EC 1.12.98	With other known acceptors
EC 1.12.99	With other acceptors
EC 1.13	**Acting on single donors with incorporation of molecular oxygen (oxygenases)**
EC 1.13.11	With incorporation of two atoms of oxygen
EC 1.13.12	With incorporation of one atom of oxygen (internal monooxygenases or internal mixed function oxidases)
EC 1.13.99	Miscellaneous
EC 1.14	**Acting on paired donors, with incorporation or reduction of molecular oxygen**
EC 1.14.11	With 2-oxoglutarate as one donor, and incorporation of one atom each of oxygen into both donors
EC 1.14.12	With NADH or NADPH as one donor, and incorporation of two atomsof oxygen into one donor
EC 1.14.13	With NADH or NADPH as one donor, and incorporation of one atom of oxygen
EC 1.14.14	With reduced flavin or flavoprotein as one donor, and incorporation of one atom of oxygen
EC 1.14.15	With reduced iron-sulphur protein as one donor, and incorporation of one atom of oxygen

(Contd.)

Table 2.1 (Continued)

Enzyme Commission Number	Classes
EC 1.14.16	With reduced pteridine as one donor, and incorporation of one atom of oxygen
EC 1.14.17	With reduced ascorbate as one donor, and incorporation of one atom of oxygen
EC 1.14.18	With another compound as one donor, and incorporation of one atom of oxygen
EC 1.14.19	With oxidation of a pair of donors resulting in the reduction of molecular oxygen to two molecules of water
EC 1.14.20	With 2-oxoglutarate as one donor, and the other dehydrogenated
EC 1.14.21	With NADH or NADPH as one donor, and the other dehydrogenated
EC 1.14.99	Miscellaneous
EC 1.15	**Acting on superoxide as acceptor**
EC 1.16	**Oxidizing metal ions**
EC 1.16.1	With NAD^+ or $NADP^+$ as acceptor
EC 1.16.3	With oxygen as acceptor
EC 1.16.8	With flavin as acceptor
EC 1.17	**Acting on CH or CH_2 groups**
EC 1.17.1	With NAD^+ or $NADP^+$ as acceptor
EC 1.17.3	With oxygen as acceptor
EC 1.17.4	With a disulphide as acceptor
EC 1.17.5	With a quinone or similar compound as acceptor
EC 1.17.99	With other acceptors
EC 1.18	**Acting on iron–sulphur proteins as donors**
EC 1.18.1	With NAD^+ or $NADP^+$ as acceptor
EC 1.18.3	With H^+ as acceptor (now EC 1.18.99)

(*Contd.*)

Table 2.1 (Continued)

Enzyme Commission Number	Classes
EC 1.18.6	With dinitrogen as acceptor
EC 1.18.96	With other known acceptors
EC 1.18.99	With H^+ as acceptor
EC 1.19	**Acting on reduced flavodoxin as donor**
EC 1.19.6	With dinitrogen as acceptor
EC 1.20	**Acting on phosphorus or arsenic in donors**
EC 1.20.1	With $NAD(P)^+$ as acceptor
EC 1.20.4	With disulphide as acceptor
EC 1.20.98	With other, known acceptors
EC 1.20.99	With other acceptors
EC 1.21	**Acting on X–H and Y–H to form an X–Y bond**
EC 1.21.3	With oxygen as acceptor
EC 1.21.4	With a disulphide as acceptor
EC 1.21.99	With other acceptors
EC 1.97	**Other oxidoreductases**

EC 2 Transferase

Transferases are enzymes transferring a group, for example, the methyl group or a glycosyl group, from one compound (generally regarded as donor) to another compound (generally regarded as acceptor). The classification is based on the scheme "donor: acceptor group transferase". The accepted names are normally formed as "acceptor group transferase" or "donor group transferase". In many cases, the donor is a cofactor (coenzyme), carrying the group to be transferred. The aminotransferases constitute a special case (subclass EC 2.6).

EC* 2.1 *Transferring one-carbon groups This subclass contains the methyltransferases (EC 2.1.1), the hydroxymethyl-, formyl- and related transferases (EC 2.1.2), the carboxy- and carbamoyl-transferases (EC 2.1.3), and the amidino-transferases (EC 2.1.4).

EC* 2.2 *Transferring aldehyde or ketone groups This single sub-subclass (EC 2.2.1) contains transketolases and transaldolases.

EC* 2.3 *Acyltransferases These enzymes transfer acyl groups, forming either esters or amides. The donor is in most cases the corresponding acyl-coenzyme A derivative (EC 2.3.1). Aminoacyltransferases form a separate sub-subclass (EC 2.3.2). Acyl groups converted into alkyl groups by a transfer reaction form sub-subclass (EC 2.3.3).

EC* 2.4 *Glycosyltransferases All enzymes transferring glycosyl groups belong to this class. Some of these enzymes also catalyse hydrolysis, which can be regarded as transfer of a glycosyl group from the donor to water. Also, inorganic phosphate can act as acceptor in the case of phosphorylases; phosphorolysis of glycogen is regarded as transfer of one sugar residue from glycogen to phosphate. However, the more general case is the transfer of a sugar from oligosaccharide or a high-energy compound to another carbohydrate molecule as acceptor. The subclass is further subdivided, according to the nature of the sugar residue being transferred, into hexosyltransferases (EC 2.4.1), pentosyltransferases (EC 2.4.2) and those transferring other glycosyl groups (EC 2.4.99).

EC* 2.5 *Transferring alkyl or aryl groups, other than methyl groups This is a somewhat heterogeneous class of enzymes transferring alkyl or related groups either substituted or unsubstituted. There is no subdivision as yet in this subclass.

EC* 2.6 *Transferring nitrogenous groups These are mainly the enzymes transferring amino groups (sub-subclass EC 2.6.1) from a donor, generally an amino acid, to an acceptor, generally a 2-oxo acid. It should be kept in mind that transamination by this reaction also involves an oxidoreduction; the donor is oxidized to a ketone, while the acceptor is reduced. Nevertheless, since the transfer of the amino group is the most prominent feature of this reaction, these enzymes have been classified as aminotransferases rather than oxidoreductases (transaminating). Most of these enzymes are pyridoxal-phosphate proteins.

EC* 2.7 *Transferring phosphorus-containing groups This is a rather large group of enzymes comprising not only those transferring phosphate but also diphosphate, nucleotidyl residues and others. The most numerous section, that of phosphotransferases, is subdivided according to the acceptor group which may be an alcohol group (EC 2.7.1), a carboxy group (EC 2.7.2), a nitrogenous group such as that of creatine (EC 2.7.3), or a phosphate group as in the case of adenylate kinase (EC 2.7.4). The diphosphotransferases are in sub-subclass EC 2.7.6, the nucleotidyltransferases in EC 2.7.7, and those with other substituted phosphate groups in EC 2.7.8. With the enzymes of sub-subclass EC 2.7.9, two phosphate groups are transferred from a donor such as ATP to two different acceptors. The protein kinases are divided into the sub-subclasses tyrosine kinases (EC 2.7.10), serine/threonine kinases (EC 2.7.11), dual-specificity kinases (EC 2.7.12), histidine kinases (EC 2.7.13) and other protein kinases (EC 2.7.99).

EC* 2.8 *Transferring sulphur-containing groups These are enzymes transferring sulphur atoms (EC 2.8.1), sulphate groups (EC 2.8.2) or coenzyme A (EC 2.8.3).

EC 2.9 Transferring selenium-containing groups This is Represented by a single enzyme which transfers a selenium-containing group.

The Enzyme Commission numbers and the classes of transferases are summarized in Table 2.2.

Table 2.2 EC 2 Transferases

Enzyme Commission Number	Classes
EC 2.1	**Transferring one-carbon groups**
EC 2.1.1	Methyltransferases
EC 2.1.2	Hydroxymethyl-, formyl- and related transferases
EC 2.1.3	Carboxy- and carbamoyltransferases
EC 2.1.4	Amidinotransferases
EC 2.2	**Transferring aldehyde or ketonic groups**
EC 2.2.1	Transketolases and transaldolases
EC 2.3	**Acyltransferases**
EC 2.3.1	Transferring groups other than amino-acyl groups
EC 2.3.2	Aminoacyltransferases
EC 2.3.3	Acyl groups converted into alkyl on transfer
EC 2.4	**Glycosyltransferases**
EC 2.4.1	Hexosyltransferases
EC 2.4.2	Pentosyltransferases
EC 2.4.99	Transferring other glycosyl groups
EC 2.5	**Transferring alkyl or aryl groups, other than methyl groups**
EC 2.5.1	Transferring alkyl or aryl groups, other than methyl groups
EC 2.6	**Transferring nitrogenous groups**

(*Contd.*)

Table 2.2 (Continued)

Enzyme Commission Number	Classes
EC 2.6.1	Transaminases
EC 2.6.2	Amidinotransferases
EC 2.6.3	Oximinotransferases
EC 2.6.99	Transferring other nitrogenous groups
EC 2.7	**Transferring phosphorus-containing groups**
EC 2.7.1	Phosphotransferases with an alcohol group as acceptor
EC 2.7.2	Phosphotransferases with a carboxy group as acceptor
EC 2.7.3	Phosphotransferases with a nitrogenous group as acceptor
EC 2.7.4	Phosphotransferases with a phosphate group as acceptor
EC 2.7.5	Phosphotransferases with regeneration of donors, apparently catalysing intramolecular transfers
EC 2.7.6	Diphosphotransferases
EC 2.7.7	Nucleotidyltransferases
EC 2.7.8	Transferases for other substituted phosphate groups
EC 2.7.9	Phosphotransferases with paired acceptors
EC 2.7.10	Protein-tyrosine kinases
EC 2.7.11	Protein-serine/threonine kinases
EC 2.7.12	Dual-specificity kinases (those acting on Ser/Thr and Tyr residues)
EC 2.7.13	Protein-histidine kinases
EC 2.7.99	Other protein kinases

(*Contd.*)

Table 2.2 (Continued)

Enzyme Commission Number	Classes
EC 2.8	**Transferring sulphur-containing groups**
EC 2.8.1	Sulphurtransferases
EC 2.8.2	Sulphotransferases
EC 2.8.3	CoA-transferases
EC 2.8.4	Transferring alkylthio groups
EC 2.9	**Transferring selenium-containing groups**
EC 2.9.1	Selenotransferases

EC 3 Hydrolase

These enzymes catalyse the hydrolysis of various bonds. Some of these enzymes pose problems because they have a wide specificity, and it is not easy to decide if two preparations described by different authors are the same, or if they should be listed under different entries. While the systematic name always includes "hydrolase", the accepted name is, in most cases, formed by the name of the substrate with the suffix -ase. It is understood that the name of the substrate with this suffix, and no other indicator, means a hydrolytic enzyme.

EC* 3.1 *Acting on ester bonds The esterases are subdivided into those acting on carboxylic esters (EC 3.1.1), thioester hydrolases (EC 3.1.2), phosphoric monoester hydrolases, the phosphatases (EC 3.1.3), phosphodiester hydrolases (EC 3.1.4), triphosphoric monoester hydrolases (EC 3.1.5), sulphatases (EC 3.1.6), diphosphoric monoesterases (EC 3.1.7) and phosphoric triester hydrolases (EC 3.1.8). The nucleases, previously included under EC 3.1.4, are now placed in a number of new sub-subclasses: the exonucleases (EC 3.1.11–16), and the endonucleases (EC 3.1.21–31).

EC 3.2 Glycosylases Glycosylases are classified under hydrolases, although some of them can also transfer glycosyl residues to oligosaccharides, polysaccharides and other alcoholic acceptors. The glycosylases are subdivided into those hydrolysing O- or S-glycosyl compounds (EC 3.2.1 and EC 3.2.3), i.e., glycosides, and those hydrolysing N-glycosyl compounds (EC 3.2.2). Accepted names for enzymes acting on D-sugars or their derivatives do not contain "D", unless ambiguity would result from the common existence of the corresponding L-sugar.

EC 3.3 Acting on ether bonds Enzymes acting on ether bonds belong to this small subclass. It is subdivided into those hydrolysing on thioether and trialkylsulphonium compounds (EC 3.3.1) and those acting on ethers (EC 3.3.2).

EC 3.4 Acting on peptide bonds (Peptidases) It is recommended that the term "peptidase" be used synonymous with "peptide hydrolase" for any enzyme that hydrolyses peptide bonds. Peptidases are recommended to be further divided into "exopeptidases" that act only near a terminus of a polypeptide chain and "endopeptidases" that act internally in polypeptide chains. The types of exopeptidases and endopeptidases are described more fully below. The usage of "peptidase" now recommended is synonymous with "protease" as it was originally used as a general term for both exopeptidases and endopeptidases, but it should be noted that previously, in Enzyme Nomenclature (1984), "peptidase" was restricted to the enzymes included in sub-subclasses EC 3.4.11–19, the exopeptidases. Also, the term "proteinase" used previously for the enzymes included in sub-subclasses EC 3.4.21–99 carried the same meaning as "endopeptidase", and has been replaced by "endopeptidase" for consistency.

The nomenclature of the peptidases is troublesome. Their specificity is commonly difficult to define, depending upon the nature of several amino acid residues around the peptide bond to be hydrolysed and also on the conformation of the substrate polypeptide chain. A classification involving the additional criterion of catalytic mechanism is therefore used.

Two sets of sub-subclasses of peptidases are recognized, those of the exopeptidases (EC 3.4.11–19) and those of the endopeptidases (EC 3.4.21–24 and EC 3.4.99). The exopeptidases act only near the ends of polypeptide chains, and those acting at a free N-terminus liberate a single amino acid residue (aminopeptidases, EC 3.4.11), or a dipeptide or a tripeptide (dipeptidyl-peptidases and tripeptidyl-peptidases, EC 3.4.14). The exopeptidases acting at a free C-terminus liberate a single residue (carboxypeptidases, EC 3.4.16–18) or a dipeptide (peptidyl-dipeptidases, EC 3.4.15). The carboxypeptidases are allocated to four groups on the basis of the catalytic mechanism the serine-type carboxypeptidases (EC 3.4.16), the metallocarboxypeptidases (EC 3.4.17) and the cysteine-type carboxypeptidases (EC 3.4.18). Other exopeptidases are specific for dipeptides (dipeptidases, EC 3.4.13), or remove terminal residues that are substituted, cyclized or linked by isopeptide bonds (peptide linkages other than those of α-carboxyl to α-amino groups) (omega peptidases, EC 3.4.19).

The endopeptidases are divided into sub-subclasses on the basis of catalytic mechanism, and specificity is used only to identify individual enzymes within the groups. These are the sub-subclasses of serine endopeptidases (EC 3.4.21), cysteine endopeptidases (EC 3.4.22), aspartic endopeptidases (EC 3.4.23), metalloendopeptidases (EC 3.4.24) and threonine endopeptidases (EC 3.4.25). Endopeptidases that could not be assigned to any of the sub-subclasses EC 3.4.21–25 were listed in sub-subclass EC 3.4.99.

There are characteristic inhibitors of the members of each catalytic type of endopeptidase; to save space these have not been listed separately for each individual enzyme. In describing the specificity of peptidases, use is made of a model in which the catalytic site is considered to be flanked on one or both sides by specificity subsites, each able to accommodate the side chain of a single amino acid residue. These sites are numbered from the catalytic site, S1... S*n* towards the N-terminus of the substrate, and S1′... S*n*′ towards the C-terminus. The residues they accommodate are numbered P1... P*n*, and P1′... P*n*′, respectively, as follows:

Substrate: - P3 - P2 - P1 + P1′ - P2′ - P3′ -

Enzyme: - S3 - S2 - S1* S1′- S2′- S3′-

In this representation, the catalytic site of the enzyme is marked '*'. The peptide bond cleaved (the scissile bond) is indicated by the symbol or a hyphen, in the structural formula of the substrate, or by a hyphen, in the name of the enzyme. In describing the specificity of endopeptidases, the term "oligopeptidase" is used to refer to those that act only on substrates smaller than proteins.

EC* 3.5 *Acting on carbon–nitrogen bonds, other than peptide bonds To this subclass belong those enzymes hydrolysing amides, amidines and other C–N bonds. The sub-subclasses are separated on the basis of the substrate: linear amides (EC 3.5.1), cyclic amides (EC 3.5.2), linear amidines (EC 3.5.3), cyclic amidines (EC 3.5.4), nitriles (EC 3.5.5) and other compounds (EC 3.5.99).

EC 3.6 Acting on acid anhydrides The enzymes acting on diphosphate bonds in compounds such as nucleoside di- and tri-phosphates (EC 3.6.1), on sulphonyl-containing anhydrides such as adenylylsulphate (EC 3.6.2) and on acid anhydrides and catalysing transmembrane movement of substances (EC 3.6.3) belong to this subclass.

EC 3.6.3 Acting on acid anhydrides; catalysing transmembrane movement of substances Several types of ATP phosphohydrolase are listed here. Entries EC 3.6.3.1 to EC 3.6.3.13 and EC 3.6.3.53 are enzymes undergoing covalent phosphorylation of an aspartate residue during the transport cycle; entries EC 3.6.3.14 and EC 3.6.3.15 refer to enzymes of complicated membrane and non-membrane location that can also serve in ATP synthesis; entry EC 3.6.3.16 is a multisubunit enzyme that is involved in arsenite transport only; entries EC 3.6.3.17 to EC 3.6.3.50 are two-domain enzymes of the ABC family; entries EC 3.6.3.51 and EC 3.6.3.52 are parts of a complex protein-transporting machinery in mitochondria and chloroplasts.

EC 3.7 Acting on carbon–carbon bonds There are relatively few carbon–carbon hydrolases; they mostly catalyse the hydrolysis of 3-oxo carboxylic acids.

EC 3.8 Acting on halide bonds These enzymes hydrolyse organic halides.

EC 3.9 Acting on phosphorus–nitrogen bonds This is the phosphoamidase group.

EC 3.10 Acting on sulphur–nitrogen bonds

EC 3.11 Acting on carbon–phosphorus bonds The enzymes in this subclass hydrolyse C-phosphono groups

EC 3.12 Acting on sulphur–sulphur bonds

EC 3.13 Acting on carbon–sulphur bonds

The Enzyme Commission numbers and the classes of hydrolases are summarized in Table 2.3.

Table 2.3 EC 3 Hydrolases

Enzyme Commission Number	Classes
EC 3.1	**Acting on ester bonds**
EC 3.1.1	Carboxylic ester hydrolases
EC 3.1.2	Thioester hydrolases
EC 3.1.3	Phosphoric monoester hydrolases
EC 3.1.4	Phosphoric diester hydrolases
EC 3.1.5	Triphosphoric monoester hydrolases
EC 3.1.6	Sulphuric ester hydrolases
EC 3.1.7	Diphosphoric monoester hydrolases
EC 3.1.8	Phosphoric triester hydrolases
EC 3.1.11	Exodeoxyribonucleases producing 5′-phosphomonoesters
EC 3.1.13	Exoribonucleases producing 5′-phosphomonoesters
EC 3.1.14	Exoribonucleases producing 3′-phosphomonoesters
EC 3.1.15	Exonucleases active with either ribo- or deoxyribonucleic acids and producing 5′-phosphomonoesters
EC 3.1.16	Exonucleases active with either ribo- or deoxyribonucleic acids and producing 3′-phosphomonoesters
EC 3.1.21	Endodeoxyribonucleases producing 5′-phosphomonoesters
EC 3.1.22	Endodeoxyribonucleases producing 3′-phosphomonoesters

(Contd.)

Table 2.3 (Continued)

Enzyme Commission Number	Classes
EC 3.1.25	Site-specific endodeoxyribonucleases specific for altered bases
EC 3.1.26	Endoribonucleases producing 5′-phosphomonoesters
EC 3.1.27	Endoribonucleases producing 3′-phosphomonoesters
EC 3.1.30	Endoribonucleases active with either ribo- or deoxyribonucleic acids and producing 5′-phosphomonoesters
EC 3.1.31	Endoribonucleases active with either ribo- or deoxyribonucleic acids and producing 3′-phosphomonoesters
EC 3.2	**Glycosylases**
EC 3.2.1	Glycosidases, i.e., enzymes hydrolysing O- and S-glycosyl compounds
EC 3.2.2	Hydrolysing N-glycosyl compounds
EC 3.2.3	Hydrolysing S-glycosyl compounds (discontinued)
EC 3.3	**Acting on ether bonds**
EC 3.3.1	Thioether and trialkylsulphonium hydrolases
EC 3.3.2	Ether hydrolases
EC 3.4	**Acting on peptide bonds (Peptidases)**
EC 3.4.11	Aminopeptidases
EC 3.4.13	Dipeptidases
EC 3.4.14	Dipeptidyl-peptidases and tripeptidyl-peptidases
EC 3.4.15	Peptidyl-dipeptidases
EC 3.4.16	Serine-type carboxypeptidases
EC 3.4.17	Metallocarboxypeptidases
EC 3.4.18	Cysteine-type carboxypeptidases

(Contd.)

Table 2.3 (Continued)

Enzyme Commission Number	**Classes**
EC 3.4.19	Omega peptidases
EC 3.4.21	Serine endopeptidases
EC 3.4.22	Cysteine endopeptidases
EC 3.4.23	Aspartic endopeptidases
EC 3.4.24	Metalloendopeptidases
EC 3.4.25	Threonine endopeptidases
EC 3.4.99	Endopeptidases of unknown catalytic mechanism
EC 3.5	**Acting on carbon–nitrogen bonds, other than peptide bonds**
EC 3.5.1	In linear amides
EC 3.5.2	In cyclic amides
EC 3.5.3	In linear amidines
EC 3.5.4	In cyclic amidines
EC 3.5.5	In nitriles
EC 3.5.99	In other compounds
EC 3.6	**Acting on acid anhydrides**
EC 3.6.1	In phosphorus-containing anhydrides
EC 3.6.2	In sulphonyl-containing anhydrides
EC 3.6.3	Acting on acid anhydrides; catalysing transmembrane movement of substances
EC 3.6.4	Acting on acid anhydrides; involved in cellular and subcellular movement
EC 3.6.5	Acting on GTP; involved in cellular and subcellular movement
EC 3.7	**Acting on carbon–carbon bonds**
EC 3.7.1	In ketonic substances

(Contd.)

Table 2.3 (Continued)

Enzyme Commission Number	Classes
EC 3.8	**Acting on halide bonds**
EC 3.8.1	In C-halide compounds
EC 3.9	**Acting on phosphorus–nitrogen bonds**
EC 3.10	**Acting on sulphur–nitrogen bonds**
EC 3.11	**Acting on carbon–phosphorus bonds**
EC 3.12	**Acting on sulphur–sulphur bonds**
EC 3.13	**Acting on carbon–sulphur bonds**

EC 4 Lyases

Lyases are enzymes cleaving C–C, C–O, C–N and other bonds by other means than by hydrolysis or oxidation. They differ from other enzymes in that two substrates are involved in one reaction direction, but only one in the other direction. When acting on the single substrate, a molecule is eliminated and this generates either a new double bond or a new ring. The systematic name is formed according to "substrate group-lyase". In common names, expressions like decarboxylase, aldolase, etc., are used. "Dehydratase" is used for those enzymes eliminating water. In cases where the reverse reaction is the more important, or the only one to be demonstrated, "synthase" may be used in the name.

EC* 4.1 *Carbon–carbon lyases This subclass contains the decarboxylases (EC 4.1.1), the aldehyde-lyases catalysing the reversal of an aldol condensation (EC 4.1.2), and the oxo acid-lyases, catalysing the cleavage of a 3-hydroxy acid (EC 4.1.3), or the reverse reactions.

EC* 4.2 *Carbon–oxygen lyases These enzymes catalyse the breakage of a carbon–oxygen bond. In the case of hydro-lyases (EC 4.2.1), it is by elimination of water; in sub-subclass EC 4.2.2 it is by the elimination of an alcohol from a polysaccharide and in sub-subclass EC 4.2.3 it is by elimination of a phosphate. A few other cases are grouped in EC 4.2.99.

EC* 4.3 *Carbon–nitrogen lyases This subclass contains the enzymes that release ammonia or one of its derivatives, with the formation of a double bond or ring. Some catalyse the actual elimination of the ammonia, amine or amide, e.g.

$$>CH{-}CH{-}({-}NH{-}R) \rightarrow >C{=}CH{-} + NH_2{-}R$$

Others, however, catalyse elimination of another component, e.g. water, which is followed by spontaneous reactions that lead to breakage of the C–N bond, e.g.

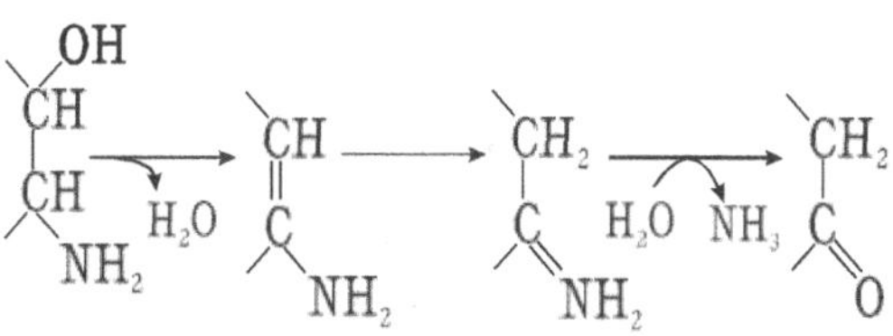

as in EC 4.3.1.17 (L-serine ammonia-lyase), so that the overall reaction is

$$>C\,({-}OH{-}){-}CH\,({-}NH_2)^- \longrightarrow >CH_2{-}CO^- + NH_3$$

i.e., an elimination with rearrangement. The sub-subclasses of EC 4.3 are the ammonia-lyases (EC 4.3.1), lyases acting on amides, amidines, etc. (EC 4.3.2), the amine-lyases (EC 4.3.3), and other carbon–nitrogen lyases (EC 4.3.99).

EC* 4.4 *Carbon–sulphur lyases Enzymes eliminating H_2S or substituted H_2S; a single sub-subclass EC 4.4.1.

EC* 4.5 *Carbon–halide lyases This subclass was originally set up on the basis of the enzyme eliminating HCl from DDT.

EC* 4.6 *Phosphorus–oxygen lyases The so-called nucleotidyl-cyclases are included here, on the basis that diphosphate is eliminated from the nucleoside triphosphate.

EC* 4.99 *Other lyases A subclass of miscellaneous enzymes.

The Enzyme Commission numbers and the classes of lyases are summarized in Table 2.4.

Table 2.4 Classes of lyases

Enzyme Commission Number	Classes
EC 4.1	**Carbon–carbon lyases**
EC 4.1.1	Carboxy-lyases
EC 4.1.2	Aldehyde-lyases
EC 4.1.3	Oxo acid-lyases
EC 4.1.99	Other carbon–carbon lyases
EC 4.2	**Carbon–oxygen lyases**
EC 4.2.1	Hydro-lyases
EC 4.2.2	Acting on polysaccharides
EC 4.2.3	Acting on phosphates
EC 4.2.99	Other carbon–oxygen lyases
EC 4.3	**Carbon–nitrogen lyases**
EC 4.3.1	Ammonia-lyases
EC 4.3.2	Lyases acting on amides, amidines, etc.
EC 4.3.3	Amine-lyases
EC 4.3.99	Other carbon–nitrogen lyases
EC 4.4	**Carbon–sulphur lyases**
EC 4.5	**Carbon–halide lyases**
EC 4.6	**Phosphorus–oxygen lyases**
EC 4.99	**Other lyases**

EC 5 Isomerases

These enzymes catalyse changes within one molecule.

EC* 5.1 *Racemases and epimerases These enzymes, catalysing either racemization or epimerization of a centre of chirality, are subdivided according to their substrates which may be amino acids (EC 5.1.1), hydroxy acids (EC 5.1.2), carbohydrates and derivatives (EC 5.1.3), or other compounds (EC 5.1.99).

EC* 5.2 *Cis–trans-isomerases These enzymes rearrange the geometry at double bonds.

EC* 5.3 *Intramolecular oxidoreductases These enzymes bring about the oxidation of one part of a molecule with a corresponding reduction of another part. They include the enzymes converting, in the sugar series, aldoses to ketoses and vice versa (sugar isomerases, EC 5.3.1), enzymes catalysing a keto-enol equilibrium (tautomerases, EC 5.3.2), enzymes shifting a carbon–carbon double bond from one position to another (EC 5.3.3), enzymes transposing S–S bonds (EC 5.3.4), and a group of miscellaneous enzymes (EC 5.3.99).

EC* 5.4 *Intramolecular transferases These enzymes transfer acyl (EC 5.4.1), phospho (EC 5.4.2), amino (EC 5.4.3), hydroxy (EC 5.4.4) or other groups (EC 5.4.99) from one position to another.

EC 5.4.2 *Phosphotransferases (Phosphomutases)* Most of these enzymes were previously listed as sub-subclass EC 2.7.5, under the heading "Phosphotransferases with regeneration of donors, apparently catalysing intramolecular transfers". The reaction for these enzymes was written in the form: $X\text{-}(P)_2 + AP = BP + X\text{-}(P)_2$.

In fact, since phosphorylation of the acceptor produces a bisphosphate identical with the donor, the overall reaction is an isomerization of A*P* into B*P*, with the bisphosphate acting catalytically. It has been shown in some cases that the enzyme has a functional phosphate group, which can act as the donor. Phosphate is transferred to the substrate, forming the intermediate bisphosphate; the other phosphate group is subsequently transferred to the enzyme:

The bisphosphate may be firmly attached to the enzyme during the catalytic cycle, or in other cases may be released so that free bisphosphate is required as an activator. Under these circumstances, it was agreed in 1983 that all of these enzymes should be listed together in the sub-subclass 5.4.2 based on the overall isomerase reaction.

EC* 5.5 *Intramolecular lyases These catalyse reactions in which a group can be regarded as eliminated from one part of a molecule, leaving a double bond, while remaining covalently attached to the molecule.

EC* 5.99 *Other isomerases A subclass of miscellaneous enzymes.

The Enzyme Commission numbers and the classes of isomerases are summarized in Table 2.5.

Table 2.5 Classes of isomerases

Enzyme Commission Number	Classes
EC 5.1	Racemases and epimerases
EC 5.1.1	Acting on amino acids and derivatives
EC 5.1.2	Acting on hydroxy acids and derivatives
EC 5.1.3	Acting on carbohydrates and derivatives
EC 5.1.99	Acting on other compounds

(*Contd.*)

Table 2.5 (Continued)

Enzyme Commission Number	Classes
EC 5.2	***cis-trans*-Isomerases**
EC 5.3	**Intramolecular oxidoreductases**
EC 5.3.1	Interconverting aldoses and ketoses
EC 5.3.2	Interconverting keto- and enol-groups
EC 5.3.3	Transposing C= C bonds
EC 5.3.4	Transposing S–S bonds
EC 5.3.99	Other intramolecular oxidoreductases
EC 5.4	**Intramolecular transferases**
EC 5.4.1	Transferring acyl groups
EC 5.4.2	Phosphotransferases (Phosphomutases)
EC 5.4.3	Transferring amino groups
EC 5.4.4	Transferring hydroxy groups
EC 5.4.99	Transferring other groups
EC 5.5	**Intramolecular lyases**
EC 5.99	**Other isomerases**

EC 6 Ligases

Ligases are enzymes that catalyse the joining of two molecules with concomitant hydrolysis of the diphosphate bond in ATP or a similar triphosphate. Ligase is commonly used for the common name, but, in a few cases, "synthase" or "carboxylase" are used. "Synthetase" may be used in the place of "synthase" for enzymes in this class.

EC 6.1 Forming carbon–oxygen bonds This subclass contains the enzymes acylating a transfer RNA with the corresponding amino acid (amino-acid-tRNA ligases (EC 6.1.1)).

EC 6.2 Forming carbon–sulphur bonds The enzymes synthesizing acyl-CoA derivatives belong to this subclass.

EC 6.3 Forming carbon–nitrogen bonds This subclass contains the amide synthases (EC 6.3.1), the peptide synthases (EC 6.3.2), enzymes forming heterocyclic rings (EC 6.3.3), enzymes using glutamine as amido N-donor (EC 6.3.5) and a few others (EC 6.3.4).

EC 6.4 Forming carbon–carbon bonds These are the carboxylating enzymes, mostly biotinyl proteins.

EC 6.5 Forming phosphoric ester bonds These include the enzymes restoring broken phosphodiester bonds in the nucleic acids (often called repair enzymes).

EC 6.6 Forming nitrogen–metal bonds This covers metal chelation of a tetrapyrrole ring system.

The Enzyme Commission numbers and the classes of ligases are summarized in Table 2.6.

Table 2.6 EC 6 Ligases

Enzyme Commission Number	Classes
EC 6.1	**Forming carbon–oxygen bonds**
EC 6.1.1	Ligases forming aminoacyl-tRNA and related compounds
EC 6.2	**Forming carbon–sulphur bonds**
EC 6.2.1	Acid-thiol ligases
EC 6.3	**Forming carbon–nitrogen bonds**

(*Contd.*)

Table 2.6 (Continued)

Enzyme Commission Number	Classes
EC 6.3.1	Acid-ammonia (or amine) ligases (amide synthases)
EC 6.3.2	Acid–amino-acid ligases (peptide synthases)
EC 6.3.3	Cyclo-ligases
EC 6.3.4	Other carbon–nitrogen ligases
EC 6.3.5	Carbon–nitrogen ligases with glutamine as Amido-N-donor
EC 6.4	**Forming carbon–carbon bonds**
EC 6.5	**Forming phosphoric ester bonds**
EC 6.6	**Forming nitrogen–metal bonds**
EC 6.6.1	Forming coordination complexes

REVIEW QUESTIONS

1. What are the six classes of enzymes?
2. Give an account of classification of enzymes.

3

CHARACTERISTICS OF ENZYMES

GENERAL CHARACTERISTICS OF ENZYMES

Enzymes are chemical catalysts that control various biochemical reactions without themselves being changed or utilized, in the living beings. An enzyme is a protein molecule that is a biological catalyst with the following important characteristics:

- All enzymes are globular proteins.
- They increase the rate of reaction without themselves being used up.
- Their presence does not affect the nature or properties of end products.
- Small amounts of an enzyme can accelerate chemical reactions.
- They are very specific in their reaction; a single enzyme catalyses a single chemical reaction or a group of related reactions.
- They are sensitive to even a minor change in pH, temperature and substrate concentration.

- Some enzymes require a cofactor for their proper functioning.
- They lower the activation energy of the reaction.

Most of the enzymes produced by the cell function within the cell and are called **endoenzymes** (intracellular) but some enzymes are liberated by the living cells and catalyse reactions in the cell's environment; such enzymes are known as **exoenzymes** (extracellular). The enzymes of the digestive tract are good examples of exoenzymes.

If the enzyme is secreted in a form that does not act upon the substrate without undergoing prior modification in its structure, it is known as **proenzyme** or **zymogen**. The zymogens are activated when they come in contact with the activator. For example, pancreatic juice contains the zymogen trypsinogen which itself is inactive but is converted to active trypsin on coming into contact with previously formed trypsin, or with the enteropeptidase (older name enterokinase). By producing zymogen, the organism is protected from the action of active enzymes in situations where their activity might prove injurious. Most cases of proenzyme activation involve the removal of inhibitory or blocking peptide moiety from the proenzyme molecule.

The activation of trypsinogen to trypsin occurs as follows:

$$\text{Trypsinogen} + \text{Enteropeptidase} \rightarrow \text{Trypsin} + \text{Octapeptide}$$

Enteropeptidase hydrolyses the peptide bond between amino acid residues 15 and 16 liberating a small peptide. The new amino terminal unit, 16, then interacts ionically with aspartic acid-194. This, in turn, alters the orientation of the lysine-145 and opens the active site on the enzyme.

Trypsin in turn is used to activate two other zymogens:

Chymotrypsinogen + Trypsin → Chymotrypsin

Procarboxypeptidase + Trypsin → Carboxypeptidase

CHEMICAL NATURE OF ENZYMES

The activity of an enzyme depends, at the minimum, on a specific protein chain. In many cases, the enzyme consists of the protein and a combination of a non-proteinaceous part. This non-protein part of a conjugated protein or enzyme is known as **prosthetic group**. The two components (protein + prosthetic group) of some conjugated proteins (enzymes) can be separated by dialysis; While this would result in the loss of activity of the enzyme, it can be regained by mixing the two separated components. This fact clearly indicates that the dialysable component is necessary for protein to act as an enzyme and hence this dialysable material (prosthetic group) is termed as **coenzyme**, the protein part of a conjugated protein as **apoenzyme** and the intact molecule or conjugated protein as **holoenzyme.**

$$\underset{\text{(Conjugated proteins)}}{\text{Holoenzyme}} \rightleftharpoons \underset{\text{(Proteins)}}{\text{Apoenzyme}} + \underset{\text{(Prosthetic group)}}{\text{Coenzyme}}$$

PROPERTIES OF ENZYMES

The two important properties of enzymes are efficiency and specificity.

Enzyme Efficiency

The most important biochemical characteristic of a living cell is its ability to achieve a large number of rapid chemical conversions at temperatures usually below 40°C, while the chemical reaction in the laboratory would require higher temperatures. Experiments with isolated enzymes from cells and tissues have shown that the efficiency of enzymes is as much as one hundred million (10^8) times greater than that of the chemical catalysis.

All catalysts including enzymes do not alter the position of the chemical equilibrium of the reaction. Usually, in the presence of an enzyme, the reaction runs in the same direction as it would without the enzyme, just more quickly. However, in the absence of the enzyme, other possible uncatalysed, spontaneous reactions might lead to different products, because in those conditions this different product is formed faster.

Furthermore, enzymes can couple two or more reactions, so that a thermodynamically favourable reaction can be used to drive a thermodynamically unfavourable one. For example, the hydrolysis of ATP is often used to drive other chemical reactions.

Energetics of enzyme action For any reaction to occur, the two or more compounds involved in the reaction must come in contact or collide. Each participant of the reaction has some amount of energy that is required to surmount the barrier separating the reactants from becoming the product. Let us consider going uphill and to the other side of the hill. The energy required for the proton on a substrate to reach an intermediate position is called the **activation energy** and the system is said to be in the transition state.

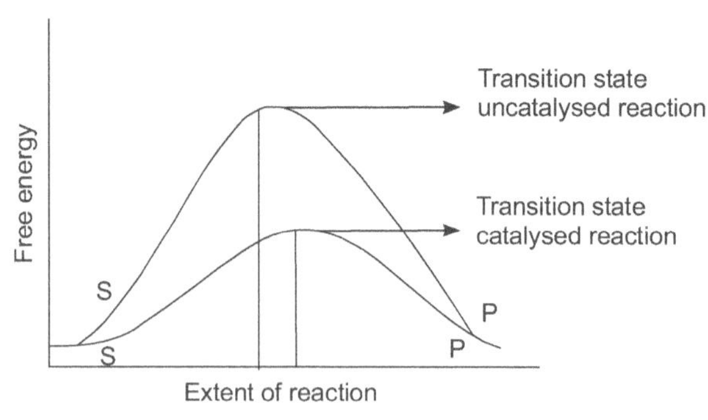

Figure 3.1 Diagrammatic representation of the free energy of activation of a chemical reaction

In Figure 3.1 note that the energy required to reach the transition state is much lower in a catalysed reaction than that required in an uncatalysed reaction (S, is the substrate and P is the product).

Enzyme Specificity

Enzymes are very specific in nature, i.e., a particular enzyme attacks only a particular substrate. There are three types of specificity, viz. stereochemical, reaction and substrate specificities.

Specificity of enzyme action is noncovalent forces through which substrates and other molecules bind to enzymes are identical in character to the forces that dictate the conformations of the proteins themselves. Both involve van der Waals, electrostatic, hydrogen-bonding and hydrophobic interactions. In general, a substrate-binding site consists of an indentation or cleft on the surface of an enzyme molecule that is complementary in shape to the substrate (geometrical complementarity). Moreover, the amino acid residues that form the binding sites are arranged to interact specifically with the substrate in an attractive manner (electronic complementarity).

Molecules that differ from the substrate in shape or functional group distribution cannot productively bind the enzyme; that is, they cannot form enzyme–substrate complexes that lead to the formation of products. The substrate-binding site may, in accordance with the lock-and-key hypothesis, exist in the absence of bound substrate or it may, as suggested by the induced fit hypothesis, form about the substrate as it binds to the enzyme. X-ray studies indicate that the substrate-binding sites of most enzymes are largely preformed but that most of them exhibit at least some degree of induced fit upon binding substrate.

With few exceptions, the enzymes are specific in their action. Their specificity lies in the fact that they may act

- on one specific type of substrate molecule
- on a group of structurally related compounds
- on only one of the two optical isomers of a compound
- on only one of the two geometrical isomers.

Accordingly, four patterns of enzyme specificity have been recognized.

1. *Absolute specificity* Some enzymes are capable of acting on only one substrate, e.g. urease acts only on urea to produce ammonia and carbon dioxide.

$$H_2N-\overset{\overset{\displaystyle O}{\|}}{C}-NH_2 + H-O-H \xrightarrow{\text{Urease}} 2NH_3 + CO_2$$

Similarly, carbonic anhydrase brings about the union of carbon dioxide with water to form carbonic acid.

$$H_2O + CO_2 \xrightarrow{\text{Carbonic anhydrase}} H_2CO_3$$

2. *Group specificity* Some other enzymes are capable of catalysing the reaction of a structurally related group of compounds. For example, lactic dehydrogenase (LDH) catalyses the interconversion of pyruvic acid and lactic acid and also of a number of other structurally related compounds.

$$\underset{\text{Pyruvic acid}}{CH_3COCOOH} + NADH + H^+ \xrightarrow[\text{dehydrogenase}]{\text{Lactic}} \underset{\text{Lactic acid}}{CH_3CHOHCOOH} + NAD^+$$

3. *Stereospecificity* Stereospecificity is about binding chiral compounds. Enzymes are highly specific both in binding chiral substrates and in catalysing their reactions. This stereospecifiity arises because enzymes, by virtue of their

inherent chirality form asymmetric active sites. For example, trypsin readily hydrolyses polypeptides that are composed of L-amino acids but not those consisting of D-amino acids. Likewise, the enzymes involved with glucose metabolism are specific for D-glucose residues.

$$\underset{\text{Ethanol}}{CH_3CH_2OH} + NAD^+ \xrightleftharpoons{\text{YADH}} \underset{\text{Acetaldehyde}}{CH_3\overset{\overset{\displaystyle O}{\|}}{C}H} + NADH + H^+$$

Enzymes are absolutely stereospecific in the reactions they catalyse. For example succinic acid is dehydrogenated by succinate dehydrogenase to give only fumaric acid not maleic acid which might have also been produced during chemical dehydrogenation.

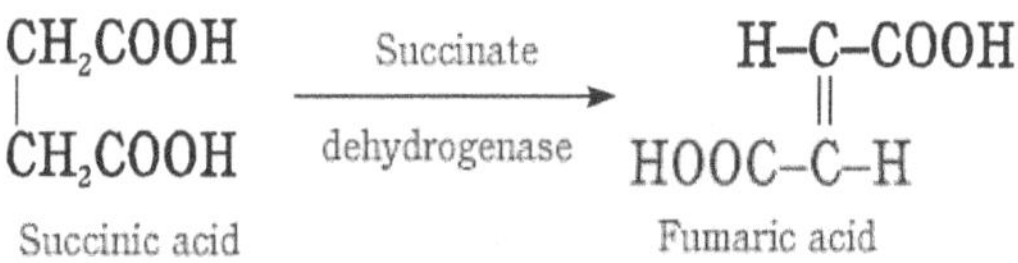

4. *Geometric specificity* The stereospecificity of enzymes is not particularly surprising, in light of the complementarity of an enzymatic binding site for its substrate. A substrate of the wrong chirality will not fit into an enzymatic binding site for much the same reasons. In addition to their stereospecificity, however, most enzymes are quite selective about the identities of the chemical groups on their substrates. Indeed such geometric specificity is a more stringent requirement than its stereospecificity. Enzymes vary considerably in their degree of geometric specificity.

A few enzymes are absolutely specific for only one compound. Most enzymes, however, catalyse the reactions of a small range of related compounds. Some enzymes particularly digestive enzymes, are so permissive in their ranges of acceptable substrates that their geometric specificities are more accurately described as preferences. For example, carboxypeptidase A catalyses the hydrolysis of C-terminal peptide bonds to all residues except arginine, lysine and proline if the preceding residue is not Proline. However, the rate of this enzymatic reaction varies with the identities of the residues in the vicinity of the C-terminus of the polypeptide. Some enzymes are not even very specific in the type of reaction they catalyse. Thus chymotrypsin, in addition to its ability to mediate peptide bond hydrolysis, also catalyses ester bond hydrolysis.

$$\underset{\text{Peptide}}{R\overset{\overset{O}{\parallel}}{C}-NHR} + H_2O \xrightarrow{\text{Chymotrypsin}} R\overset{\overset{O}{\parallel}}{C}-O^- + H_3\overset{+}{N}R$$

$$\underset{\text{Ester}}{R\overset{\overset{O}{\parallel}}{C}-OR} + H_2O \xrightarrow[H^+]{\text{Chymotrypsin}} R\overset{\overset{O}{\parallel}}{C}-O^- + HOR$$

Moreover, the acyl group acceptor in the chymotrypsin reaction need not be water; amino acids, alcohols or ammonia can also act in this capacity. It should be realized that such permissiveness is much more the exception than the rule. Indeed most intracellular enzymes function *in vivo* to catalyse a particular reaction on a specific substrate.

Enzymes show this kind of specificity due to a particular shape of the enzyme, specific arrangement of amino acid in their active site and the structure of the substrate. The active site has two regions:

1. **Binding site** This includes the region of the active site which comes in contact with the substrate or where the substrate binds to the enzyme.
2. **Catalytic site** This is the region within the active site and is responsible for catalysis.

The active site of an enzyme may be clefts or crevices with three-dimensional entity. The substrates are bound to the binding site of the enzyme by relatively weak forces.

MODES OF ENZYME ACTION

Two theories have been proposed to explain the mode of action of enzymes.

Lock-and-Key Theory

In 1890, Emil Fisher proposed a model to explain the great specificity of enzymes. He explained the interaction between substrate and enzymes in terms of "Lock and Key". According to this concept, the catalytic site of the enzyme by itself is complementary in shape to that of the substrate. That means they fit each other as shown in the Figure 3.2.

Figure 3.2 Lock-and-Key hypothesis

The enzymic reaction is possible if the substrate matches the active centre as the key fits the lock. If the substrate ("Key") becomes slightly modified, it no longer fits the active centre ("Lock") and no reaction takes place. Also, the presence of a substrate at the active site may exclude water molecules and thus make the region more non-polar. Both of these factors could be responsible for some degree of change in the tertiary structure. Therefore, in the lock-and-key mechanism, the active site is always structurally intact, with the catalytic sites aligned and freely accessible. Thus, a suitable reacting group, whether part of an appropriately bound substrate or not, can come into contact with the region of catalytic activity and some degree of reaction takes place.

Induced Fit Hypothesis

In order to account for the above observation, Koshland (1958) proposed the induced fit hypothesis. According to this hypothesis, the structure of the substrate may be complementary to that of the active site in the enzyme–substrate complex, but not in the free enzyme. A conformational change takes place in the enzyme during the binding of the substrate, which results in the required matching of structures.

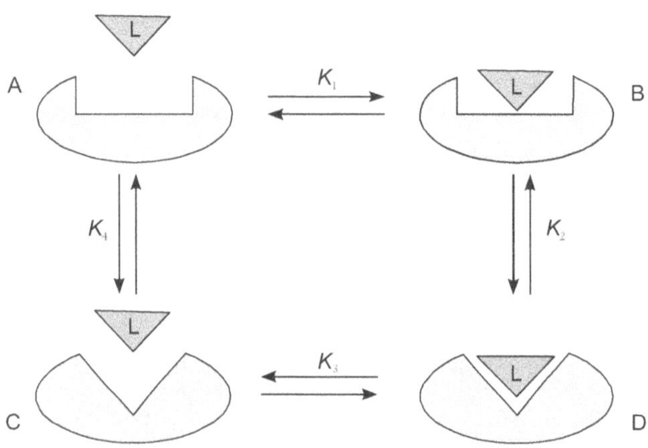

Figure 3.3 Induced fit hypothesis

The induced fit hypothesis (Figure 3.3) requires that the active site be flexible and the substrate be rigid, allowing the enzyme to wrap itself around the substrate, thereby bringing together the corresponding catalytic sites and reacting groups. Also, in this mechanism, different catalytic components might be separated by a considerable margin in the free enzyme, minimizing the risk of a chance collision of a reactive group with both of them. It is also possible that access to the catalytic groups of the free enzyme might be blocked. Only when the binding group of the substrate is recognized by the corresponding site of the enzyme and the binding process proceeds, does conformational change take place, which results in all the relevant groups in the substrate and enzyme coming together.

FACTORS AFFECTING ENZYME ACTIVITY

Effect of pH

Enzymes are affected by changes in pH. Enzyme action is greatest within a narrow range of pH because all the enzymes are active. Changing the pH changes the H bonds, and thus the shape of the active site. Therefore, the substrate can no longer bind to the active site and so enzyme action decreases. The most favourable pH value—the point where the enzyme is most active—is known as the **optimum pH**. The optimum pH of few enzymes are given in Table 3.1. Enzymes have an optimum pH at which their activity is maximal; at higher or lower pH, activity decreases (Figure 3.4). This is because the amino acid side chains in the active site may act as weak acids and bases with critical functions that depend on their maintaining a certain state of ionization.

The pH range over which an enzyme undergoes changes in activity can provide a clue to what amino acid is involved. For example, a change in the activity near pH 7.0 often reflects titration of a histidine residue. The effects of pH must be interpreted carefully because in a closely packed environment of a protein, the pKa of amino acid side chains can be significantly altered, e.g. a nearby positive charge can lower the pKa of a lysine residue and a nearby negative charge can increase it. Such effects sometimes result in a pKa that is shifted by two or more pH units from its normal (free amino acid) value.

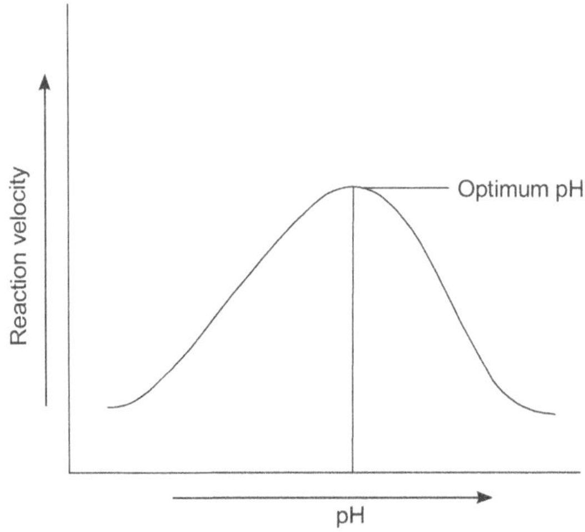

Figure 3.4 Effect of pH on enzyme activity

Table 3.1 pH for optimum activity

Enzyme	pH optimum
Lipase (pancreas)	8.0
Lipase (stomach)	4.0–5.0
Lipase (castor oil)	4.7
Pepsin	1.5–1.6
Trypsin	7.8–8.7
Urease	7.0
Invertase	4.5
Maltase	6.1–6.8
Amylase (pancreas)	6.7–7.0
Amylase (malt)	4.6–5.2
Catalase	7.0

The explanation of this pH effect is mainly based on two factors:

1. *Ionization of the enzyme, particularly at the active site* Like all proteins, enzyme molecules possess numerous ionizable groups whose state of ionizability depends on pH. For an enzyme molecule to be active as a catalyst, certain of these groups must be ionized while others must remain unionized. This state of affairs would obviously prevail only within a limited pH range which would depend on the pKa values of the groups concerned.
2. *Ionization of the substrate, or the coenzyme* In some cases, substrate, like the enzyme, is also capable of being ionized. Thus in such cases also it would be reasonable to assume that only one ionic form of the substrate molecules might be capable of undergoing the reaction. This specific ionic substrate would also exist over a limited pH range.

Thus owing to the above two facts, the optimum pH for the reaction would be the pH at which the ionization states of the substrate and enzyme were most favourable for the reaction. The influence of pH on enzyme activity explains why simple organisms like bacteria flourish only within a restricted pH range and why complicated mechanisms for maintaining the required pH of the body fluids operate in complex organisms like humans.

Effect of Temperature

Like most chemical reactions, the rate of an enzyme-catalysed reaction increases as the temperature is raised. However, this happens only up to a certain temperature commonly known as optimum temperature. (usually in the range of 37°–45°C) above which enzymes become denatured and lose activity which in turn is due to loss of the secondary and tertiary structure of the protein (Figure 3.5). As the enzyme is inactivated, the reaction which it catalyses slows down and ultimately stops. At 0°C, the enzyme action is low because the movement of molecules is low. Storage of enzymes at 5°C or below is generally the most suitable. Some enzymes lose their activity when frozen. Thus optimum temperature of an enzyme may be defined as the temperature at which its activity is maximum.

The ratio of the reaction rate at temperature t + 10°C to that at t° C is known as temperature coefficient Q_{10}. The value of temperature coefficient is usually 2 which means the rate of reaction is double the initial rate when the temperature is increased by 10°C.

$$Q_{10} = \frac{\text{Rate of reaction at } t + 10^\circ\text{C}}{\text{Rate of reaction at } t^\circ\text{C}}$$

A 10°C rise in temperature will increase the activity of most enzymes by 50 to 100%. Over a period of time, enzymes will be deactivated at even moderate temperatures.

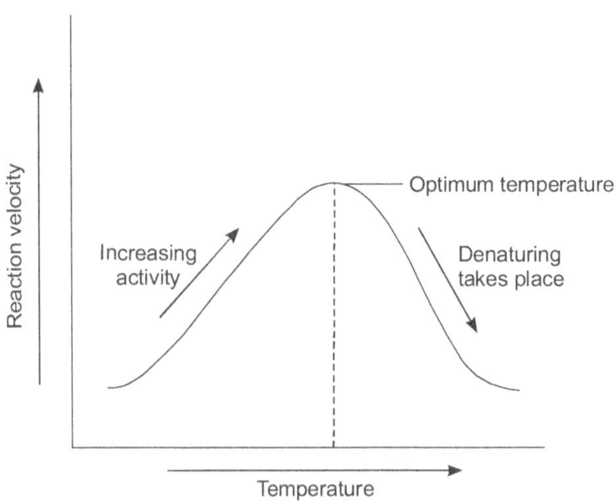

Figure 3.5 Effect of temperature on enzyme action

Effect of Enzyme Concentration

Under given conditions, two molecules of enzyme acting independently in solution will transform twice as much substrate in a given time as one molecule; the velocity should therefore be proportional to the enzyme concentration.

$$V = K\,[E]$$

This is the normal case, when deviations are found they can be explained in one of the following ways.

1. It is sometimes found that if the enzyme concentration is greatly increased, there is a falling-off from the linear relationship as shown in Figure 3.6.

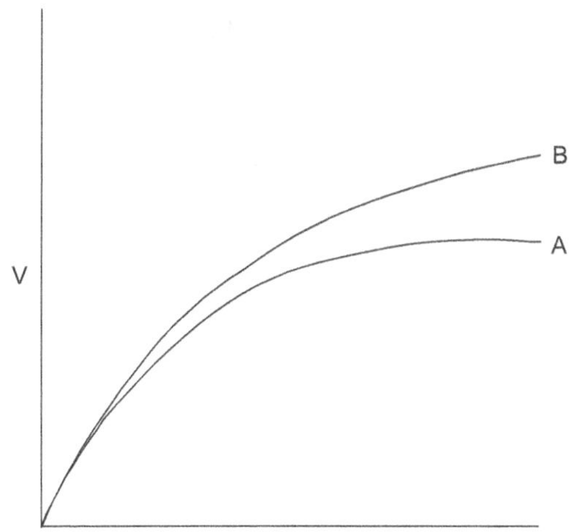

Figure 3.6 Limitation of velocity by test method

This is due to a limitation in the capacity of the method of estimation and does not indicate a decrease in the activity of the enzyme. In cases where the test method depends on a second (added) enzyme, the activity of the latter will set a definite limit to the velocity which can be measured so that a curve similar to curve A will be obtained and an increase in the amount of the second enzyme added will raise the limit as in curve B. Thus, the catalytic efficiency of the enzyme will be constant as long as the other enzymes are in excess, but will fall off at higher concentrations and this limit will vary inversely with its concentration.

2. When an enzyme is available in the pure state it becomes possible to add relatively enormous amounts. If it has an appreciable affinity for one of the components of the system, e.g. a coenzyme, the whole of this component may be bound by the enzyme and thus will not be available to react with the other enzymes. In this case, the velocity will fall as the amount of enzyme is increased (Figure 3.7).

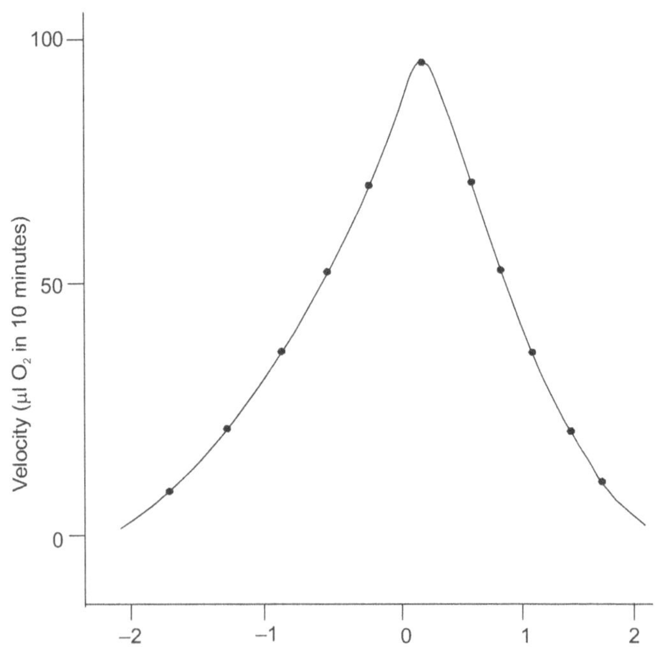

Figure 3.7 Effect of large amount of enzyme in a complex system

3. Very rarely a curve like the one shown in Figure 3.8 will be obtained. This indicates the presence of small amounts of some highly toxic impurity in one of the components of the incubation mixture other than the enzyme solution. This will poison the first amounts of enzyme added and it is only when enough of the enzyme is added to combine with the whole of the impurity that further amounts of enzyme will remain active.

   Thus the normal type of linear curve is obtained, but this is displaced from the origin by an amount which is proportional to the amount of the toxic impurity.

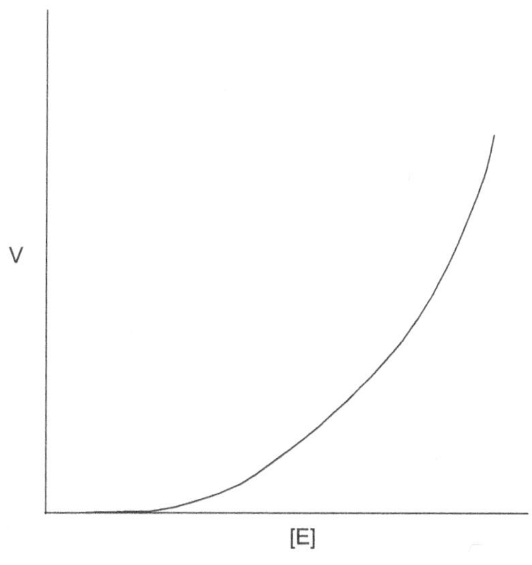

Figure 3.8 Effect of toxic impurities in reagents

4. If the enzyme preparation contains a reversible inhibitor which combines with the enzyme to give an inactive complex EI,

$$E + I \rightleftharpoons EI$$

   Then the percentage of the enzyme which is in the inactive form increases as the concentration of inhibitor in the mixture increases. But since the inhibitor is being added with the enzyme, the inhibitor concentration increases with the enzyme concentration, so the observed activity at higher enzyme concentrations falls below the levels that would be expected from a linear relationship.

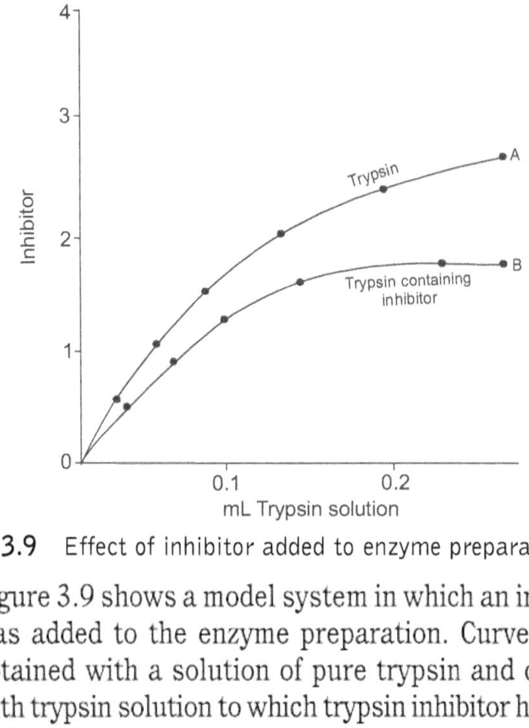

Figure 3.9 Effect of inhibitor added to enzyme preparation

Figure 3.9 shows a model system in which an inhibitor was added to the enzyme preparation. Curve A was obtained with a solution of pure trypsin and curve B with trypsin solution to which trypsin inhibitor had been added. Whenever a plot of velocity against enzyme concentration is found to curve downwards in this way, the presence of a dissociable inhibitor should be suspected and an effort must be made to remove it by dialysis or other means.

5. Proteinases acting on proteins have been repeatedly shown to give non-linear enzyme concentration curves, usually showing a downward curvature. According to "Schütz law" developed by J.Schütz (1866), the velocity

is proportional to the square root of the enzyme concentration.

$$V = K[E]^2$$

$$x = K\sqrt{e}$$

6. If the enzyme preparation contains a dissociable activator or coenzyme, the reverse effect to that described under reversible inhibitors (point 4) will be obtained. As the concentration increases, an increasing proportion of the enzyme will be in the activated form and an upward curvature of the enzyme concentration curve will be obtained. At high dilution, a velocity proportional to the square of the enzyme concentration would be expected while at high concentrations the velocity will become proportional to the first power of the enzyme concentration, when the enzyme becomes saturated with the activator. Such an effect is shown in Figure 3.10.

$$V = K[E]^2$$

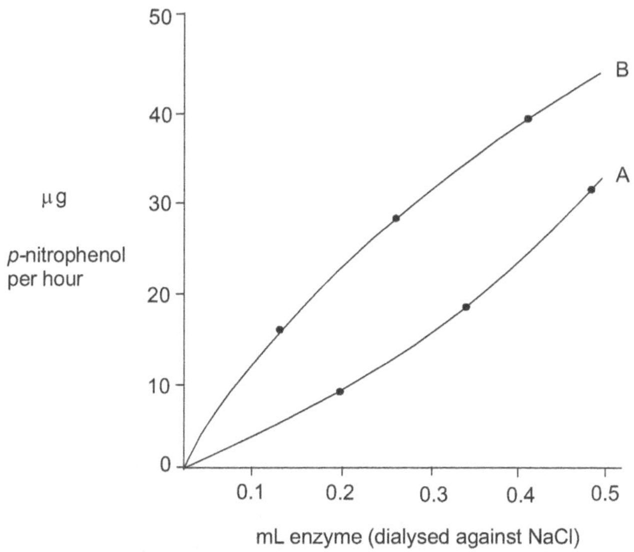

Figure 3.10 Effect of activator (arylsulphatase) in enzyme preparation

Curve A shows the activity of a purified arylsulphatase which had been dialysed against sodium chloride. The enzyme is activated by chloride and if the activities are determined in constant and optimal chloride concentration, a straight line is obtained as shown in curve B.

Effect of Substrate Concentration

Substrate concentration is one of the most important factors, which determines the velocity of enzyme reactions. In nearly all cases, when initial velocity is plotted against substrate concentration, a section of a rectangular hyperbola is obtained as shown in Figure 3.11.

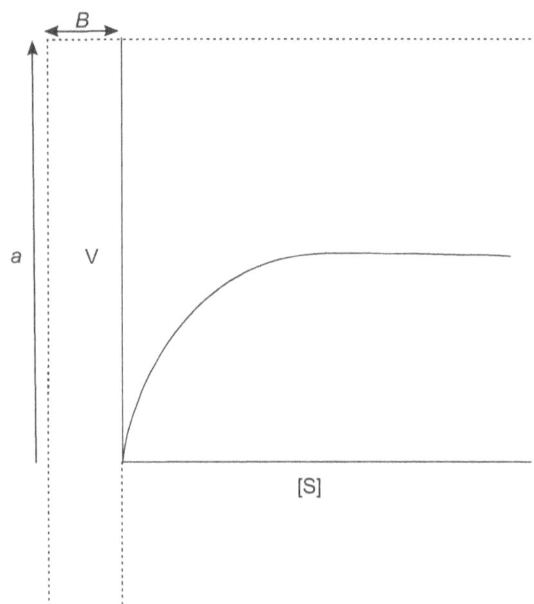

Figure 3.11 Hyperbolic form of typical substrate concentration curve

Michaelis theory A theory involving a dissociation of this type was put forward in 1913 by Michaelis and Menten and has been the foundation of a greater part of the enzyme kinetics. It adopts the earlier suggestion that the enzyme first forms a complex with its substrate and this subsequently breaks down giving the free enzyme and the products of the reaction.

The reaction rate is proportional to the substrate concentration. But this is true up to a certain concentration, after which the increase in concentration of substrate does not further increase the velocity of the reaction.

Effect of Inhibitors

The rates of enzyme-catalysed reactions are decreased by specific inhibitor compounds that combine with the enzyme and prevent enzyme and substrate from combining normally. The toxicity of many compounds such as HCN, H_2S results from their action as enzyme inhibitors. Many drugs also act to inhibit specific enzymes. Thus knowledge of enzyme inhibitors is vital to understanding drug action and toxic agents. Enzyme inhibitors are molecular agents that interfere with catalysis of enzymatic reactions.

Enzymes catalyse virtually all cellular processes, so it should not be surprising that enzyme inhibitors are among the most important pharmaceutical agents. Moreover, information about enzymes themselves is revealed by studying enzyme inhibition.

Enzyme inhibitions may be competitive, noncompetitive and irreversible in nature.

Competitive inhibition In this type of inhibition the inhibitor is a compound whose structure closely resembles the substrate. Therefore, the inhibitor competes with the substrate for the active site of the enzyme. In this type of inhibition both the enzyme–substrate (ES) and the enzyme–inhibitor (EI) complexes are formed. While the inhibitor (I) occupies the active site, it prevents binding of the substrate to the enzyme. The relative amounts of the two complexes formed during the reaction depend partly upon the affinity of the enzyme towards the substrate and inhibitor and partly on the concentrations of the two. Thus, if the inhibitor is present in high concentrations EI complexes will be favoured. An increase in substrate concentration can reverse the inhibition; therefore it is a reversible inhibition (Figure 3.12).

$$\begin{array}{l} E + S \rightleftharpoons ES \longrightarrow E + P \\ + \\ I \\ \quad \Big\Updownarrow K_i \\ EI \end{array}$$

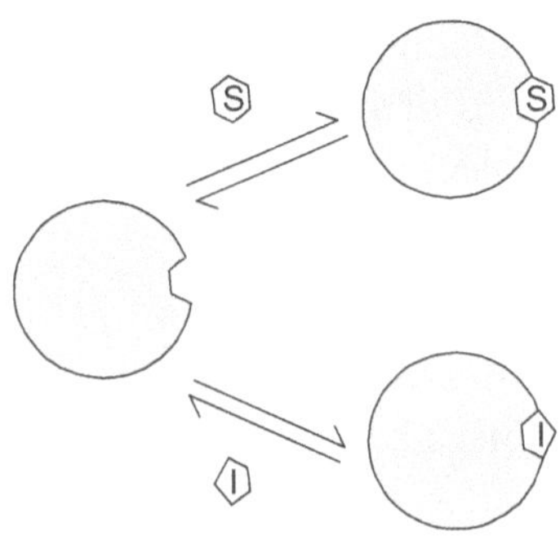

Figure 3.12 Competitive inhibition

Some examples of competitive inhibitors are listed below.

Substrate	**Competitive inhibitor**
Succinic acid	Malonic acid
p-aminobenzoic acid	Sulphanilamide
Hypoxanthine	Allopurinol

Noncompetitive inhibition It is an inhibition where inhibitor is a substance that interacts with the enzyme, but usually not at the active site. The noncompetitive inhibitor reacts either far off from or very close to the active site. The net effect of a noncompetitive inhibitor is to change the shape of the enzyme and thus the active site, so that the substrate can no longer interact with the enzyme to give a reaction. Noncompetitive inhibitors are usually reversible, but are not influenced by concentrations of the substrate as is the case for a reversible competitive inhibitor (Figure 3.13).

$$\begin{array}{ccc} E + S & \rightleftharpoons ES & \longrightarrow E + P \\ & + & \\ & I & \\ & \Big\updownarrow K_i & \\ & ESI & \end{array}$$

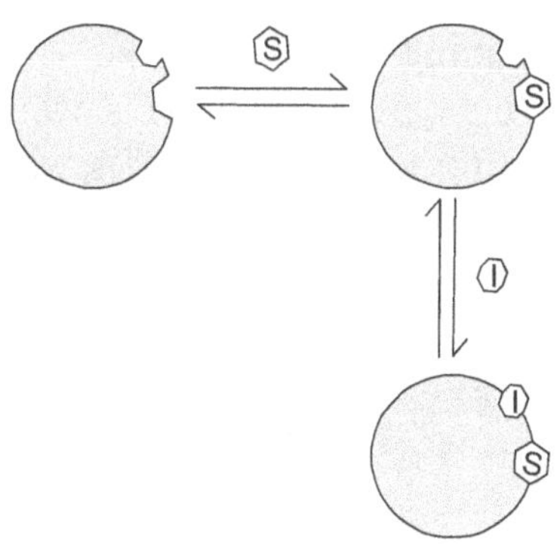

Figure 3.13 Noncompetitive inhibition

Irreversible inhibition Irreversible inhibitors form strong covalent bonds with an enzyme. These inhibitors may act at, near, or far off from the active site. Consequently, they may not be displaced by the addition of excess substrate. In any case, the basic structure of the enzyme is modified to the degree that makes it inactive. Many enzymes can be poisoned more or less specifically by substances, in a way that the enzyme is completely and irreversibly inactivated. Since the inhibitor bears no structural relationship to the substrate it is called noncompetitive inhibition, and the inhibitors show such inhibition are called noncompetitive inhibitors. Such compounds are often reagents which are capable of reacting covalently with functional groups (–OH, –SH, $-NH_2$) of the active site of the enzyme and thus make the active site inactive for the substrate.

The reaction of diisopropylfluorophosphate with a hydroxyl group of the serine residue in the enzyme, acetylcholine esterase, results in the complete loss of activity of the enzyme. Thus diisopropylfluorophosphate and similar phosphate and thiophosphate esters act as nerve poisons by inhibiting acetylcholine esterase. This type of inhibition may also occur when anions interact with a cation that is associated with the active site, e.g. anions like CN^-, F^-, S^{-2}, and oxalate inhibit enzymes containing cations like Fe^{2+}, Mg^{2+} Cu^{2+} and Ca^{2+} respectively. This type of inhibition is overcome by addition of powerful chelating agents, viz., EDTA, which form complexes with the metal ions.

REVIEW QUESTIONS

1. List the general characteristics of enzymes.
2. Define
   i. Endoenzymes
   ii. Exoenzymes
   iii. Zymogens
   iv. Apoenzymes

   v. Holoenzymes
   vi. Active site

3. What do you mean by enzyme efficiency?
4. What is activation energy?
5. Write notes on specificity of enzyme.
6. Describe the modes of enzyme action—lock-and-key hypothesis and induced fit model.
7. Discuss how the following affect enzyme activity:
   i. pH
   ii. Temperature
   iii. Enzyme concentration
   iv. Substrate concentration
   v. Inhibitors

4

ISOENZYMES AND MULTIENZYME COMPLEXES

ISOENZYMES

Many enzymes occur in more than one molecular form in the same species, in the same tissue or even in the same cell. In such cases, the different forms of the enzyme catalyse the same reaction but they differ from each other in their kinetic properties and in amino acid composition and so they can be separated by suitable techniques such as electrophoresis. Such multiple forms of the enzymes are called **isoenzymes** or **isozymes**.

Isozymes are enzymes that differ in amino acid sequence but catalyse the same chemical reaction. These enzymes usually display different kinetic parameters (i.e., different K_m values), or different regulatory properties. The existence of isozymes permits the fine-tuning of metabolism to meet the particular needs of a given tissue or developmental stage (for example, lactate dehydrogenase (LDH)).

Isozymes were first described by R.L. Hunter and Clement Markert in 1957. They defined Isozymes as different variants of the same enzyme having identical functions and present in the same individual. This definition encompasses enzyme variants that are the products of different genes and thus represent different loci (described as isozymes), and enzymes that are the products of different alleles of the same gene (Table 4.1).

Table 4.1 Multiple forms of enzymes*

Group	Reason of multiplicity	Example
1	Genetically independent proteins**	Malate dehydrogenase in mitochondria and cytosol
2	Heteropolymers (hybrids) of two or more polypeptide hydrogenase chains, noncovalently bound**	Hybrid forms of lactate dehydrogenases
3	Genetic variants (allelozymes)**	Glucose 6-phosphate dehydrogenase in man
4	Conjugated or derived proteins	
	a. Proteins conjugated with other groups	Phosphorylase *b*, glycogen synthase *a*
	b. Proteins derived from single polypeptide chains	The family of chymotrypsins arising from chymotrypsinogen
5	Polymers of a single subunit	Glutamate dehydrogenase of molecular weight 1,000,000 and 250,000
6	Conformationally different forms	All allosteric modifications of enzymes

* Artifacts occurring during preparation are outside the scope of this chapter.

** These classes fall into the category of isozymes.

Isozymes are usually the result of gene duplication, but can also arise from polyploidization or nucleic acid hybridization. Over evolutionary time, if the function of the new variant remains identical to the original, then it is likely that one or the other will be lost as mutations accumulate, resulting in a pseudogene. However, if the mutations do not immediately prevent the enzyme from functioning, but instead modifies either its function or its pattern of gene expression, then the two variants may both be favoured by natural selection and become specialized for different functions. For example, they may be expressed at different stages of development or in different tissues.

Isozymes

Enzymes can be regulated by their quaternary structures. In particular, a given enzyme may have more than one arrangement of different polypeptide subunits, each of which gives rise to a different level of activity. The arrangement can be controlled by the level of expression of each subunit. These are called isozymes.

Nomenclature of Isozymes or Isoenzymes

The 1964 report recommended that the individual isoenzymes (isozyme) should be distinguished and numbered on the basis of their electrophoretic mobility, with the number 1 being assigned to the form having the highest mobility towards the anode.

Of all the means available to indicate the different properties of isoenzymes (electrophoresis, chromatography, kinetics criteria, chemical structure, etc.), electrophoresis is the most widely used for the following reasons:

- It is still important in the study of enzyme heterogeneity to avoid any artifactual consequences of the handling or "purification" of enzymes, and zone electrophoresis is one procedure in which the resolution of individual protein species is not unduly influenced by their application in the original state, as tissue homogenates, or the like.

- The degree of resolution is another factor of prime importance in this field, and resolution by electrophoretic procedures is generally more effective than the other methods of protein separation.
- Electrophoresis offers the advantages of rapidity and broad applicability.

The recognition of these facts, and the consequent, wide utilization of electrophoretic procedures by workers in this field, has built up a substantial literature on the subject and has facilitated communication and reference.

Other choices suggested for the distinction of isoenzyme forms include kinetic criteria and structural data. While kinetic criteria are valuable as an adjunct to investigations of enzyme multiplicity, they cannot provide information on the extent of heterogeneity. With regard to structural criteria, the ultimate goal is to define the interrelationships of multiple forms of enzymes in chemical terms. However, it is not useful at this stage to provide general recommendations for isoenzyme nomenclature based on structural considerations, because such information is available only for a very few enzyme systems. When structural details become available, notations in upper case Roman letters, as presently used for lactate dehydrogenase and aldolase, may be used. A system similar to that used for haemoglobin, in which the polypeptide chains are represented by Greek letters, would also be acceptable.

Isozymes (isoenzymes) or their subunits should not be labelled on the basis of tissue distribution (brain type, heart type) since confusion may arise on account of species variation; homologous forms may occur in altogether different tissues in other species.

It is therefore recommended that:

1. In naming isozymes (isoenzymes), the normal enzyme name (either systematic or trivial) should be used, followed by a number. The numbers should be allotted consecutively, preferably on the basis of electrophoretic mobility under defined conditions, with lower numbers given to the forms with higher mobility towards the anode. In photographs or diagrams of electrophoretic results, the anode and cathode should be clearly identified.
2. In case of isozymes (isoenzymes) where complex patterns occur, with major groups each composed of several different electrophoretic zones, the numbers may be used to designate the major groups, with subscript lower case letters applied consecutively for the individual subzones (1_a, 1_b, 1_c, 2_a, 2_b, etc.).
3. For unambiguous identification of particular isozymes (isoenzymes), additional characteristics such as molecular weight, stability, or subunit structure should be given where available. Subunits may be denoted by upper case Roman letters or lower case Greek letters, but not by terms based on tissue distribution.

Lactate Dehydrogenase (LDH)

Isozymes are present in the serusm and tissues of mammals, amphibians, birds, insects, plants and unicellular organisms. A well-known example for isozymes is lactate dehydrogenase (LDH) which catalyses the oxidation of lactate to pyruvate with NAD acting as a coenzyme.

This enzyme (LDH) is found in five electrophoretically distinct forms (Figure 4.1). These five isozymic forms of LDH are tetramers each made from two types of units, H and M. H refers to the heart LDH and M refers to the muscle LDH.

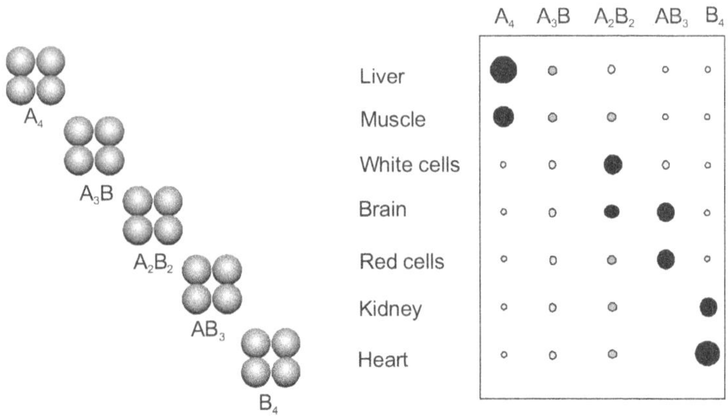

Figure 4.1 The five isomers of LDH

Heart LDH usually predominates in the heart and is active at low levels of pyruvate. This has four identical subunits (HHHH). Muscle LDH is the characteristic of skeletal muscle and maintains its activity in much higher concentration of pyruvate. This also has four identical subunits (MMMM). The two types of subunits, i.e., H and M have the same molecular weight (35,000) but differ in amino acid composition and also differ immunologically. Synthesis of these two subunits H and M is controlled by distinct genetic loci.

LDH can be formed from H and M subunits to yield a pure H tetramer and a pure M tetramer. Only the tetrameric molecule possesses catalytic activity. Combinations of H and M subunits may produce three additional types of hybrid enzymes, thus making the total number of possible forms as five.

HHHH	LD1
HHHM	LD2
HHMM	LD3
HMMM	LD4
MMMM	LD5

This is confirmed by the fact that when the two subunits are mixed in equal proportions, a sequence of five bands is obtained by electrophoresis. The various isomeric forms of LDH differ in their V_{max} and K_m, especially for pyruvate and in the degree of their allosteric inhibition by pyruvate.

LDH catalyses the transfer of two electrons and one hydrogen ion from lactate in NAD^+.

$$\text{Lactate} \xrightleftharpoons{NAD^+ \;\rightarrow\; NADH + H^+} \text{Pyruvate}$$

This reaction occurs at a measurable rate only in the presence of the enzyme catalyst, LDH.

Pyruvate may be produced from carbohydrates, by glycolysis or from amino acid; under anaerobic conditions it may undergo LDH-mediated conversion to lactate but when oxygen is freely available, pyruvate is metabolized to enter the tricarboxylic acid cycle. The tricarboxylic acid cycle and the pathway of glycolysis are important from the point of view of making energy available in a suitable form within the cell; both lead to the synthesis of ATP, an important intermediate in energy metabolism. Lactate can be produced from only pyruvate and can be metabolized back only to pyruvate.

Most of the pyruvate so formed is channelled into the TCA cycle under aerobic conditions to ensure maximum ATP production. Thus, the tissues that have sufficient oxygen supply do not require lactate production to take place and tend to be rich in the H4 isoenzyme; this converts pyruvate into lactate at a relatively low rate.

Under anaerobic conditions, the TCA cycle cannot operate and the cell depends on glycolysis for energy production. Without a constant supply of NAD^+, this too would break down but the LDH-mediated conversion of pyruvate to lactate can ensure that the NAD^+ levels are maintained. Therefore, in tissues which can become oxygen-starved, an LDH isoenzyme M4 with a high capacity for converting pyruvate to lactate is required. The lactate produced eventually finds its way to the heart or liver, via the bloodstream and pyruvate is re-formed.

However, we can conclude that the requirement of the different forms of LDH is dependent on the oxygen status on the cell. It can also be suggested that five different active monomeric forms of LDH may require five different genes.

Assay method

i. Serum sample is subjected to electrophoresis at pH 8.6 using starch, agar or polyacrylamide gel medium.
ii. The five different isozymic forms of LDH have different charges at this pH and migrate to five regions of the electrophoretogram.
iii. A typical dehydrogenase assay reagent contains:
   - reduced substrate (lactate)
   - coenzyme (NAD)
   - oxidized dye—nitro blue tetrazolium salt (NBT)
   - an intermediate electron carrier to transport electrons between NADH and the dye, phenazine methosulphate (PMS)
   - buffer
iv. Isozymes are then localized by means of their ability to catalyse reduction of a colourless dye to a coloured form.

Stains with specific activity have been developed for several isozymes. These allow all of the enzyme zones that contain particular type of activity, for example LDH activity, to be located after gel electrophoresis. The activity stain distinguishes the enzyme of interest from other activities that may have been resolved in the gel, even when very crude mixtures such as cell extracts have been analysed. This technology has been widely applied in genetic studies because large populations of individuals of a given species can be analysed for a given isozyme or set of isozymes with relatively little effort. Most of the enzymes of glycolysis have either two or four identical polypeptide chains, and isozymes of these enzymes frequently have been observed.

Diagnostic importance The estimation of lactate dehydrogenase activity in blood serum and the separation of the isoenzymes are of considerable diagnostic importance in medicine. In certain solid tumours, there is an increase in the serum LD1 and LD2. These isoenzymes are also present in the blood of patients with acute leukaemia.

Hexokinase

Mammals have several forms of hexokinase, all of which catalyse the conversion of glucose into glucose 6-phosphate. The predominant hexokinase isoenzyme in liver is hexokinase D, also called glucokinase, which differs in two important aspects from the hexokinase isoenzymes in muscles.

1. Firstly, when the glucose concentration in the blood is high, as it is after a meal rich in carbohydrates, excess blood glucose is transported into hepatocytes, where glucokinase converts it into glucose 6-phosphate.
2. Secondly, glucokinase is inhibited not by its reaction product glucose 6-phosphate but by its isomer, fructose 6-phosphate, which is always in equilibrium with glucose 6-phosphate because of the action of phosphoglucoisomerase.

Glucokinase is variant of hexokinase which is not inhibited by glucose 6-phosphate. Its different regulatory features and lower affinity for glucose (compared to other hexokinases), allows it to serve different functions in cells of specific organs, such as control of insulin release by the beta cells of the pancreas, or initiation of glycogen synthesis by the liver cells. Both of these processes must occur only when glucose is abundant, or problems occur.

Specificity in Hexokinase

- Hexokinase has to discriminate between the normal substrate, glucose, and water.
- The enzyme has a molecular weight of 100,000 and contains two domains that basically form a "pac man" like structure, in which the active site resides at the cleft.
- Water has access to the active site, and can compete with the normal substrate to hydrolyse ATP into ADP and inorganic phosphate.
- Upon binding of a sugar, however, the jaws of the two lobes of the enzyme collapse to form a properly oriented active site, with amino acids positioned to facilitate phosphoryl transfer.
- This allows the enzyme to discriminate against water by a factor of 106.

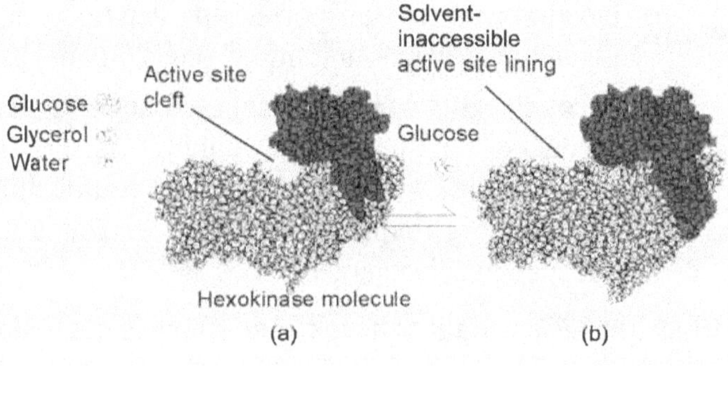

Other examples of isoenzymes are malate dehydrogenase, isocitrate dehydrogenase, glucose 6-phosphate dehydrogenase, aspartate aminotransferase, creatine kinase, phosphatases, esterases, transaminases, transphosphorylases and proteolytic enzymes.

Isozymes and Allozymes as Molecular Markers

Population genetics is essentially a study of the causes and effects of genetic variation within and between populations, and in the past isozymes have been amongst the most widely used molecular markers for this purpose. Although they have now been largely superseded by more informative DNA-based approaches (such as direct DNA sequencing, single-nucleotide polymorphisms and microsatellites), they are still amongst the quickest and the cheapest marker systems to develop, and remain (as of 2005) an excellent choice for projects that only need to identify low levels of genetic variation, e.g. quantifying mating systems.

MULTIENZYME COMPLEX SYSTEM

These are not single independent enzymes which catalyse a single reaction but are aggregates of several enzymes catalysing a sequence of many steps.

Pyruvate Dehydrogenase

The pyruvate dehydrogenase complex consists of multiple copies of each of the three enzymes, which are

i. pyruvate dehydrogenase (E1),
ii. dihydrolipoyl transacetylase (E2), and
iii. dihydrolipoyl dehydrogenase (E3).

The number of copies of each subunit and therefore the size of the complex varies from one organism to another (Table 4.2). The pyruvate dehydrogenase complex isolated from *E. coli* is about 45 nm in diameter.

Each component of this complex enzyme is so arranged as to provide efficient coupling of the individual reactions catalysed by these enzymes. In other words, the product of the first enzyme becomes the substrate of the second and so on.

Table 4.2 Subunits of pyruvate dehydrogenase complex

Enzyme	Coenzyme	Mol. wt. of subunit	No. of subunits per complex
Pyruvate dehydrogenase	TPP	96,000	24
Dihydrolipoyl transacetylase	Lipoate, CoA	65,000 70,000	24
Dihydrolipoyl dehydrogenase	FAD, NAD	56,000	12

The overall reaction catalysed by pyruvate dehydrogenase complex is oxidative decarboxylation of pyruvate to acetyl-CoA, which is an irreversible reaction.

$$\text{Pyruvate} + \text{CoASH} + \text{NAD}^+ \rightarrow \text{Acetyl-CoA} + \text{CO}_2 + \text{NADH}$$

E. coli enzyme consists of some 60 polypeptide chains and has a molecular weight of about 4,60,000. Three separate catalytic activities are present and are as follows:

1. Pyruvate dehydrogenase (E1), a TPP-requiring (TPP—thiamine pyrophosphate) enzyme, decarboxylates the pyruvate with an intermediate formation of hydroxyethyl-TPP.
2. The second step is also carried out by pyruvate dehydrogenase (E1) itself. During the reaction, acetyl

group and two electrons are transferred from TPP to the oxidized form of lipoyllysyl group of the core enzyme, dihydrolipoyl transacetylase (E2). This results in the formation of acetyl dihydrolipoamide.

3. E2 catalyses the third step of the reaction. During this step, acetyl group is transferred to CoA resulting in the formation of acetyl-CoA and dihydrolipoamide-E2.

   The fourth and fifth steps are carried out for the regeneration of disulphide form of the lipoyl group of E2 to prepare the enzyme complex for another round of oxidation.

4. The third enzyme of this complex, dihydrolipoyl dehydrogenase (E3) catalyses the transfer of two hydrogen atoms from the dihydrolipoamide-E2 to the FAD prosthetic group of E3 which results in the oxidation of dihydrolipoamide and the reduction of FAD to $FADH_2$.
5. In this step, reduced E3 is re-oxidized by NAD^+. Thus the E2 enzyme's active sulphydryl groups are re-oxidized by the E3 enzyme-bound FAD, which is thereby reduced to $FADH_2$. $FADH_2$ is then re-oxidized to FAD by NAD^+, producing NADH.

The enzyme complex is about 300 Å in diameter and its features have been observed by electron microscopy. It has a polyhedral structure, in which each of the subunits appears approximately spherical. The complex is held together by noncovalent interactions and they may easily undergo dissociation at alkaline pH. The subunits of E1 protein can be separated from those of E2 and E3 proteins at neutral pH and high urea concentration, the E2 and E3 proteins can be separated from each other. If the various subunits are mixed together at neutral pH in the absence of urea, the multienzyme complex will spontaneously reform, but E1 and E3 subunits will not re-associate unless E2 is present. It appears that 24 subunits of E2 form a core to the complex, with a symmetrical arrangement of E1 and E3 subunits around this core; along each of the 12 edges of a cube is a molecule of E1, probably consisting of two subunits, and on each of the six faces of the cube is a molecule of E3, again probably a dimer. It is possible that the flexible side chain of each lipoamide cofactor enables the lipoyl head to make contact with the active groups on adjacent enzymes and thus link the various processes taking place.

Fatty Acid Synthase

It is a multienzyme complex. It is an ellipsoid dimer of two identical polypeptide monomers. It is made up of six enzymes and one acyl carrier protein (ACP). The acyl carrier protein contains 4´-phosphopantetheine as its prosthetic group which has an –SH group.

The cysteine –SH group of one monomer lies close to the phosphopantetheine SH (Pan–SH) and vice versa.

Reactions of fatty acid synthesis are as follows:

1. Acetyl-CoA is first carboxylated to malonyl-CoA by acetyl-CoA carboxylase with the help of ATP. Acetyl-CoA carboxylase has a requirement for biotin and it is a multienzyme protein.

   This reaction takes place in two steps:

   i. Carboxylation of biotin
   ii. Transfer of the carboxyl to acetyl-CoA to form malonyl-CoA

   Acetyl-CoA carboxylase is a rate-limiting enzyme, it is an allosteric enzyme with citrate as its activator and palmityl-CoA as its inhibitor. Once malonyl-CoA is synthesized, the remaining reactions are catalysed by fatty acid synthetase as given below:

$$\text{Acetyl-CoA} \xrightarrow{\text{Acetyl-CoA carboxylase}} \text{Malonyl-CoA}$$

2. Acetyl-CoA combines with cysteine's –SH group catalysed by acetyl transacylase.
3. Malonyl-CoA combines with adjacent –SH on the 4′ phosphopantetheine of ACP of the other monomer catalysed by malonyl transacylase. There is a formation of acyl (acetyl) malonyl enzyme.
4. The acetyl group attacks the methylene group of malonyl residue catalysed by 3-ketoacyl synthase to form 3-ketoacyl enzyme.
5. 3-ketoacyl enzyme is reduced by 3-ketoacyl reductase to form a hydroxyacyl enzyme with the help of reduced NADP.
6. 3-hydroxyacyl enzyme is dehydrated by the hydratase enzyme to form 2,3-unsaturated acyl enzyme.
7. This is again reacted by enoyl reductase to form acyl enzyme with the help of reduced NADP.
8. This acyl enzyme is acted upon by thioesterase with the addition of water to form palmitate after recycling seven times through steps 1 to 4.

$$\text{Acetyl-CoA} + 7\text{Malonyl-CoA} + 14\text{NADPH} + 14\text{H}^+ \rightarrow \text{Palmitate} + 7\text{CO}_2 + 6\text{H}_2\text{O} + 8\text{CoASH} + 14\text{NADP}^+$$

Tryptophan Synthetase

Tyrptophan synthetase of *E. coli* is an isoenzyme which contains two different functional subunits. The enzyme catalyses the reaction:

$$\text{Indole 3-glycerol phosphate} + \text{L-serine} \xrightarrow{\text{Pyridoxal phosphate}} \text{L-tryptophan} + \text{Glyceraldehyde 3-phosphate}$$

It can be dissociated into two α subunits, each with a molecular weight of 29,000 and a β2 subunit of molecular weight 90,000. The β2 subunit further dissociates in the presence of 4M urea to give two β subunits, each of which has a binding site for the coenzyme pyridoxal phosphate.

The isolated α subunits will catalyse the following reaction.

Indole 3-glycerol phosphate □ Indole + Glyceraldehyde 3-phosphate

The isolated β2 subunit also has catalytic activity, and catalyses the following reaction.

$$\text{Indole} + \text{L-serine} \xrightarrow{\text{Pyridoxal phosphate}} \text{L tryptophan}$$

Therefore, the different subunits of tryptophan synthetase can catalyse two separate halves of the overall reaction. However, the rates of these partial reactions are 5% less than the rate of the reaction catalysed by the intact α2β2 enzyme. And indole is not released from the intact enzyme. Therefore, it is apparent that the oligomer has a degree of organization not possessed by the isolated subunits. The intermediate compound, indole, is passed directly from the active site of the α subunit to that of a β subunit, which is presumably in close proximity to increase the efficiency of the overall reaction.

Multienzyme complexes are often allosteric, enabling their activities to be regulated by feedback inhibition.

REVIEW QUESTIONS

1. Define isoenzyme.
2. Discuss the nomenclature of isoenzymes.
3. Give an account of LDH and hexokinase as isoenzymes.
4. Discuss how isozymes and allozymes can be used as molecular markers.
5. What are multienzymes? Describe with a suitable example.

5

COENZYMES

Coenzymes are small, organic, non-protein molecules that carry chemical groups between enzymes. These molecules are not bound tightly by enzymes and are released as a normal part of the catalytic cycle. Coenzyme molecules are often vitamins or are made from vitamins.

The oxidized form of coenzyme I is the cationic form of the molecule, in which the positive charge on the nitrogen atom may be regarded as balanced by some anion, e.g. Cl^-, in the solution. At pH 7.5 the acidic groups of the two phosphate residues in coenzyme I are in the ionized state, giving a net single negative charge on the molecule, which may be regarded as a dipolar ion with an additional negative charge; in coenzyme II these two groups and also the two acidic groups of the extra phosphate residue are ionized, giving a net charge of –3. The D-ribose residues are linked to the adenine and nicotinamide nitrogen atoms respectively by β-glycosidic links. Both coenzymes give normal adenylic acid on hydrolysis by nucleotide pyrophosphatase and therefore must contain β-riboside-linked adenine.

SIGNIFICANCE OF NUCLEOTIDE STRUCTURE IN COENZYMES

A striking feature of the structure of coenzymes is that the majority are either actual nucleotides or have some structural analogy with nucleotides. Many have a nitrogenous base at one end of the molecule and a phosphate group at the other, with or without a carbohydrate between, or have two such structures joined in the form of a dinucleotide. This is true of the phosphate carriers shown in Figure 5.1, which are actual

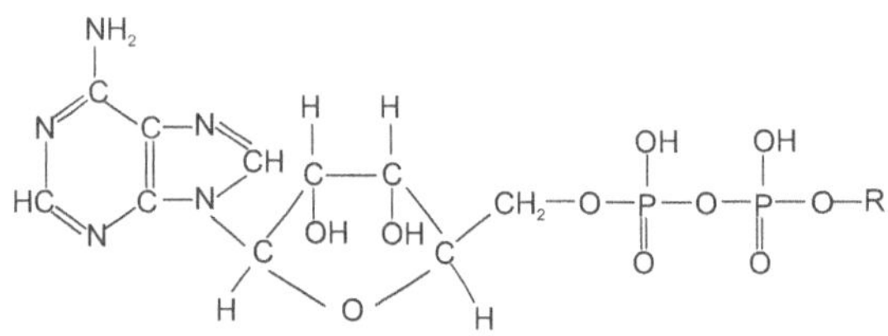

Adenosine di(tri)phosphate

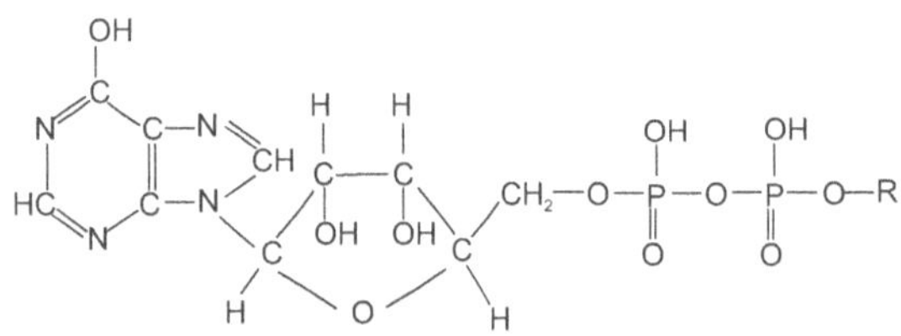

Inosine di(tri)phosphate

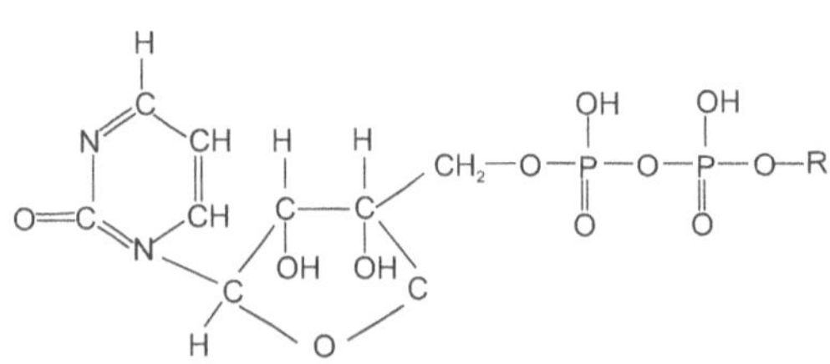

Cytidine di(tri)phosphate

Figure 5.1 Structure of nucleoside di- and triphosphates

nucleotide phosphates, of coenzymes I and II, which are true dinucleotides but contain nicotinamide instead of a purine or pyrimidine and also of the flavin nucleotides, in which one ribose is replaced by ribitol and the base is dimethylisoalloxazine.

Further, one-half of the coenzyme A molecule is a nucleotide, although the functional group is not, and pyrophosphothiamine may be regarded as a pyrimidine nucleotide phosphate in which the sugar is replaced by a thiazolinium ring, while in phosphopyridoxal the phosphate is directly attached to the base.

Five of these coenzymes A, I, II, III and FAD contain in addition to the relative moiety, an adenylic acid residue. It may not be without significance that three of the hydrogen carriers contain as part of their structure, a grouping which when free is associated with phosphate transport, in view of the fact that the oxidations in which these carriers take part are coupled with phosphorylation *in vivo*. Moreover, coenzyme I can itself accept a phosphate residue from ATP, becoming coenzyme II. No doubt the significance of these interrelationships will become clearer as a result of future research.

MODE OF REDUCTION

The coenzymes can be readily and reversibly reduced, either by chemical reducing agents such as dithionite or by those dehydrogenases for which they are specific; the reduced form is the same in either case. In the reduction, two equivalents of hydrogen per molecule are required. Warburg and Christian showed that it is the nicotinamide ring which is involved in the reduction.

When the coenzymes are reduced, the ultraviolet absorption spectrum undergoes a change; the oxidized form shows only a band at 260 nm, due to the purine and pyrimidine rings, but the reduced form shows in addition a band at 340 nm.

This band is due to the quinonoid bond structure of the reduced nicotinamide ring and is shown also by a number of simple derivatives of nicotinamide (e.g. nicotinamide methiodide) in the reduced state. Nucleotides which do not contain nicotinamide do not give this band. Pullman, SanPietro and Colowick made use of the fact that nicotinamide methiodide can be oxidized by alkaline ferricyanide to a mixture of the 2- and 6-pyridones (Figure 5.2); thus the two pyridones can be separated. If the nicotinamide had contained a deuterium atom attached to the 2-carbon atom, the 6-pyridone would contain a deuterium atom, while the 2-pyridone would contain none, since it would have been removed in the oxidation. Conversely, if the deuterium atom had been attached to the 6-carbon atom, only the 2-pyridone would contain deuterium.

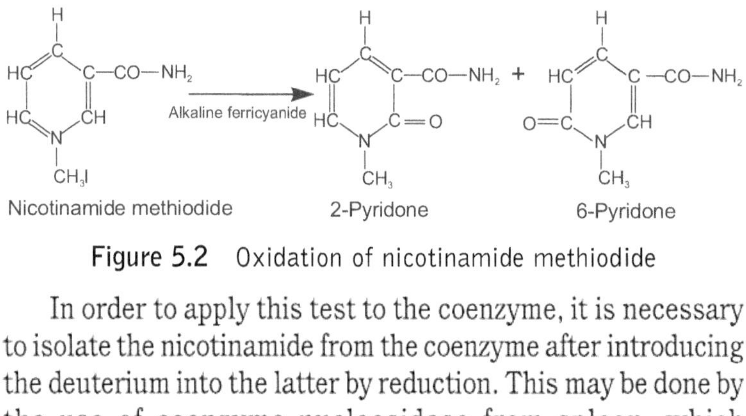

Figure 5.2 Oxidation of nicotinamide methiodide

In order to apply this test to the coenzyme, it is necessary to isolate the nicotinamide from the coenzyme after introducing the deuterium into the latter by reduction. This may be done by the use of coenzyme nucleosidase from spleen, which hydrolyses the nicotinamide–riboside bond. However, this enzyme does not act on the reduced form of the coenzyme; consequently, this must first be reoxidized. If both the original reduction and the reoxidation are carried out enzymatically, the deuterium will all be removed again; it is therefore necessary either to reduce chemically and oxidize enzymatically, or to reduce enzymatically and oxidize chemically.

Final proof that a 1, 4-reduction of the nicotinamide ring of the coenzyme, and not a 1,2 or a 1,6-reduction, takes place in dehydrogenase reactions was given by Loewus, Vennesland and Harris. By chemical synthesis they prepared three nicotinamides, each containing a deuterium ion attached to carbon atoms 2, 4 and 6 respectively. These three labelled nicotinamides were separately introduced into coenzyme I by an enzymic exchange using coenzyme nucleosidase, giving preparations of oxidized coenzyme labelled in the three different positions. These were then reduced by dithionite in H_2O, each giving a mixture of the two isotopic stereoisomers. When these three preparations of reduced coenzyme I were oxidized separately with pyruvate and lactate dehydrogenase, and the lactate formed was isolated and analysed for deuterium, it was found that no deuterium at all was transferred from either the 2nd or the 6th position, whereas an amount approximating to the theoretical value was transferred from the 4th position. This shows convincingly that the 2nd and the 6th positions are not concerned in the dehydrogenase reactions and that these reactions are exclusively, 1, 4-reductions. The reduction is represented in Figure 5.3.

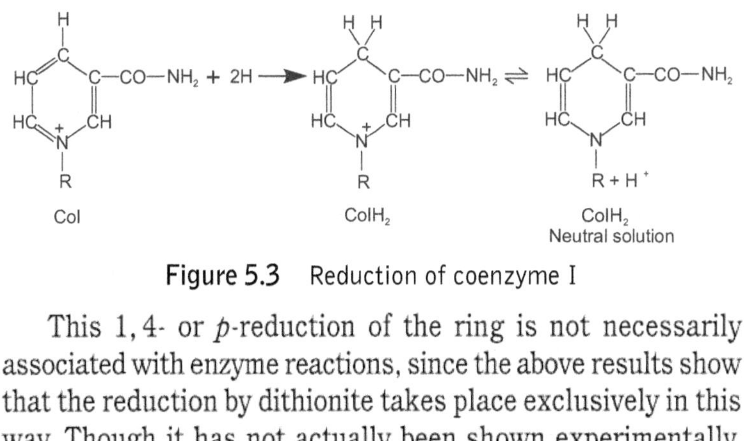

Figure 5.3 Reduction of coenzyme I

This 1, 4- or *p*-reduction of the ring is not necessarily associated with enzyme reactions, since the above results show that the reduction by dithionite takes place exclusively in this way. Though it has not actually been shown experimentally, there is little doubt that coenzymes II and III are reduced in a similar way.

PROPERTIES AND DISTRIBUTION

Coenzyme I is reduced by many different substrates in the presence of the appropriate dehydrogenases. Coenzyme II seems to be reduced by fewer enzymes. The enzymes which reduce coenzyme I do not usually reduce coenzyme II and vice versa, though there are exceptions.

The coenzymes are also reduced by certain chemical reducers, of which the best known is dithionite. The reduced forms are not oxidized at significant rates by O_2, or by dyes such as methylene blue, or by cytochrome *c*. They are oxidized by many dehydrogenases acting in reverse in the catalysis of anaerobic fermentations, by dyes in the presence of diaphorases, by cytochrome *c* or b_5 in the presence of the respective cytochrome reductases, by cystine or glutathione in the presence of their reductases, and non-enzymatically by chemical oxidizers like ferricyanide and (more slowly) by quinones, the latter reaction being greatly accelerated by specific enzymes. The stability of the oxidized and reduced forms vary with pH in opposite directions; the reduced form is extremely unstable in acid but relatively stable in alkaline solutions, while the oxidized form is fairly stable in acid but rather less stable in alkali.

FUNCTION OF COENZYMES

The main function of coenzymes is undoubtedly to act as hydrogen carriers in anaerobic and aerobic oxidations and fermentations. The reduced forms of coenzymes I and II appear to act as cofactors without undergoing oxidation in the hydroxylation of aromatic and steroid rings.

The functional role of coenzymes is to act as transporters of chemical groups from one reactant to another. The chemical groups carried can be as simple as the hydride ion (H^++2e^-) carried by NAD or the mole of hydrogen carried by FAD; or they can be even more complex than the amine ($-NH_2$) carried by pyridoxal phosphate.

Since coenzymes are chemically changed as a consequence of enzyme action, it is often useful to consider enzymes to be a special class of substrates, or second substrates, which are common to many different holoenzymes. In all cases, the coenzymes donate the carried chemical group to an acceptor molecule and are thus regenerated to their original form. This regeneration of coenzyme and holoenzyme fulfils the definition of an enzyme as a chemical catalyst, since (unlike the usual substrates, which are used up during the course of a reaction) coenzymes are generally regenerated.

ROLE OF COENZYMES IN CLINICAL ENZYMOLOGY

Coenzymes participate in many of the enzyme analyses performed in the clinical laboratory. Supplementation of assay systems with optimal levels of coenzymes has recently been recommended as part of efforts to achieve interlaboratory standardization of enzyme measurements. Aspartate aminotransferase and alanine aminotransferase require pyridoxal phosphate for expression of enzyme activity. The role of this coenzyme in enzymatic transamination and the effects of its supplementation on the clinical estimation of these two enzymes is reviewed.

Other coenzymes are flavins, coenzymes for glutathione reductase, glucose oxidase, cholesterol oxidase and diaphorase, as well as thiamine pyrophosphate, coenzyme for transketolase.

Catalase and peroxidase are used as examples of haemoproteins utilized in clinical measurements. Two peptide coenzymes, colipase and glutathione are also considered. Measurement of apoenzyme stimulation upon supplementation with specific coenzymes is a valuable technique for quantitative coenzyme measurements or assessment of vitamin nutritional status.

COENZYME I (NAD^+)

The coenzyme NAD^+ was first discovered by the British biochemist Arthur Harden and William Youndin in 1906. The structure of coenzyme I is given in Figure 5.4. It consists of two nucleotides joined through their phosphate groups: one nucleotide containing an adenine base, and the other containing nicotinamide. NAD^+ is involved in redox reactions. The coenzyme is therefore found in two forms in cells—**NAD^+** is an oxidizing agent that accepts electrons from other molecules and becomes reduced. This reaction forms **NADH**, which can then be used as a reducing agent to donate electrons.

$$RH_2 + NAD^+ \rightarrow NADH + H^+ + R$$

These electron transfer reactions are the main function of NAD^+. Due to the importance of these functions, the enzymes involved in NAD^+ metabolism are targets for drug discovery. Some NAD^+ is converted into nicotinamide adenine dinucleotide phosphate ($NADP^+$); the chemistry of this related coenzyme is similar to that of NAD^+, but it has different roles in metabolism.

In appearance, all forms of this coenzyme are white amorphous powders that are hygroscopic and highly water-soluble. Both NAD^+ and NADH absorb strongly in the ultraviolet due to the adenine base. NAD^+ and NADH differ in their fluorescence. NAD^+ is synthesized through two metabolic pathways. It is produced either in a de novo pathway from amino acids, or in salvage pathways by recycling preformed components such as nicotinamide back to NAD^+. It acts as a coenzyme in redox reactions, as a donor of ADP-ribose groups in ADP-ribosylation reactions, as a precursor of the second messenger molecule cyclic ADP-ribose as well as acting as a substrate for bacterial DNA ligases and a group of enzymes called sirtuins that use NAD^+ to remove acetyl groups from proteins. The coenzyme NAD^+ is not itself currently used as a treatment for any disease. However, it is potentially useful in the therapy of neurodegenerative diseases such as Alzheimer's and Parkinson's disease.

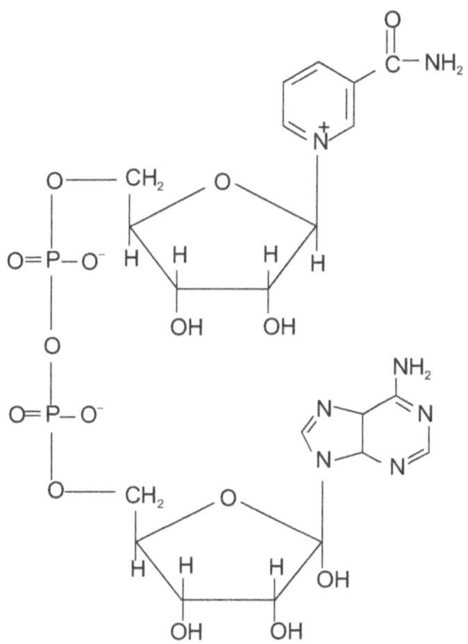

Figure 5.4 Structure of coenzyme I (NAD^+)

COENZYME II ($NADP^+$)

It is an important coenzyme, functioning as a hydrogen carrier in a wide range of redox reactions; the H is carried on the nicotinamide residue. The oxidized form of the coenzyme is

NADP+, the reduced form is NADPH. It is an important component of the enzymatic systems concerned with biological oxidation–reduction systems. It is also known as NAD, triphosphopyridine nucleotide (TPN), coenzyme II and codehydrogenase II.

The compound is similar in structure (Figure 5.5) and function to nicotinamide adenine dinucleotide (NAD). It differs structurally from NAD in having an additional phosphoric acid group esterified at the 2′ position of the ribose moiety of the adenylic acid portion.

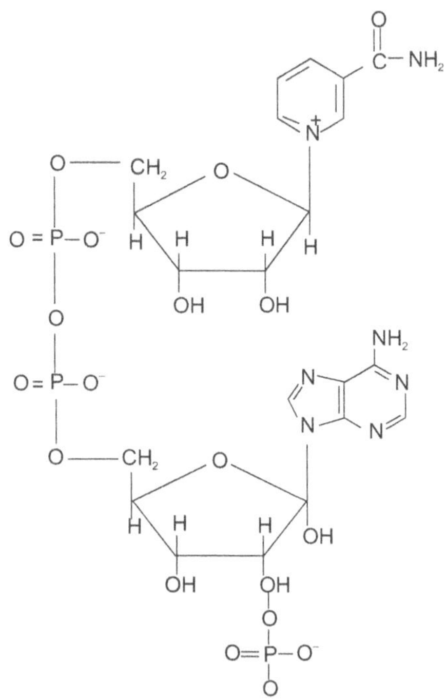

Figure 5.5 Structure of coenzyme II (NADP+)

In biological oxidation–reduction reactions, the NADP molecule becomes alternately reduced to its hydrogenated form (NADPH) and reoxidized to its initial state. It is used in anabolic reactions such as lipid and nucleic acid synthesis, which require NADPH as a reducing agent. $NADP^+$ differs from NAD^+ by the presence of an additional phosphate group in $NADP^+$ on the 2´ position of the ribose ring that carries the adenine moiety. In chloroplasts, NADP is reduced by ferredoxin $NADP^+$ reductase in the last step of the electron chain of the light reactions of photosynthesis. The NADPH produced is then used as reducing power for the biosynthetic reactions in the Calvin cycle of photosynthesis. It is the source of reducing equivalents for cytochrome 450 hydroxylation of aromatic compounds, steroids, alcohols and drugs.

COENZYME A

This important coenzyme (CoA, CoASH or HSCoA) was originally discovered as a coenzyme for acetylations in liver and microorganisms by Lipmann in 1947. An enzymic pathway for the biosynthesis of coenzyme A has been completely elucidated. The structure is shown in Figure 5.6 , and will be seen to be 3´-phospho ADP pantoylalanylcystamine. It thus has some similarity to an adenine dinucleotide in which the second nucleoside is replaced by pantetheine (i.e., pantothenlyl-cystamine). Like coenzyme II, it has a third phosphate group on the ribose residue of the adenylic acid, but in this case it is in the 3rd position instead of the 2nd position.

Acyl groups are added to the thiol group of coenzyme A by a number of enzymes, which fall into two categories: certain transacylases transfer acyl groups to coenzyme A from other thiol groups or from β-ketoacids; These reactions are reversible, because the two acylated compounds involved have roughly equivalent free energies of hydrolysis; a number of synthetases also acylate coenzyme A, using free acids and making use of ATP energy to form the thioester bond. The acyl group may be removed from acyl-CoA compounds by the

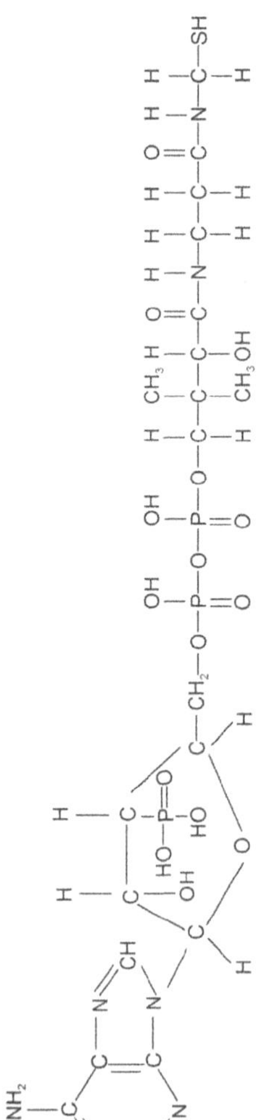

Figure 5.6 Structure of coenzyme A

reversal of any of the enzyme reactions just mentioned, or by a number of transacylases which transfer the acyl group to other molecules.

The molecules which are formed by these transacylases are "energy-poor", and the reactions are therefore effectively irreversible. In four cases, the acyl group may also be removed by direct hydrolysis.

Coenzyme A is chemically a thiol, it can react with carboxylic acids to form thioesters, thus functioning as an acyl group carrier. It assists in transferring fatty acids from the cytoplasm to the mitochondria. A molecule of coenzyme A carrying an acetyl group is also referred to as acetyl-CoA. When it is not attached to an acyl group, it is usually referred to as "CoASH or HSCoA".

Coenzyme A is synthesized in a five-step process from pantothenate:

1. Pantothenate (Vit B_5) is phosphorylated to 4´-phosphopantothenate by the enzyme pantothenate kinase.
2. A cysteine is added to 4´-phosphopantothenate by the enzyme phosphopantothenoyl cysteine synthetase to form 4´-phospho-N-pantothenoyl cysteine (PPC).
3. PPC is decarboxylated to 4´-phosphopantetheine by phosphopantothenoyl cysteine decarboxylase (CoAC).
4. 4´-phosphopantetheine is adenylylated to form dephospho-CoA by the enzyme phosphopantetheine adenylyl transferase (CoAD).
5. Dephospho-CoA is phosphorylated using ATP to coenzyme A by the enzyme dephosphocoenzyme A kinase (CoAE).

Acetyl-CoA

Acetyl-CoA is the thioester between coenzyme A (a thiol) and acetic acid (an acyl group carrier) (Figure 5.7). Two acetyl-CoA can be condensed to create acetoacetyl-CoA, the first step in the HMG-CoA/mevalonic acid pathway leading to synthesis of isoprenoids. In animals, HMG-CoA is a vital precursor to cholesterol and ketone synthesis. Its main function is to convey the carbon atoms within the acetyl group to the citric acid cycle to be oxidized for energy production.

Acetyl-CoA is produced during the second step of aerobic cellular respiration, pyruvate decarboxylation, which occurs in the matrix of the mitochondria. Acetyl-CoA then enters the citric acid cycle. The oxidative conversion of pyruvate into acetyl-CoA is referred to as the pyruvate dehydrogenase reaction. It is catalysed by the pyruvate dehydrogenase complex. In animals, acetyl-CoA is central to the balance between carbohydrate metabolism and fat metabolism. In normal circumstances, acetyl-CoA from fatty acid metabolism feeds into the citric acid cycle, contributing to the cell's energy supply.

Acetyl-CoA is also the source of the acetyl group incorporated onto certain lysine residues of histidine and non-histone proteins in the post-translational modification acetylation, a reaction catalysed by acetyltransferases. In plants and animals, cytosolic acetyl-CoA is synthesized by ATP citrate lyase.

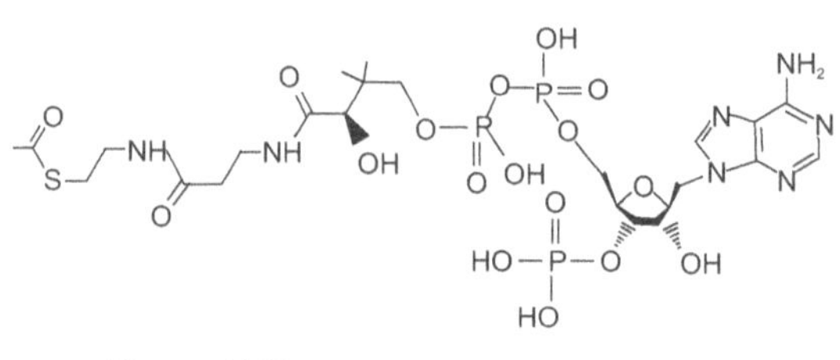

Figure 5.7 Structure of acetyl-CoA

$$\text{Citrate} + \text{CoA} + \text{ATP} \xrightarrow{\text{ATP citrate lyase}} \text{Acetyl-CoA} + \text{Oxaloacetate} + \text{ADP} + \text{Pi}$$

When glucose is abundant in the blood of animals, it is converted via glycolysis in the cytosol to pyruvate, and then to acetyl-CoA in the mitochondria. The excess of acetyl-CoA results in the production of excess citrate, which is exported into the cytosol to give rise to cytosolic acetyl-CoA. Acetyl-CoA can be carboxylated in the cytosol by acetyl-CoA carboxylase, giving rise to malonyl-CoA, a substrate required for synthesis of flavonoids and related polyketides, for elongation of fatty acids to produce waxes, cuticle and seed oils in members of the Brassica family, and for malonation of proteins and other phytochemicals. In plants, these include sesquiterpenes, brassinosteroids (hormones) and membrane sterols.

Acyl-CoA

Acyl-CoA is a coenzyme involved in the metabolism of fatty acids. It is formed when coenzyme A attaches to the end of a long-chain fatty acid (Figure 5.8) inside living cells. The CoA is then removed from the chain, carrying two carbons from the chain with it, forming acetyl-CoA. This then enters the citric acid cycle, eventually forming adenosine triphosphate.

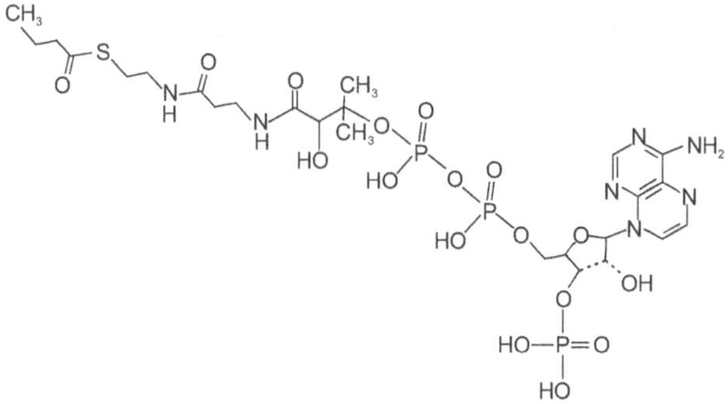

Figure 5.8 Structure of acyl-CoA

To be oxidatively degraded, a fatty acid must first be activated in a two-step reaction catalysed by acyl-CoA synthetase. First, the fatty acid displaces the diphosphate group of ATP, then coenzyme A (HSCoA) displaces the AMP group to form an acyl-CoA. The second step, transfer of the acyl group to CoA, conserves free energy in the formation of a thioester bond. Consequently, the overall reaction

$$\text{Fatty acid} + \text{CoA} + \text{ATP} \xrightarrow{\text{Acyl-CoA synthetase}} \text{Acyl-CoA} + \text{AMP} + \text{PPi}$$

has a free energy change near zero. Subsequent hydrolysis of the product PPi (by the enzyme inorganic pyrophosphate) is highly exergonic, and this reaction makes the formation of acyl-CoA spontaneous and irreversible. Fatty acids are activated in the cytosol, but oxidation occurs in the mitochondria. Because there is no transport protein for CoA adducts, acyl group must enter the mitochondria via a shuttle system involving the small molecule carnitine.

COENZYME M

Coenzyme M is required for methyl transfer reactions in the metabolism of methanogens. The coenzyme is an anion with formula $HSCH_2CH_2SO_3$. The cation is unimportant, but the sodium salt is most available. The coenzyme is the C1 carrier in methanogenesis. It is converted to methyl coenzyme M, the thioether $CH_3SCH_2CH_2SO_3^-$, in the penultimate step to methane formation. Coenzyme M reacts with coenzyme B, 7-thioheptanoylthreoninephosphate, to give a heterodisulphide, releasing methane:

$$CH_3\text{-S-CoM} + \text{HS-CoB} \xrightarrow{\text{Methyl coenzyme M reductase}} CH_4 + \text{CoB-S-S-CoM}$$

This conversion is catalysed by the enzyme methyl coenzyme M reductase, which contains cofactor F_{430} as the prosthetic group.

COENZYME B

Coenzyme B (7-mercaptoheptanoyl threonine phosphate) is required for redox reactions in methanogens. The molecule contains a thiol, which is its principal site of reaction. Coenzyme B reacts with 2-methylthioethane sulphonate (methyl coenzyme M) to release methane in methanogenesis.

$$\underset{\text{Methyl coenzyme M}}{CH_3\text{-S-CoM}} + \underset{\text{Coenzyme B}}{\text{HS-CoB}} \xrightarrow[\text{reductase}]{\text{Methyl coenzyme M}} \underset{\text{Methane}}{CH_4} + \text{CoB-S-S-CoM}$$

This conversion is catalysed by the enzyme methyl coenzyme M reductase, which contains cofactor F_{430} as the prosthetic group. A related conversion that utilizes both HS-CoM is the reduction of fumarate to succinate, catalysed by fumarate reductase.

$$CH_3\text{-S-CoM} + \text{HS-CoB-}\underset{\text{Fumarate}}{O_2CCH = CHCO_2^-} \rightarrow \underset{\text{Succinate}}{O_2^- CCH_2^- CH_2CO_2^-} + \text{CoB-S-S-CoM}$$

COENZYME B_{12}

In 1926, two American physicians, George Minot and George William Murphy discovered that patients suffering from pernicious anaemia could be cured by feeding them about half a pound of liver a day. This anti-pernicious factor (AP factor) was later isolated in crystalline form in 1948 independently by E. Lester Smith in England and Edward Rickes in the United States. It was then named as vitamin B_{12} or cyanocobalamine. It is the last B vitamin to be isolated and is also known as factor X. The coenzyme form of this vitamin (deoxyadenosyl cobalamine or cobamide coenzyme) was first isolated by Barker of California. Coenzyme B_{12} has been called a **biologic Grignard reagent**.

Many compounds with vitamin B_{12} activity have been isolated from natural sources. Cyanocobalamine is the most common form and is sometimes also written as vitamin B_{12a}. In other forms, cyanide ion is replaced by other ions, e.g. hydroxyl ion in hydroxocobalamine (vitamin B_{12b}), nitrite ion in nitrocobalamine (vitamin B_{12c}). The latter two, B_{12b} and B_{12c} can be converted to vitamin B_{12a} by treatment with cyanide.

The structure of vitamin B_{12} coenzyme (5´-deoxyadenosyl cobalamine) is similar to that of cyanocobalamine except that here the CN group is replaced by adenosine and the linking with cobalt atom taking place at 5´ carbon atom of the ribose of adenosine. Vitamin B_{12} coenzyme is the only known example of a carbon metal bond in a biomolecule. Vitamin B_{12} is converted into coenzyme B_{12} by extracts from microorganisms supplemented with ATP.

Coenzyme B_{12} is associated with many biochemical reactions, which are as follows:

1. **1,2 shift of a hydrogen atom** Coenzyme B_{12} catalyses 1,2 shift of a hydrogen atom from one carbon of the substrate to the next with a concomitant 2,1 (reverse) shift of some other group, e.g. hydroxyl, alkyl, etc. Conversion of methylmalonyl-CoA to succinyl-CoA is an example of 1,2-shift.
2. **Carrier of a methyl group** Coenzyme B_{12} also serves as a carrier of a methyl group, obtained from N^5-methyltetrahydrofolate to the appropriate acceptor molecule. In the reaction, a methyl group occupies the 5-deoxyadenosyl coordination position of coenzyme B_{12}. Methylation of homocysteine to produce methionine is an example of such reaction.

3. **Isomerization of decarboxylic acids** Coenzyme B_{12} is associated with isomerization of dicarboxylic acids, e.g. glutamic acid into β-methylaspartic acid.
4. **Dismutation of vicinal diols** coenzyme B_{12} also catalyses dismutation of vicinal diols to the corresponding aldehydes, e.g. propane 1,2-diol into propionaldehyde.

COENZYME F_{420}

Coenzyme F_{420} (8-hydroxy 5-deazaflavin) is a flavin derivative. This coenzyme is involved in redox reactions in methanogens. The coenzyme is a substrate for 5, 10 methylene-tetrahydromethanopterin reductase, methylene-tetrahydromethanopterin dehydrogenase and coenzyme F_{420} hydrogenase. Coenzyme F_{420} hydsrogenase is an enzyme that catalyses the chemical reaction.

$$H_2 + \text{Coenzyme } F_{420} \rightarrow \text{Reduced coenzyme } F_{420}$$

Thus, the two substrates of this enzyme are H_2 and coenzyme F_{420} whereas its product is reduced coenzyme F_{420}. This enzyme belongs to the family of oxidoreductases, acting on hydrogen as donor with other, known acceptors. This enzyme participates in folate biosynthesis and has 3 cofactors: iron, nickel and deazaflavin.

ADENOSINE TRIPHOSPHATE

Adenosine 5´triphosphate (ATP) was discovered by Karl Lohmann in 1929 and was proposed to be the "molecular unit of currency" of intracellular energy transfer. ATP is a multifunctional nucleotide, made from adenosine diphosphate (ADP) or adenosine monophosphate (AMP). ATP is the main energy source for the majority of cellular functions such as photosynthesis, cellular respiration, motility, cell division and biosynthetic reactions such as the synthesis of macromolecules including DNA, RNA and proteins. ATP also plays a critical role in the transport of macromolecules across cell membranes, e.g. exocytosis and endocytosis.

The structure of this molecule (Figure 5.9) consists of a purine base (adenine) attached to the 1´carbon atom of a pentose sugar (ribose). Three phosphate groups are attached at the 5´ carbon atom of the pentose sugar. It is the addition and removal of these phosphate groups that interconvert ATP, ADP and AMP. ATP is highly soluble in water and rapidly hydrolysed at extreme pH. ATP can be produced by redox reactions using carbohydrates or lipids as energy source. The overall process of cellular respiration can produce 30 molecules of ATP from a single molecule of glucose.

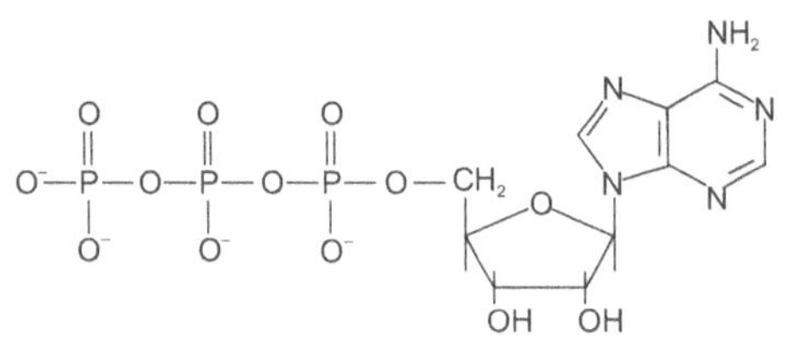

Figure 5.9 Structure of ATP

ATP can also be produced by a number of distinct cellular processes; the three main pathways used to generate energy in eukaryotic organisms are glycolysis, the citric acid cycle/ oxidative phosphorylation and beta oxidation. ATP is critically involved in maintaining cell structure by facilitating assembly and disassembly of elements of the cytoskeleton. Some proteins that bind ATP do so in a characteristic protein fold known as the Rossmann fold, which is a nucleotide binding structural domain that can also bind the cofactor NAD. Enzyme inhibitors of ATP-dependent enzymes such as kinases are needed to examine the binding sites and transition states involved in ATP-dependent reactions. ATP analogs are also used in X-ray crystallography to determine a protein structure in complex with ATP, often together with other substrates.

FLAVIN ADENINE DINUCLEOTIDE (FAD)

FAD is derived from riboflavin (vitamin B_2). Many oxidoreductases, called flavoenzymes or flavoproteins, require FAD as a prosthetic group which functions in electron transfers. It is a redox cofactor involved in metabolism. The structure of FAD is given in Figure 5.10.

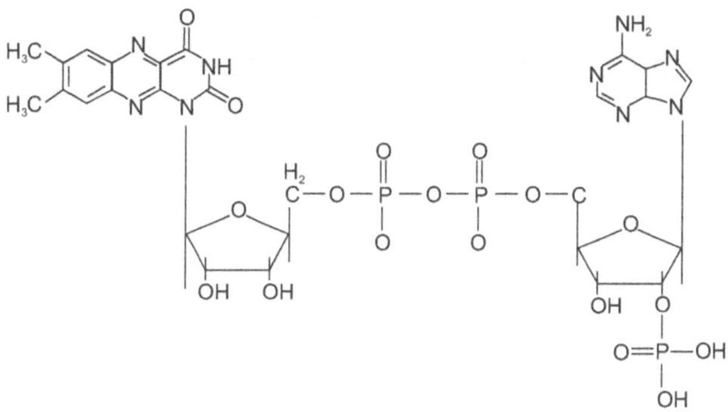

Figure 5.10 Structure of FAD

FAD can exist in two different redox states. FAD can be reduced to the $FADH_2$, whereby it accepts two hydrogen atoms: $FADH_2$ is an energy-carrying molecule, and the reduced coenzyme can be used as substrate for oxidative phosphorylation in the mitochondria. $FADH_2$ is reoxidized to FAD, which produces enough of a proton gradient across the inner mitochondrial membrane for the enzyme, ATP synthase, to produce 1.5 equivalents of the high-energy compound ATP. The primary sources of reduced FAD in eukaryotic metabolism are the citric acid cycle and the beta oxidation reaction pathways. In the citric acid cycle, FAD is a prosthetic group in the enzyme succinate dehydrogenase that oxidizes succinate to fumarate, whereas in beta oxidation it serves as a coenzyme in the reaction of acyl-CoA dehydrogenase.

FLAVIN MONONUCLEOTIDE

Flavin mononucleotide (riboflavin 5´-phosphate) (Figure 5.11) is a coenzyme synthesized from riboflavin by the enzyme riboflavin kinase and functions as prosthetic group of various oxidoreductases including NADH dehydrogenase. FMN acts as a component of complex I of the electron transport chain. During the catalytic cycle, the reversible interconversion of the oxidized form (FMN), semiquinone (FMNH) and the reduced form ($FMNH_2$) occur. FMN is a stronger oxidizing agent than NAD and is involved in electron transfers. It is the principal form in which riboflavin is found in cells and tissues.

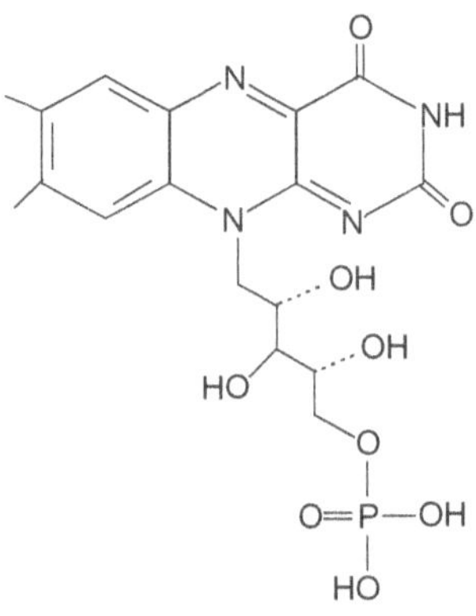

Figure 5.11 Structure of FMN

LIPOATE

Lipoate is a widely distributed hydrogen carrier which could replace the requirement for acetate in the growth of certain bacteria, which they called the acetate-replacing factor. The structures of the oxidized and reduced forms of lipoate are shown in Figure 5.12.

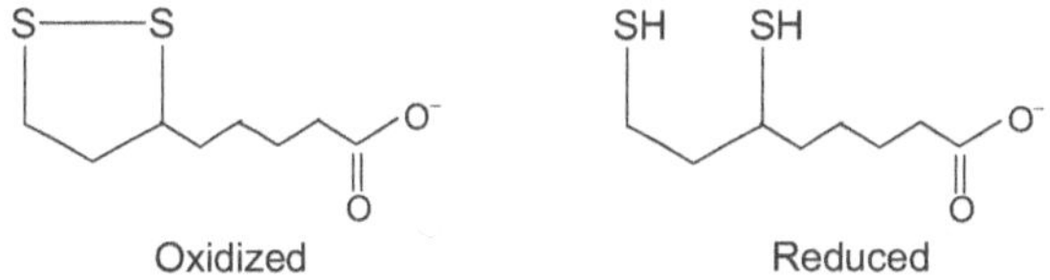

Figure 5.12 Oxidized and reduced forms of lipoate

The essential change is the reduction of a disulphide group to two thiol groups. It will be seen that one of the carbon atoms is asymmetric, so that two optical isomers exist. The natural isomer of the oxidized form is dextrorotatory, but gives rise on reduction to the laevorotatory reduced form. Only the natural isomers are active in the pyruvate oxidase system.

In the oxidized form, the substance is yellowish, having an absorption band at about 335 nm this is due to the five-membered ring of this form. Powerful reducing agents such as zinc in hydrochloric acid (not borohydride), open the ring by the reduction of the disulphide bond.

On irradiation of the oxidized form with long-wave ultraviolet light, it readily becomes reduced and this is regarded as being great importance in photosynthesis, in which it may act as the primary acceptor of hydrogen produced in the photolysis of water.

Apart from this, it is reduced by pyruvate and by α-ketoglutarate in the presence of their respective dehydrogenases and pyrophosphothiamine. In these cases, the reduction is accompanied by the transfer of an acyl group from the substrate to the lipoate, giving acetyl-hydrolipoate and succinyl-hydrolipoate respectively. The acetyl group has been shown to be attached to the thiol in the 6th position (Figure 5.13). The reduced lipoate after removal of any attached acyl group is oxidized by coenzyme in the presence of lipoate dehydrogenase. It may also be oxidized by mild chemical oxidizing agents such as iodine.

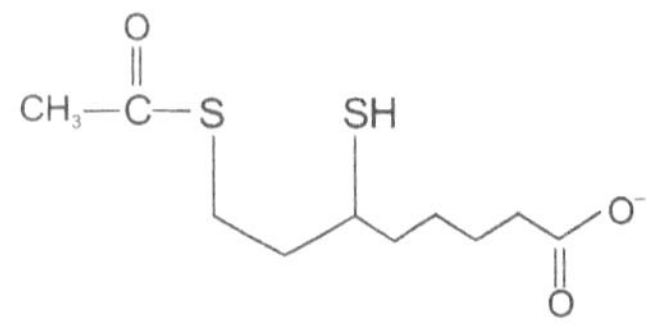

Figure 5.13 Lipoic acid functions in acyl transfers

PHOSPHOPYRIDOXAL

This forms the prosthetic group of a number of enzymes which catalyse a wide variety of reactions involving amino acids. In some of its reactions which are transaminations it acts as an amino-carrier; in the other cases it forms a reactive compound with the amino acid and thus may act as a "handle." The structure of phosphopyridoxal was finally established in 1952 by Baddiley and Mathas by total synthesis as shown in Figure 5.14.

Figure 5.14 Structures of phosphopyridoxal and phosphopyridoxamine

The mechanism of the action of phosphopyridoxal has been believed to act in all cases by forming an azomethine (Schiff's base) by combination of its aldehyde group with the amino group of the substrate. This can undergo tautomerism in the following way:

The fate of this substance depends on the nature of the enzyme protein to which the phosphopyridoxal is attached and

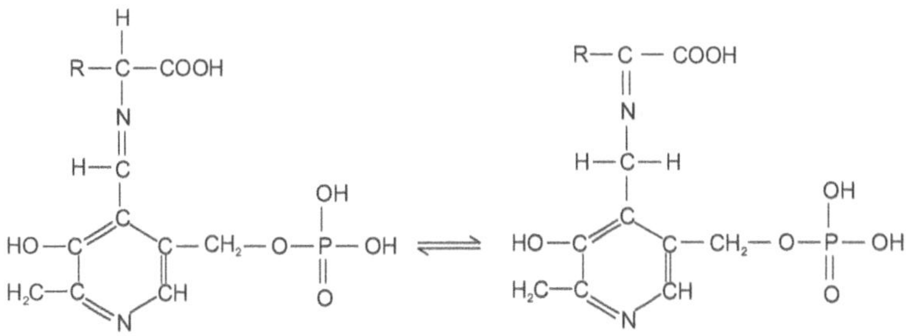

on the group R. In the amino acid decarboxylases, the carboxyl group becomes labile and is readily given off, leaving combined amine which is then readily liberated by hydrolysis. In yet other cases, water or some other molecule is eliminated from the substrate, leaving an unsaturated residue which is then hydrolysed off from the phosphopyridoxal, for example, with serine and threonine dehydrases and alliinase.

COENZYME Q

Q (also known as ubiquinone) is a benzoquinone, where Q refers to the quinone. It is a component of the electron transport chain and participates in aerobic cellular respiration, generating energy in the form of ATP. The benzoquinone portion of coenzyme Q10 is synthesized from tyrosine, whereas the isoprene side chain is synthesized from acetyl-CoA through the mevalonate pathway. The mevalonate pathway is also used for the first step of cholesterol biosynthesis.

Coenzyme Q (Figure 5.15) was first discovered by Fred L Crane in 1957. The various kinds of coenzyme Q can be distinguished by the number of isoprenoid side chains they have. The most common CoQ in human mitochondria is Q10. The 10 refers to the number of isoprene repeats. Coenzyme Q has the ability to transfer electrons and therefore act as an antioxidant. Coenzyme Q10 is a treatment for some of the very rare and serious mitochondrial disorders and other metabolic disorders, where patients are not capable of producing enough coenzyme Q10 because of their disorder. Coenzyme Q is beneficial in the treatment of congestive heart failure, cancer and blood pressure.

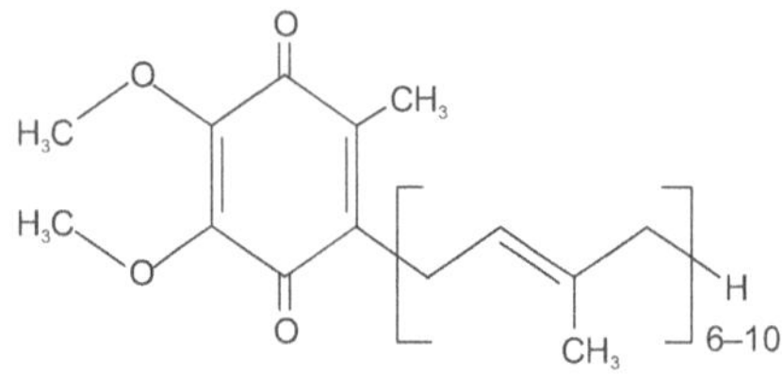

Figure 5.15 Structure of coenzyme Q

TETRAHYDROFOLIC ACID

Tetrahydrofolic acid (THF) (Figure 5.16) is a folic acid derivative. It is produced from dihydrofolic acid by dihydrofolate reductase. This reaction is inhibited by methotrexate. It is converted into 5,10 methylenetetrahydrofolate by serine hydroxymethyltransferase. It is a coenzyme in many reactions, especially in the metabolism of amino acids and nucleic acids. It acts as a donor of a group with one carbon atom. It gets this carbon atom by sequestering formaldehyde produced in other processes. A shortage in THF can cause anaemia. Tetrahydrofolic acid is reduced by the drug methotrexate, which is used to impair nucleotide synthesis. This is a potent drug used in chemotherapy and in treating rheumatism.

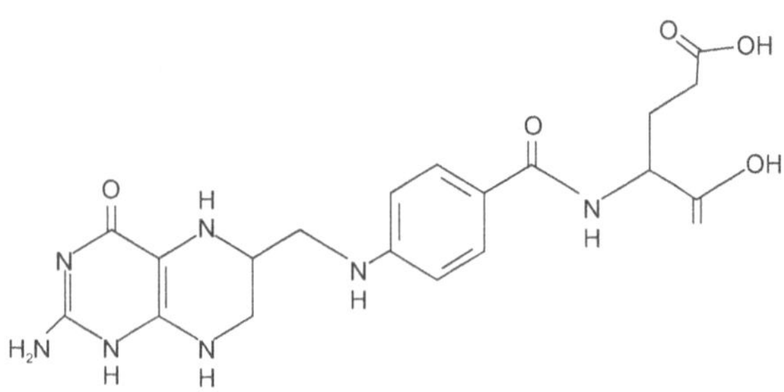

Figure 5.16 Structure of tetrahydrofolic acid

GLUTATHIONE

Glutathione is a sulphur-containing peptide, discovered in 1921 by Hopkins. The reduced form of glutathione (Figure 5.17) is a dipeptide of cysteine and glutamic acid. The structure was established as α-glutamyl L-cysteinylglycine by synthesis. The functional group in the molecule is the thiol group and it is customary to represent reduced glutathione.

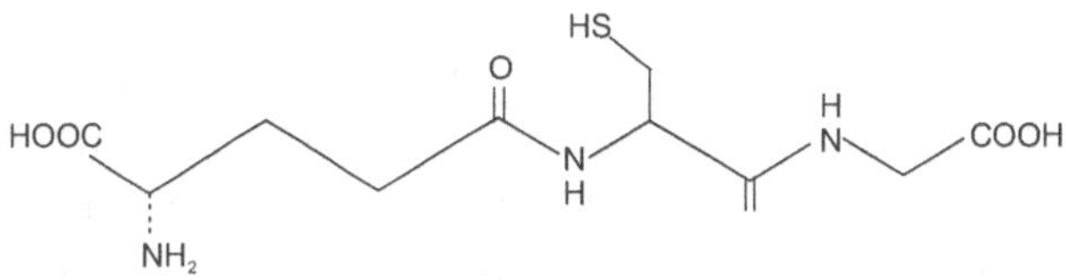

Figure 5.17 Structure of reduced glutathione

Reduced glutathione is oxidized to the disulphide by mild oxidizing agents, e.g. iodine or ferricyanide. It is also oxidized by molecular oxygen under suitable conditions in the presence of traces of catalytic metals and by cytochrome *c*. It is oxidized enzymatically by glutathione dehydrogenase in the presence of dehydroascorbate. Powerful reducing agents are needed to reduce the oxidized form back to the thiol form but enzymatically it can be reduced by coenzymes I and II in the presence of glutathione reductase. Since glutathione undergoes enzymic oxidation and reduction, it can act as a biological hydrogen carrier and it was the first such carrier to be discovered.

Glutathione can provide a path for the oxidation of the coenzymes through ascorbate and either ascorbate oxidase in plants or cytochrome oxidase in animals. Glutathione acts as a specific coenzyme for the glyoxalase system, for maleylacetoacetate isomerase and for formaldehyde dehydrogenase. The action of the glutathione in the glyoxalase system has been much studied, but the metabolic significance of this system is still completely unknown. Many enzymes are still enzymes, active only in the thiol state, and it has been suggested that a very important function of glutathione is to keep these enzymes in the reduced form.

ASCORBATE

Ascorbate exists in the oxidized and reduced forms. The name "ascorbic acid" is given to the reduced form only (Figure 5.18) and the oxidized form is known as "dehydroascorbic acid".

Ascorbate is an effective reducing agent, rapidly reducing many dyes and even reducing neutral silver nitrate to metallic silver. It is also oxidized by the usual oxidizing agents (e.g. iodine) and by oxygen in the presence of traces of catalytic metals. In plants a specific copper-containing ascorbate oxidase brings about this reaction; this enzyme is not present in animal tissues, where ascorbate may be oxidized through the cytochrome system.

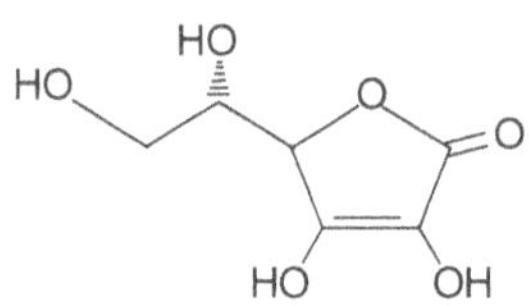

Figure 5.18 Structure of ascorbic acid

Both forms of ascorbic acid are lactones. The reduced form is acidic because of ionization of the ene-diol structure, while the oxidized form is neutral. The lactone ring is stable in ascorbic acid, but readily hydrolyses in dehydroascorbic acid to give an open-chain acid.

$$CH_2OH.CHOH.CHOH.CO.CO.COOH$$

The lactone form of dehydroascorbic acid can readily be reduced to ascorbic acid by H_2S or by glutathione in the presence of glutathione dehydrogenase. The open chain form is not reduced back to ascorbic acid by H_2S or by glutathione in the presence of glutathione dehydrogenase. This can undergo further irreversible reaction if the product is not reduced within a short time.

An oxidized form of ascorbic acid, which appears not to be dehydroascorbic acid, but to be fairly rapidly converted into it, can bring about the oxidation of reduced coenzymes I and II in the presence of the coenzyme, ascorbate reductase. This oxidized form is produced, at least transitorily, by the action of ascorbate oxidase. It may be concluded from these observations that ascorbate can act as a biological hydrogen carrier, although its significance in this respect is not yet entirely clear.

THIAMINE PYROPHOSPHATE (TPP)

Thiamine pyrophosphate (Figure 5.19) is the active form of thiamine (vitamin B_1). Thiamine pyrophosphate or thiamine diphosphate (ThDP) is a thiamine derivative, cleaved by thiamine pyrophosphate. The part of TPP that is most commonly involved in reactions is its thiazolium ring. The C2 of this ring is capable of acting as an acid and donating its proton and forming carbanion. Normally reactions that form carbanions are highly unfavourable, but the positive charge on the tetravalent nitrogen just adjacent to the carbanion stabilizes the negative charge, making the reaction more favourable. In several reactions, including that of pyruvate decarboxylase, alphaketoglutarate dehydrogenase and transketolase, TPP catalyses the reversible cleavage of a substrate compound at a carbon–carbon bond connecting a carbonyl group to an adjacent reactive group (usually a carboxylic acid or an alcohol). It achieves this in four basic steps:

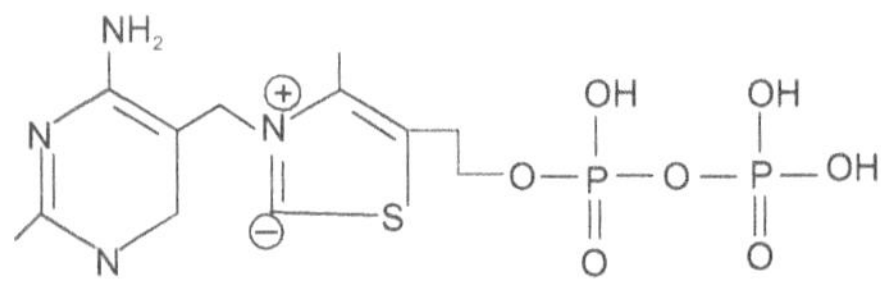

Figure 5.19 Structure of TPP

1. The carbanion of the TPP ylid nucleophilically attacks the carbonyl group on the substrate. (This forms a single bond between the TPP and the substrate).
2. The target bond on the substrate is broken, and its electrons are pushed towards the TPP. This creates a double bond between the substrate carbon and the TPP carbon and pushes the electrons in the N–C double bond in TPP entirely to the nitrogen atom, reducing it from a cation to a neutral atom.
3. The electrons push back in the opposite direction forming a new bond between the substrate carbon and another atom.
4. The TPP substrate bond is broken, reforming the TPP ylid and the substrate carbonyl.

PYROPHOSPHOTHIAMINE

Pyrophosphothiamine (Figure 5.20) was first recognized as a cofactor by Lohmann and Schuster in 1937. In addition to the decarboxylation of α-keto acids, pyrophosphothiamine is involved in the α-keto acid oxidation system and in acetoin formation. Its exact mechanism of action is still uncertain, but it acts in the first stage of these systems, giving CO_2 and a compound with aldehyde. The aldehyde group may be liberated from this compound as such, or transferred to an acceptor, either without oxidation (acetoin formation) or accompanied by oxidation to an acyl group, or liberated in the oxidized form as a free acid. This is therefore another case in which the group transferred may undergo oxidation while combined with the carrier, like the tetrahydrofolate system. It is possible that the intermediate may be a Schiff's base, as in the case of phosphopyridoxal, but now the aldehyde group is supplied by the substrate and the amino group by the cofactor.

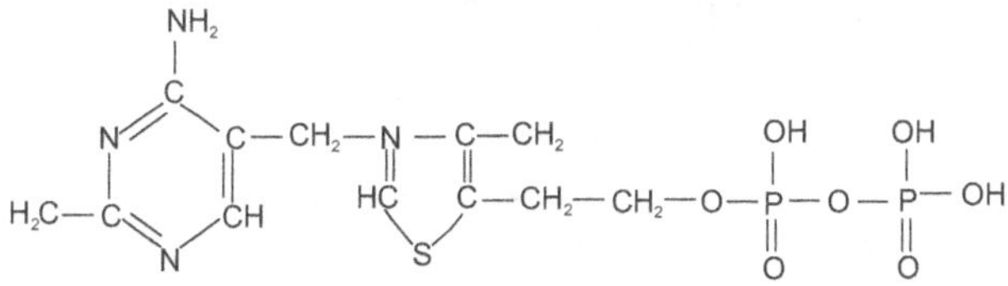

Figure 5.20 Structure of pyrophosphothiamine

URIDINE PHOSPHATE

Uridine phosphate (UDP) is a nucleotide. It is an ester of pyrophosphoric acid with the nucleoside uridine. UDP consists of the pyrophosphate group, the pentose sugar ribose and the nucleobase uracil.

PHOSPHOMUTASE COENZYMES

The phosphomutase coenzyme is converted into product and an equivalent amount of substrate is converted into coenzyme by the same transfer of a single phosphate residue. In nearly all cases, if the phosphate is regarded as being transferred from position α to position β, the coenzyme is the α, β-diphosphate. In one case, however, an α, β-diphosphate is converted into α, β, γ-diphosphate, the coenzyme being the β-monophosphate, thus,

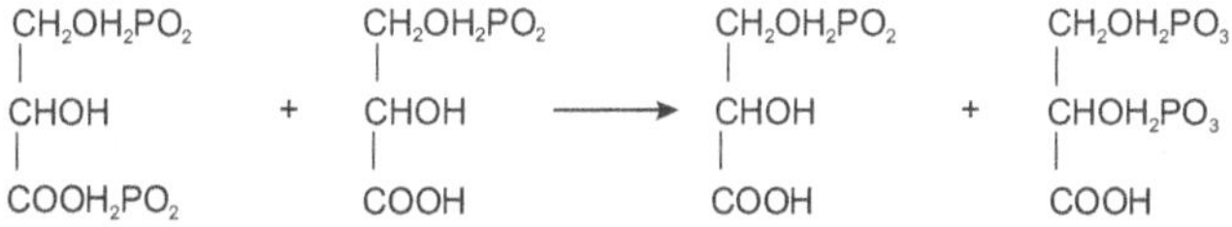

These phosphomutase reactions are special cases of the general transfer reactions.

$$AP + B \rightleftharpoons A + BP$$

In the first type AP is identical with BP; in the second, A is identical with B.

GLUTATHIONE COENZYMES

Glutathione has already been mentioned as a hydrogen carrier, but reduced glutathione also acts as a specific coenzyme for a few enzyme systems. In these cases, cysteine cannot replace glutathione. The glutathione probably forms a covalent compound with the substrate in all these cases. The identification of reduced glutathione as the coenzyme of glyoxylase was made by Lohmann in 1932.

Kuhnau had shown the spontaneous formation of a compound of glutathione and methylglyoxal. This was confirmed by Jowett and Quastel who suggested that this compound was the true substrate of the enzyme. Yamazoye showed the formation of another compound in the presence of the enzyme; this compound slowly breaks down to lactic acid and glutathione at above pH 7, and the breakdown was acclerated by crude glyoxylase preparations. Hopkins and Margan showed that when methylglyoxal was the substrate, the system consisted of two protein components in addition to glutathione; they partially purified these and called them "enzyme" and "factor". The factor played no part when phenylglyoxal was the substrate. Reduced glutathione is also the specific coenzyme of maleylacetoacetate isomerizes. It is probable that the *cis–trans* isomerization catalysed by this enzyme takes place through the intermediate formation of an additional compound which does not possess *cis–trans* isomerism, by addition of the thiol group of glutathione to the double bond of the maleic residue, thus

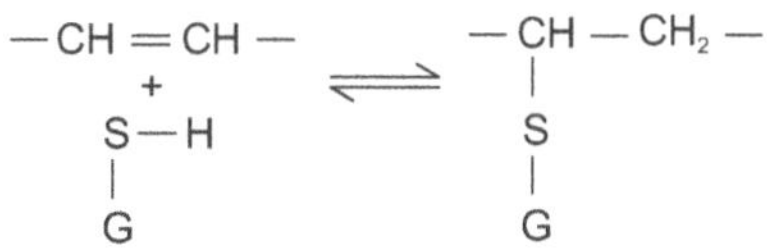

Cis-trans isomerization

CYTOCHROMES

The name "cytochrome" was given by Keilin in 1925. The name at present used appears to include all intracellular haemoproteins with the exception of haemoglobin, myoglobin, peroxidase and catalase. The group includes substances with many different functions. The functions of the cytochromes are unknown, but they all appear to act by undergoing oxidation and reduction. The cytochromes fall into three groups, differing in the nature of the haem prosthetic group. The types are designated by the letters *a*, *b* and *c*. These types can be distinguished on the basis of the nature of the pyridine haemochromogen which they yield: the '*b*' types yield normal protohaemochromogen, whereas the '*a*' and '*c*' types give special haemochromogens. A possible arrangement of these groups is shown in the Figure 5.21.

The spectra of cytochrome *c* may be taken as typical of those of the oxidized and reduced forms of cytochromes (Figure 22). On oxidation, which is a one-equivalent reaction involving a change of state of the iron atom from ferrous to ferric, the sharp α- and β-bands disappear, but the γ-band remains, although it is shifted towards the ultraviolet. However, the reduced 4 cytochromes show little or no β-band.

As a general rule, the *b* cytochromes are oxidized directly by molecular oxygen, whereas in general the others do not have this property of autoxidizability. Reduced cytochrome *c* in neutral solution is not oxidized by molecular oxygen, but it is oxidized by H_2O_2, by ferricyanide, and by copper salts. The reaction with H_2O_2 is greatly catalysed by cytochrome *c* peroxidase in some microorganisms.

Reduced cytochrome *c* is very rapidly oxidized by oxygen when it is brought into contact with tissue preparations, owing to the presence in them of the enzyme cytochrome oxidase and

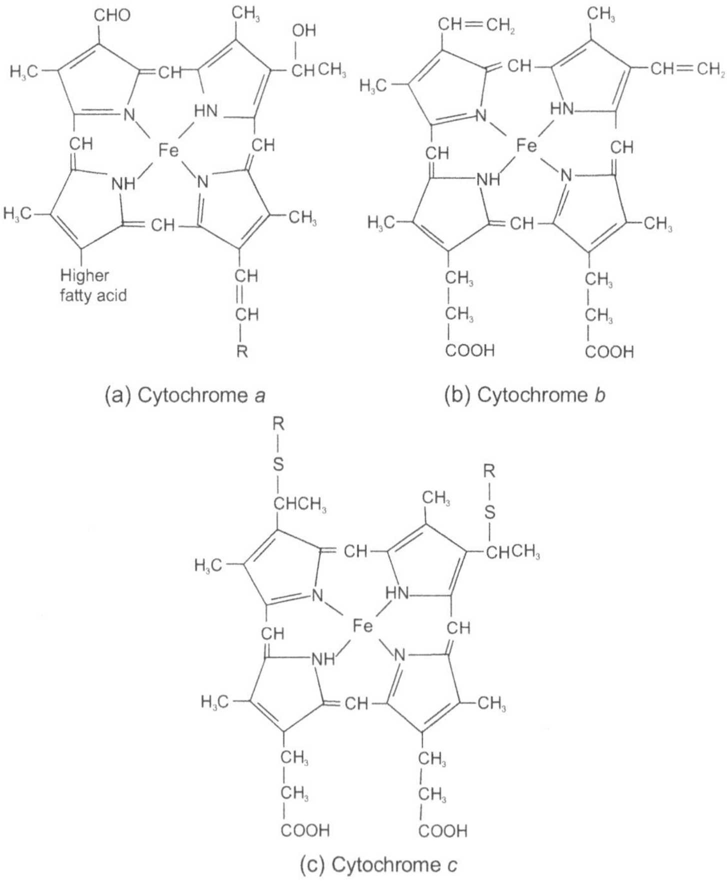

Figure 5.21 Cytochromes

this reaction forms the terminal step of the system. Oxidized cytochrome *c* is reduced by such chemical reducing agents such as dithionite, cysteine, polyphenols and ascorbate. It is also reduced by cytochromes b_3 and c_1 and by a number of enzymes, including the cytochrome *c* and the systems involving the succinate and acyl-CoA dehydrogenases. Since its enzymic oxidation and reduction are both rapid in tissues, it is a very effective carrier in the oxidation of the relevant substrates by molecular oxygen. It will be seen that its mode of action closely resembles that of the coenzymes, and that, it acts as a true cofactor in these systems.

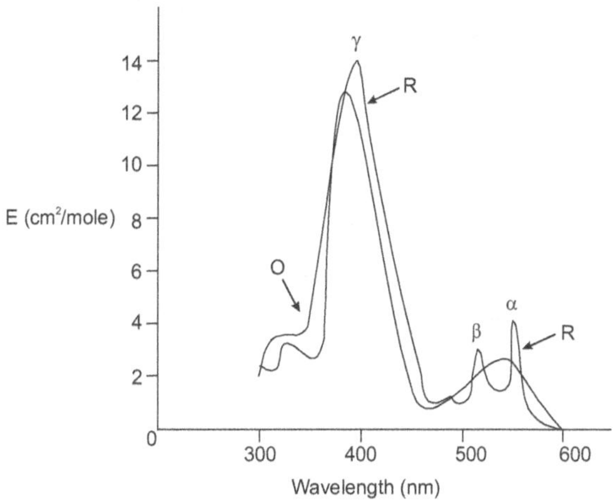

Figure 5.22 Spectra of oxidized (O) and reduced (R) forms of cytochrome *c*

RELATIONSHIP BETWEEN COENZYMES AND VITAMINS

It will have been noticed that many of the coenzymes to which we have referred are derivatives of vitamins. Taking them in the order in which they are given, coenzymes I, II and III contain nicotinamide (vitamin B group), lipoate is a growth factor for some microorganisms, ascorbate (vitamin C) is a vitamin for the primates and the guinea pig, coenzyme A contains pantothenic acid (vitamin B group), tetrahydrofolate is derived from folic acid (vitamin B group, contains the simpler vitamin *p*-aminobenzoic acid), the flavin prosthetic groups contain riboflavin (vitamin B_{12}), phosphopyridoxal is a derivative of pyridoxine (vitamin B_6) and pyrophosphothiamine contains thiamine (vitamin B_1).

It may be well that the main function of the vitamins is to serve as the operative part of specialized coenzymes and prosthetic groups, (Table 5.1) in other words, organisms are often unable to synthesize important parts of many of the their essential enzyme systems, which have to be supplied. There are almost certainly other cases (Table 5.2) which are not mentioned in the Table 5.1. Other vitamins which may well have coenzyme functions are biotin and vitamin B_{12}.

Table 5.1 Coenzymes in group transfer reactions

Coenzyme	**Chemical group transferred**
Adenosine triphosphate	Phosphate group
NAD^+ and NADP	Electrons
S-adenosyl methionine	Methyl group
Coenzyme A	Acetyl group and other acyl group
Coenzyme Q	Electrons
Cytidine triphosphate	Diacylglycerols
Tetrahydrofolic acid	Methyl, formyl, methylene and formimino groups
Menaquinone	Carbonyl group
Ascorbic acid	Electrons
Coenzyme F_{420}	Electrons
3´-phosphoadenosine-5´-phosphosulphate	Sulphate group

Substances closely related to vitamins, but incapable of working in the same way, would be expected to combine with the enzymes in the place of the active group and thereby acts as a special type of competitive inhibitor. Such substances have

Table 5.2 Vitamins and coenzymes

Vitamin	Coenzyme	Reaction type	Coenzyme class
B_1 (Thiamine)	TPP	Oxidative decarboxylation	Prosthetic group
B_2 (Riboflavin)	FAD	Oxidation/ reduction	Prosthetic group
B_3 (Pantothenate)	CoA	Acyl group transfer	Cosubstrate
B_4 (Pyridoxine)	PLP	Transfer of groups to and from amino acids	Prosthetic group
B_{12} (Cobalamine)	5-deoxyadenosyl cobalamine	Intramolecular rearrangements	Prosthetic group
Niacin	NAD^+	Oxidation/ reduction	Cosubstrate
Folic acid	Tetrahydrofolate	One-carbon group transfer	Prosthetic group
Biotin	Biotin	Carboxylation	Prosthetic group

been called antivitamins. When administered, often in very small quantities, they produce the symptoms of a deficiency of the corresponding vitamin, and may be highly toxic. For example, aminopterin antagonizes the action of tetrahydrofolate and also competitively inhibits its formation from folate. Other antivitamins are pyridine 3-sulphonic acid, which competes with nicotinic acid and 6, 7-didemethylriboflavin and galactoflavin, both of which compete with riboflavin.

REVIEW QUESTIONS

1. What are metalloenzymes?
2. Differentiate metalloenzymes from metal-activated enzymes.
3. Discuss metalloenzymes with suitable examples.

6

ACTIVE SITE AND ITS DETERMINATION

In an enzyme-catalysed reaction, the first step is the formation of an enzyme–substrate complex in which the substrate or substrates bind to the active site, usually noncovalently. Specificity of binding comes from the close fit of the substrate within the active site pocket, which is due primarily to van der Waals interactions between the substrate and non-polar groups on the enzyme, combined with complementary arrangements of polar and charged groups around the bound molecule. This fit is so specific that even a small change in the chemical composition of the substrate will abolish the binding. This portion of the enzyme which is in contact with or in very close proximity to the substrate during the reaction is known as **"active site"**.

The enzymes are therefore large molecules which catalyse the reactions of molecules much smaller than themselves. Even in those cases where the substrate of an enzyme is itself a very large molecule, such as the case of ribonuclease, which catalyses the depolymerization of ribonucleic acid, it is clear that at any one time the enzyme is involved with only a very small portion of the substrate. Similarly proteolytic enzymes interact with only a small portion of the protein substrates and are also able to hydrolyse small peptides and similar molecules. The great difference in size between enzyme and substrate therefore means that during the catalytic process only a very small portion of the enzyme is in contact with the substrate.

The active site comprises only a small portion of the total enzyme molecule and is usually at or near the surface, since it must be accessible to the substrate molecules. It binds with the substrate molecule by relatively weak forces like hydrogen bonds, hydrophobic interactions, temporary covalent bonds (van der Waals) or a combination of all of these to form the enzyme–substrate complex. Residues of the active site will act as donors or acceptors of protons or other groups on the substrate to facilitate the reaction. In other words, the active site modifies the reaction mechanism in order to decrease the activation energy of the reaction. The product is usually unstable in the active site due to steric hindrances that force it to be released and return the enzyme to its initial unbound state.

It has a three-dimensional structure and consists of portions of a polypeptide chain. The active site in the enzyme molecules are grooves or crevices from which water is largely excluded. The active site contains amino acids such as aspartic acid, glutamic acid, lysine, serine, aspartate, cysteine, arginine, tyrosine, etc. Among these amino acids, serine is the most frequently found.

The amino acid residues involved may be widely separated in the primary structure, being brought together in space because of the twists and turns within the molecule. The amino acid residues in the active site which do not have a binding or catalytic function may not contribute to the specificity of the enzyme. The side chain groups (–COOH, $-NH_2$, $-CH_2OH$, etc.) present in the active site serve as catalytic groups in the active site. They might not interfere with the binding of the substrate but they might interfere with the binding of other chemically similar structures.

DETERMINATION OF AMINO ACIDS AT THE ACTIVE SITE

Horecker *et al.* in 1962 showed that if the enzyme–substrate reaction mixture was treated with sodium borohydride, a strong reducing agent, then an inactive complex was produced. On hydrolysis of this complex, ε *N*-glyceryl lysine was found among the products. From this it was concluded that dihydroxyacetone phosphate normally binds to the ε (i.e., side chain) amino group of a lysine residue in the enzyme by a Schiff's base (–N=CH–) linkage. The presumed sequences for the normal reaction and for the procedures used to identify the substrate-binding site are as follows:

Trinitrobenzylsulphonate modifies the lysine residue at the active site of the enzyme, for example, myosin ATPase.

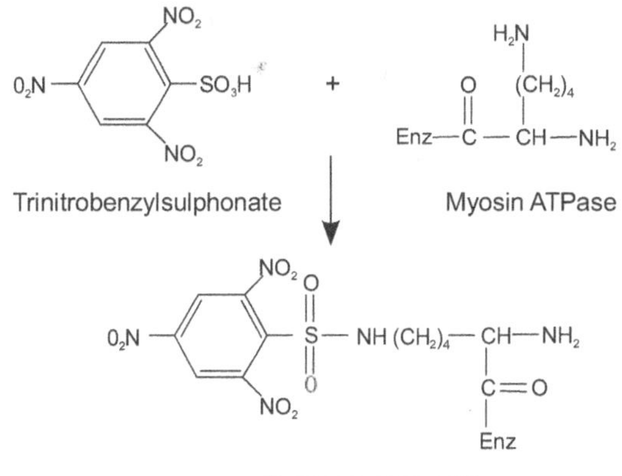

α-ε-diaminocaproic acid
(*p*-dinitrobenzyl sulphonyl lysine derivative)

In general, once an enzyme–substrate complex has been trapped as an inactive complex, it may be subjected to partial hydrolysis and the amino acid sequence for a few residues on each side of the binding site is determined. It may also be possible to determine the complete primary structure of the inactivated complex and hence the substrate-binding site may be precisely located.

CHEMICAL MODIFICATION OF AMINO ACIDS IN THE ACTIVE SITE

A number of chemical studies have been carried out in order to determine which amino acid residues may be present at the active site of the enzyme. For most part, these have consisted of treating the enzyme with chemical reagents which react with certain amino acid side chains and testing the modified enzyme for activity.

If the modification has destroyed or decreased the activity of the enzyme then it has usually been deduced that the amino acid residue was present at the active site. Such a deduction is not always justified. Destruction or modification of an amino acid residue of enzyme may cause loss of activity for several reasons. In order to eliminate this possibility, chemical modification should also be carried out in the presence of the substrate or of some other compound which binds to the active site (e.g. a competitive inhibitor).

If the substrate or inhibitor protects the enzyme from inactivation, relative to the situation in the absence of such a compound, then it may be assumed with more justification that the reaction occurred at the active site. Even when this has been established, it must be borne in mind that an amino acid residue shown to be present at the active site and also shown to be essential for enzymatic activity may play one of two roles in the process. First, it may be recognized for the binding of the substrate to the enzyme, if this is the case, then modification may not totally destroy activity. The dissociation constant k_S may be affected but in the presence of a large concentration of substrate, the maximum velocity V may be unchanged.

Secondly, the amino acid residue may be involved in the actual catalytic process. After modification, the enzyme would be expected to be totally inactive, but may still bind the substrate. The wide variety of chemically reactive side groups in amino acid allows the chemist to use a number of chemical reagents to bring about modification of an enzyme. For the most meaningful results, a highly selective reagent is to be preferred, so that each type of amino acid can be studied individually.

Photo-oxidation

Another technique which may be used to inactivate enzymes by modification of amino acid side chains is **photo-oxidation**, i.e., oxidation by activated oxygen in the presence of a photosensitizer such as methylene blue or rose bengal. This method is non-specific and may oxidize histidine, tryptophan, methionine and cysteine residues. However, some degree of specificity may be obtained by careful choice of the photosensitizing dye and pH.

Photo-oxidation of tryptophan residues in lysozyme inactivates the enzyme, suggesting the presence of these residues at the active site.

Modification by Substrate Analogue

An alternative way of producing a complex more stable than the normal enzyme–substrate complex is to replace the natural substrate by an analogue which binds to the same site on the enzyme but is then less readily removed. In the case of chymotrypsin, enzymes form stronger linkages with irreversible inhibitors, since in this case there is no subsequent reaction at all. Thus diisopropylphosphofluoridate (DPF) binds to the **serine** residue at the active site in chymotrypsin and other serine proteases to form very stable complexes.

In chymotrypsin, DPF selectively phosphorylates the serine residue at position 195 thus indicating that serine residue at position 195 is essential for the catalysis. Partial hydrolysis of each enzyme–inhibitor complex gives a series of peptide fragments, which can be separated from each other and analysed. These phosphorylated peptides have given important information on the amino acid sequences near the active site in this group of enzymes, sometimes called the "serine enzymes".

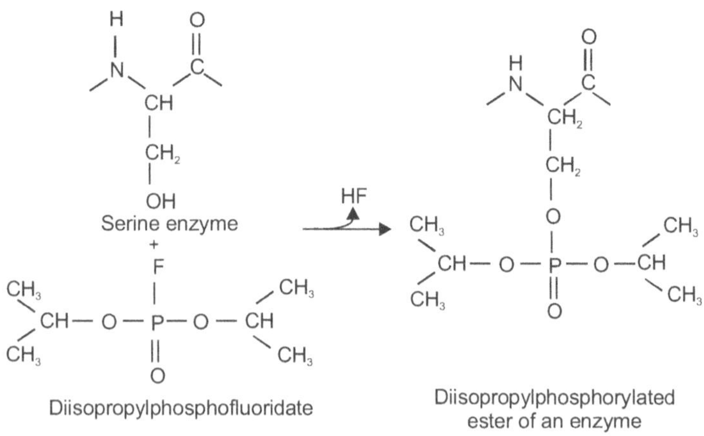

The amino acid sequence of any fragment containing DPF must be the primary active structure. In this way it can be shown that the amino acid sequence around the essential serine residue of chymotrypsin is **–Gly–Asp–Ser–Gly–Gly–Pro–**. A similar sequence is found for trypsin.

The assumption that an irreversible inhibitor binds at the active site of an enzyme is particularly valid when the inhibitor resembles a substrate. For example, tosyl (*N*-toluenesulphonyl) L-phenylalanine chloromethyl ketone (TPCK) resembles esters which are hydrolysed by chymotrypsin, but TPCK itself acts as an irreversible inhibitor of this enzyme by alkylating the histidine-57 residue. Thus there is evidence that serine-195 and histidine-57 are present at the active site of chymotrypsin.

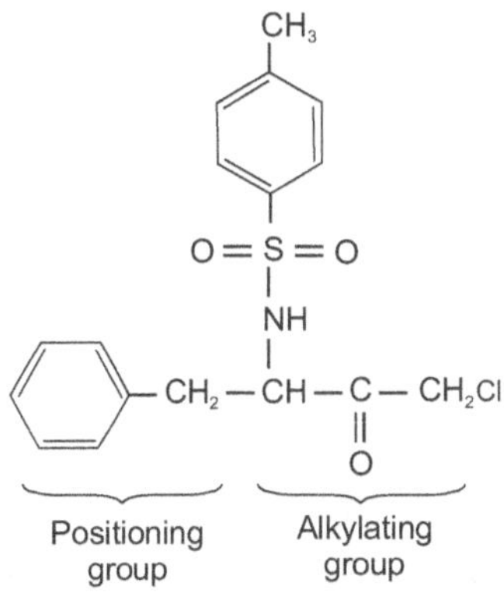

TPCK (*N*-tosyl L-phenylalanine chloromethyl ketone)

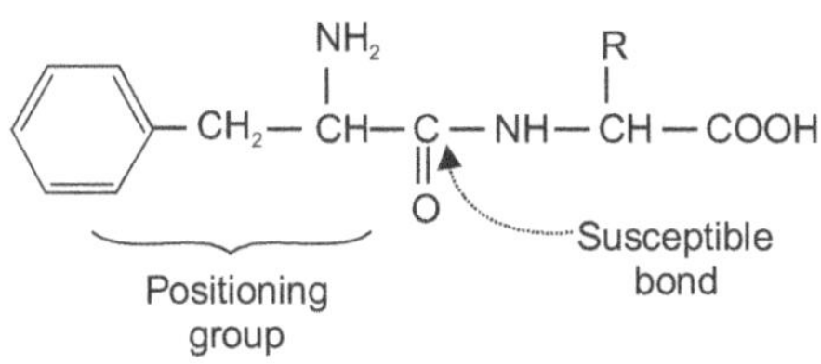

Structure of normal substrate

Tosyl (toluene sulphonic acid) reacts with serine to give the following:

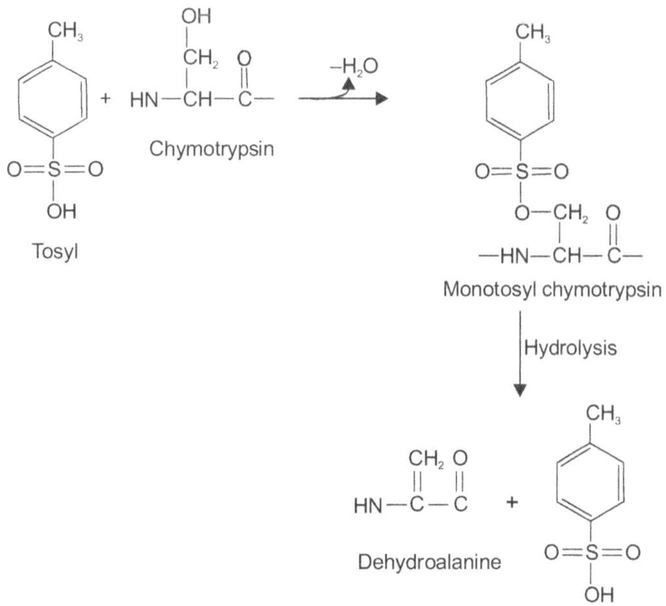

Reaction of TPCK with His-57 of chymotrypsin is as follows:

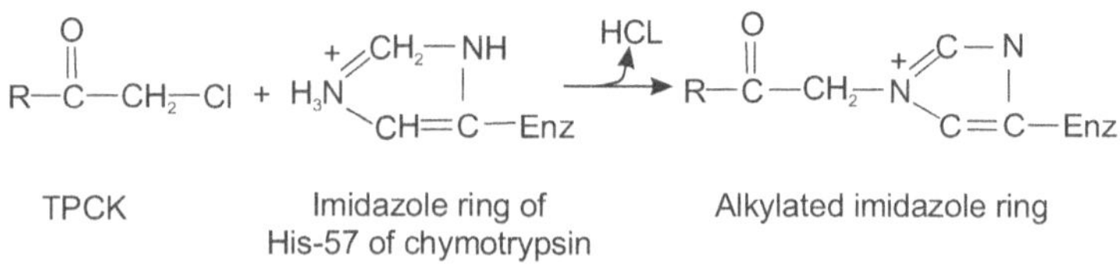

Enzymes containing serine at the active site and that are inactivated in this manner can sometimes be slowly reactivated by reagents with a higher reactivity for the phosphorylating agent than the enzyme itself. It is highly significant that those functional groups of an enzyme required for catalytic activity are usually more accessible or reactive than similar groups elsewhere in the molecule that are not directly involved in catalysis.

For example, ribonuclease contains many functional groups capable of reaction with iodoacetate, but the imidazole group of histidine residue 12 and 119 are far more relative than all the others. Similarly, chymotrypsin contains many serine residues but only that at position 195 is phosphorylated by DPF under mild conditions.

Amino acid sequences around inactive serine residues of some enzymes are:

Enzymes	**Amino acid sequence in active site**
Chymotrypsin	–Gly–Asp–Ser–Gly–Gly
Trypsin	–Gly–Asp–Ser–Gly–Pro
Thrombin	–Asp–Asp–Ser–Gly
Esterase	–Gly–Asp–Ser–Gly
Phosphoglucomutase	–Thr–Asp–Ser–His–Asp
Phospholyase	–Gly–Ile–Ser–Val–Arg

Alkylation

The thiol group of **cysteine** residues may be alkylated by halogeno-compounds of the type which also alkylate histidine residues. For example, cysteine-25 of the proteolytic enzyme, papain, is alkylated by iodoacetamide, which results in inactivation of the enzyme.

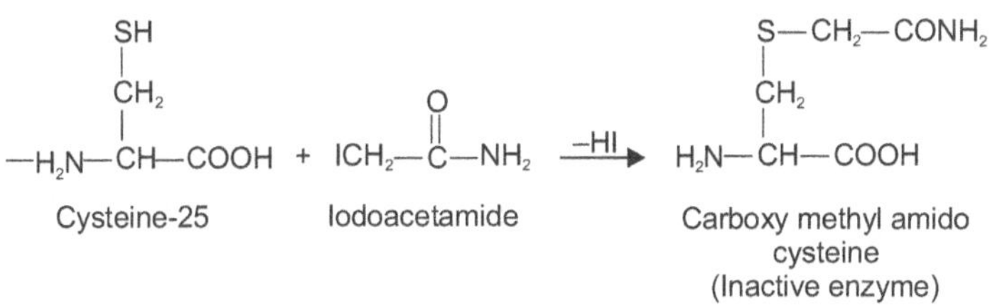

Enzymes containing cysteine residues may also be inhibited by unsaturated compounds such as *N*-ethylmaleimide.

For example, reaction with unsaturated compounds like *N*-ethylmaleimide inhibits myosin ATPase which contains cysteine at the active site.

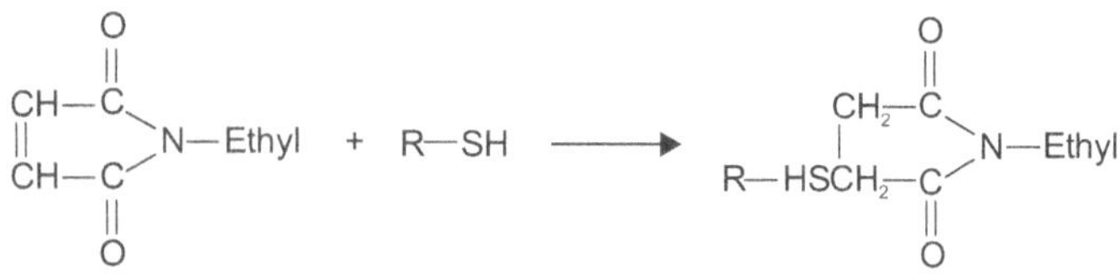

Methionine residues, in common with those of other amino acids, could be modified by alkylation or photo-oxidation. Hence, it was important to determine precisely which amino acids were affected in each modification experiment. Koshland and his colleagues showed that photo-oxidation of chymotrypsin under certain conditions led to the oxidation of only a single methionine residue (methionine-192), which formed the corresponding sulphoxide; this modification led to only partial inactivation of the enzyme, as did alkylation of the same methionine residue. On the basis of these studies, it was concluded that methionine-192 is present at the active site of chymotrypsin and may play a part in substrate-binding, but is not otherwise involved in catalytic activity.

An interesting example of the inactivation of an enzyme by alkylation at **histidine** is afforded by ribonuclease. In this enzyme, two histidine residues appear to be present at the active site and either but not both of them can react with iodoacetate. Iodoacetamide has no effect. 1 or 3 carboxymethyl histidine could be isolated from the product. The inactivation of ribonuclease by iodoacetate was prevented by the presence of PO_4 which binds to the active site.

The most important halogen-alkylating agents used for the modification of enzymes at histidine are those which are closely related to substrates of the enzymes. Schoellmann and Shaw used chloroketones related to ester substrates of chymotrypsin and trypsin.

TPCK (tosyl phenylalanine chloroketone) totally inactivated chymotrypsin by alkylation of the histidine at position 57 in the primary sequence.

The chloroketone inactivates efficiently because due to its similarity to a substrate, it is first bound to the active site of the enzyme by the usual noncovalent forces. This brings the chloroketone groups into close proximity with the histidine residue which is present at the active site. TPCK does not inactivate trypsin, whose specificity is not for derivation of phenylalanine but for lysine derivatives. The corresponding chloroketone, TLCK (tosyl lysine chloroketone), inactivates trypsin, also by alkylation at an essential histidine.

The importance of specific inhibitors of this type based on substrate analogues are accomplished not only for the study of enzymes but mainly for the pharmaceutical sciences since inhibitors may have great therapeutic value. A histidine residue at the active site of carbonic anhydrase may be alkylated by a reagent not derived from a substrate but for a specific inhibitor. Acetazolamide binds strongly to the enzyme at the zinc ion which is at the active site. Chloroacetazolamide, which is an active halogen compound derived from it, irreversibly inhibits the enzyme by first binding at the zinc and then reacting with a histidine residue.

The histidine side chain also reacts with diazo compounds such as diazosulphonic acid. This couples with the imidazole ring giving a chromophore which absorbs at 450 nm, allowing the nature of the modified protein to be identified. Diazosulphonic acid inactivates fructose diphosphate aldolase, the inactivators being prevented by the presence of the substrate fructose diphosphate.

Therefore, it is probable that this inactivation is due to reaction at a histidine although diazosulphonic acid can also react with tyrosine and lysine side chains. Photo-oxidation which is oxidation by activated oxygen, light-induced in the presence of a photosensitizer such as methylene blue or rose bengal, is another method of modifying proteins at histidine residues.

The affinity label TPCK was designed to resemble a normal substrate for chymotrypsin. However instead of usual susceptible structure, TPCK contains a group of potent alkylation agent. Incubation of TPCK with chymotrypsin causes it to bind to the active site in the same way as the normal substrate.

However, instead of hydration of the substance, alkylation of essential His-57 residue occurs indicating that the latter is close to the susceptible bond normally undergoing hydration. The alkylated His residue can be isolated following complete hydration of all the peptide bonds of the enzyme. Thus His-57, serine-195 of chymotrypsin has been identified as participants in its catalytic activity.

When ribonuclease is treated with iodoacetate at pH 5.5 the enzyme undergoes alkylation and loses its catalytic activity. Two different inactive forms are found in one:

1. The imidazole ring of histidine residue 119 is alkylated.
2. His-12 is alkylated.

Since no other functional groups in the ribonuclease molecule is alkylated under these conditions, the conclusion is that the residues 12 and 119 of ribonuclease are necessary for catalytic activity.

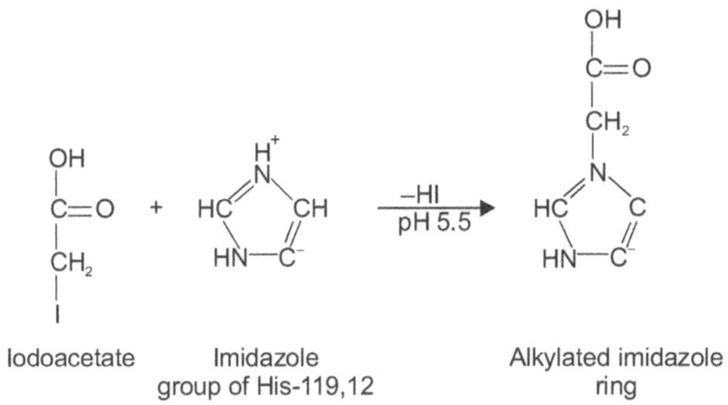

Iodination, Nitration and Acylation

A **tyrosine** residue (tyrosine-248) was found to be present at the active site of carboxylase A by a variety of modification techniques. Vallee and colleagues showed that iodination or nitration of the benzene ring, or acylation of the phenolic –OH group, all inactivate the enzyme.

The presence of **aspartate** and **glutamate** at the active site of certain enzymes could be demonstrated by the production of esters of the side chain's carboxyl groups. For example, Erlanger (1966) showed that *p*-bromophenacyl bromide can inactivate pepsin by forming an ester linkage with an aspartate residue.

This modification can be prevented by the presence of excess substrate, showing that the aspartate residue is indeed at the activity site.

Other chemical reagents also have been used to identify specific amino acid residues of various enzymes involved in catalytic activity, e.g. reactions of enzyme with substrate coupled between fructose diphosphate aldolase and dihydroxyacetone phosphate.

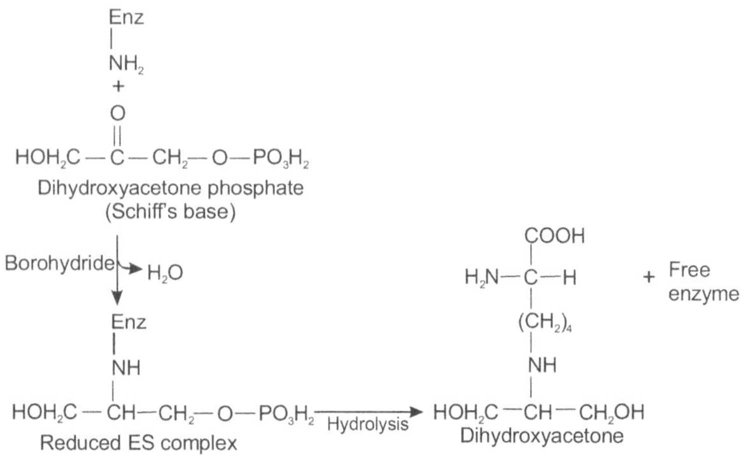

Addition of strong reducing agent sodium borohydride and mixture of aldolase (fructose diphosphate aldolase and isotopically labelled, dihydroxyacetone phosphate) results in the formation of isotopically labelled covalent aldol dihydroxyl acetone phosphate compound which is catalytically inactive.

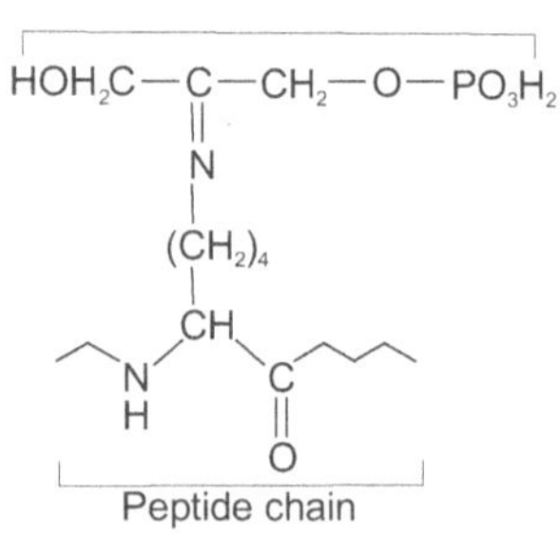

Catalytically inactive

After complete hydration of this compound with acid, a stable ε-*N* glycerol derivative of single lysine residues was found in the hydrolysate. This derivative did not form in the absence of the reducing agent. From the structure of the labelled product, it was concluded that aldolase combines covalently but reversibly with dihydroxyacetone phosphate to form a labile "Schiff's base" between the amino group of lysine residue in the active site of the enzyme and the carbonyl group of the substrate. Treatment with borohydride reduces this very labile intermediate to a stable but inactive covalent derivative. In this way very labile, covalent enzyme–substrate compounds of a number of enzymes acting on substrate with either an amino or carboxyl end have been trapped successfully in their stable reduced forms. In this way one can find out the presence of lysine residue in the active or catalytic site and participation of lysine residue in the catalytic activity. Thus lysine residue (lysine-41) has been demonstrated to be present at the active site of ribonuclease by modification with dinitrofluorobenzene.

ENZYME MODIFICATION BY SITE-DIRECTED MUTAGENESIS

Site-directed mutagenesis involves recombinant DNA technology, enzymes with specific amino acid substitution may be synthesized and the effect on activity observed. Aspartate-102 was replaced with arginine in trypsin, and it was found that at neutral pH, activity of the ester substrate was reduced 10,000-fold. Similarly in triose phosphate isomerase, glutamate-165 is found to be present at the active site and its replacement with aspartate reduced its activity 1000-fold.

Therefore, the active site, a small portion of the enzyme molecule determines the specificity of the binding. This fit is very specific in that a small change in the chemical composition of the substrate will abolish the binding. Also the side chains in the active site which serve as catalytic groups in them might interfere with the binding of the other chemically related structures.

REVIEW QUESTIONS

1. What are coenzymes?
2. Discuss how NAD and NADP participate in oxidation–reduction reactions.
3. Write the structure of CoA and list the reactions it participates in.
4. Differentiate acetyl-CoA from acyl-CoA.
5. Write notes on coenzyme B_{12}.
6. How do FAD and FMN act as coenzymes?
7. Write notes on:
   i. TPP
   ii. Lipoate
   iii. Glutathione
   iv. Ascorbate
8. Bring out the relationship between coenzymes and vitamins?

7

ENZYME CATALYSIS

MECHANISM OF ENZYME ACTION

An enzyme (E) molecule has a highly specific binding site or active site to which its substrate (S) binds to produce the enzyme–substrate complex (ES). The reaction proceeds at the binding site to produce the products (P), which remain associated with the enzyme (enzyme–product complex). The product is then liberated and the enzyme molecule is freed in an active site to initiate another round of catalysis. Apparently, the affinity of the binding site for the product is much lower than that for the substrate.

Enzymes reduce the overall level of activation energy or Δ G. Even a modest reduction in the value of ΔG leads to a very large increase in the reaction rate. The various mechanisms used by enzymes to lower the activation energy are briefly described below (Figure 7.1):

1. Stabilization of the transition state of the substrate reduces the activation energy of the overall reaction. This is the most important feature of enzymatic catalysis. The binding sites of enzymes bind more

strongly to the substrate molecule in the transition state than to those in the ground or stable state. In addition, when a substrate molecule in ground state binds to the binding site, it is forced into a configuration closer to that of transition state; this lowers the energy needed for reaching the transition state.

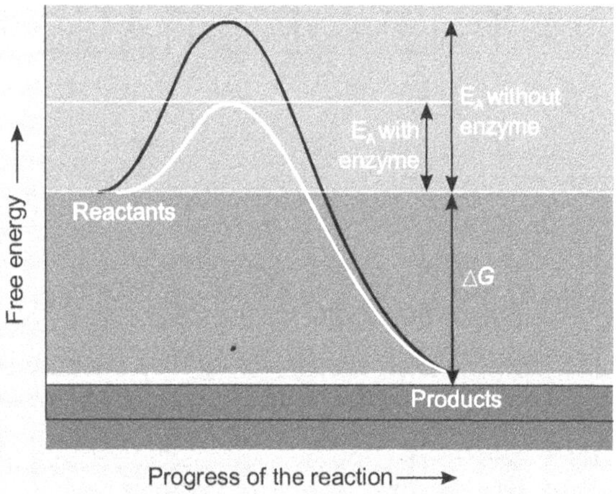

Figure 7.1 Lowering of activation energy

2. When the substrate binds to an enzyme, it is held close to another chemical group and in the optimal condition for the reaction to proceed; this is called catalysis by approximation. In contrast, substrate molecules in the transition state must collide with each other, a random process, for the reaction to occur in uncatalysed systems.
3. The chemical groups such as amino groups in the side chain of amino acids present in the active site become covalently bound to a group in the substrate molecule; such substrate molecules are much more reactive than the non-bound molecules. Another substrate may then

react with the substrate that had been bounded to the enzyme molecule to form the product. Then the product will become free from the enzyme. This strategy of enzyme action is called covalent catalysis.

4. The amino acid side chains may act as acids, (donors of protons) or bases (acceptors of protons) and thereby catalyse reactions. In addition, electrons may also be transferred during the course of a reaction; such reactions are called redox reactions.

CATALYTIC MECHANISMS

The rate of a reaction is a function of its free energy of activation. A catalyst acts by lowering the kinetic barrier, that is, catalyst stabilizes the transition state with respect to the uncatalysed reaction. Therefore, catalysis is a process that increases the rate at which a reaction approaches equilibrium. Enzymes are powerful catalysts because of two related properties: their specificity of substrate-binding combined with the optimal arrangement of their catalytic groups. The types of catalytic mechanisms that enzymes employ have been classified as:

1. Acid–base catalysis
2. Covalent catalysis
3. Metal ion catalysis
4. Electrostatic catalysis
5. Proximity and orientation effects
6. Preferential binding of the transition state complex

ACID–BASE CATALYSIS

Acid catalysis is a process in which partial proton transfer from a Bronsted acid (donate protons) lowers the free energy of a reaction's transition state. For example, an uncatalysed keto-enol tautomerization reaction occurs quite slowly as a result of the high energy of its carbanion-like transition state. Proton donation to the oxygen atom, however, reduces the carbanion character of the transition state thereby catalysing the reaction. A reaction may be general base catalysis if its rate is increased by partial proton abstraction by a Bronsted base (Figure 7.2).

Mechanism of Keto-enol Tautomerization

Uncatalysed

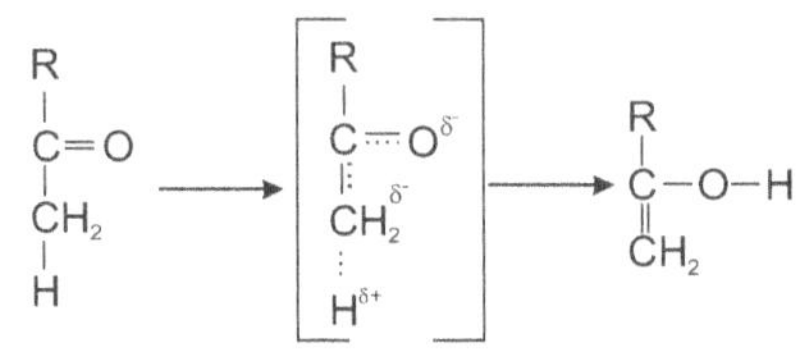

Keto Transition state Enol

Acid catalysis

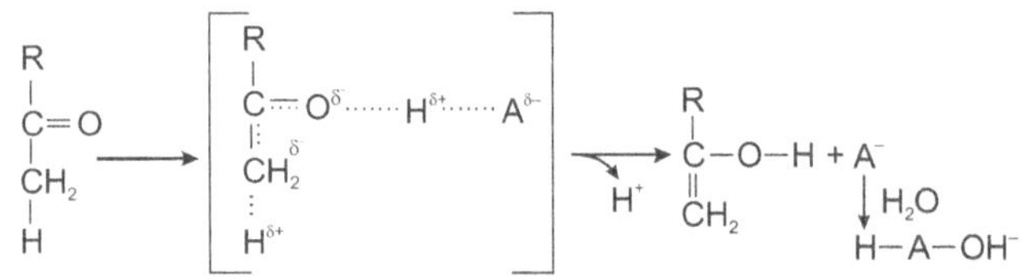

Keto Transition state Enol

Base catalysis

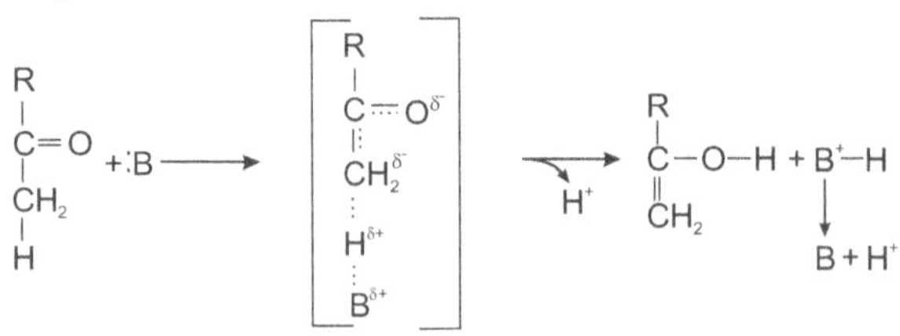

Keto Transition state Enol

Some reactions may be simultaneously subject to both processes a concerted general acid–base catalysed reaction. Mutarotation of glucose is the best example for acid–base catalysis. A glucose molecule can assume either of two anomeric cyclic forms through an intermediary of its linear form.

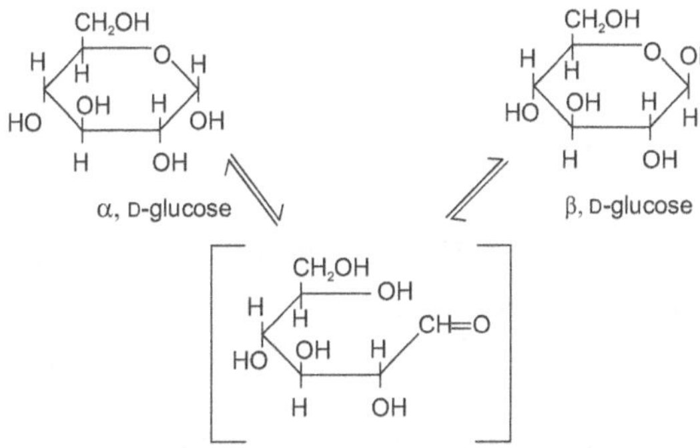

In aqueous solvents, the initial rate of mutarotation of α-D-glucose, follows the relationship

$$v = \frac{d(\alpha\text{-D-glucose})}{dt} = k_{obs}[\alpha\text{-D-glucose}]$$

where, k_{obs} is the reaction's apparent first-order rate constant. The mutarotation rate increases with the concentrations of general acids and general bases.

This model is consistent with the observation that in aprotic solvents such as benzene, 2, 3, 4, 6 *o*-tetramethyl α-D-glucose does not undergo mutarotation. Yet the reaction is catalysed by the addition of phenol, a weak benzene-soluble acid, together with pyridine, a weak benzene-soluble base according to the rate equation.

$$v = k[\text{phenol}][\text{pyridine}][\text{tetramethyl}][\alpha\text{-D-glucose}]$$

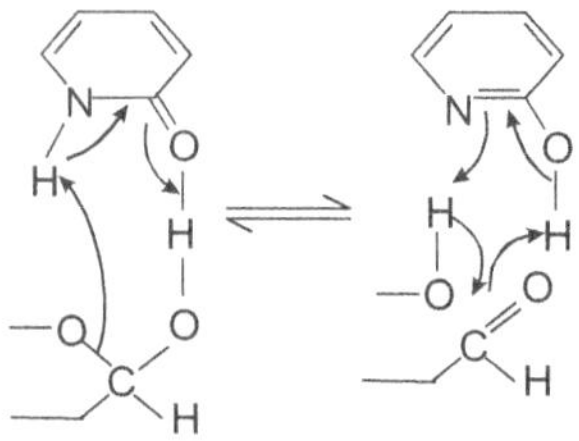

The reaction follows the rate law

$$v = k'[\alpha\text{-pyridone}]\frac{n!}{r!(n-r)!}[\text{Tetramethyl-}\alpha\text{-D-glucose}]$$

where $k' = 7000\ Mk$. This increased rate constant indicates that α-pyridone catalyses mutarotation in a concerted fashion since 1 M α-pyridone has the same catalytic effect at impossibly high concentrations of phenol and pyridine (e.g. 70 M phenol and 100 M pyridine).

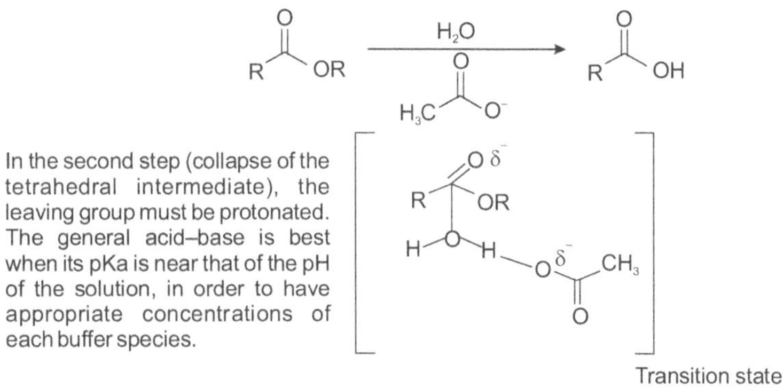

Figure 7.2 General base catalysis and ester hydrolysis

Many types of biochemically significant reactions are susceptible to acid or base catalysis. These include the hydrolysis of peptides and esters, the reactions of phosphate groups, tautomerization and additions of carbonyl groups. The side chains of the amino acid residues Asp, Glu, His, Cys,

Tyr and Lys have pK's in or near the physiological pH range, which permits them to act in the enzymatic capacity of general acid and/or base catalysts in analogy with known organic mechanisms.

Indeed, the ability of enzymes to arrange several catalytic groups about their substrates makes concerted acid–base catalysis a common enzymatic mechanism (Figure 7.2).

COVALENT CATALYSIS

Covalent catalysis involves rate acceleration through the transient formation of a catalyst substrate covalent bond. The decarboxylation of acetoacetate, as chemically catalysed by primary amines, is an example of such a process. In the first stage of the reaction, the amine nucleophilically attacks the carbonyl group of acetoacetate to form a Schiff base (imine bond).

The protonated nitrogen atom of the covalent intermediate then acts as an electron sink so as to reduce the high-energy enolate character of the transition state. The formation and decomposition of the Schiff base occurs quite rapidly so that it is not rate-determining in this reaction sequence (Figure 7.3). Covalent catalysis may be conceptually divided into two stages:

1. The nucleophilic reaction between the catalyst and the substrate to form a covalent bond.
2. The withdrawal of electrons from the reaction centre by the electrophilic catalyst.

Reaction mechanisms are classified as either electrophilic catalysis or nucleophilic catalysis depending on which of these effects provides the greater driving force for the reaction, that is, which catalyses its rate-determining step. The primary amine-catalysed decarboxylation of acetoacetate is an electrophilically catalysed reaction since its nucleophilic phase, Schiff base formation, is not its rate-determining step. In other covalently catalysed reactions, however, the nucleophilic phase may be rate-determining.

An important aspect of covalent catalysis is that the more stable the covalent bond formed, the less facilely it will decompose in the final steps of a reaction. A good covalent catalyst must therefore combine the seemingly contradictory properties of high nucleophilicity and the ability to form a good leaving group, that is, to easily reverse the bond formation step.

Groups with high polarizabilities such as imidazole and thiol functions have these properties and make good covalent catalysis. Other enzyme functional groups that participate in covalent catalysis include the imidazole moiety of His, the thiol group of Cys, the carboxyl function of Asp and the hydroxyl group of Ser. In addition, several coenzymes such as thiamine pyrophosphate and pyridoxal phosphate function in association with their holoenzymes mainly as covalent catalysts.

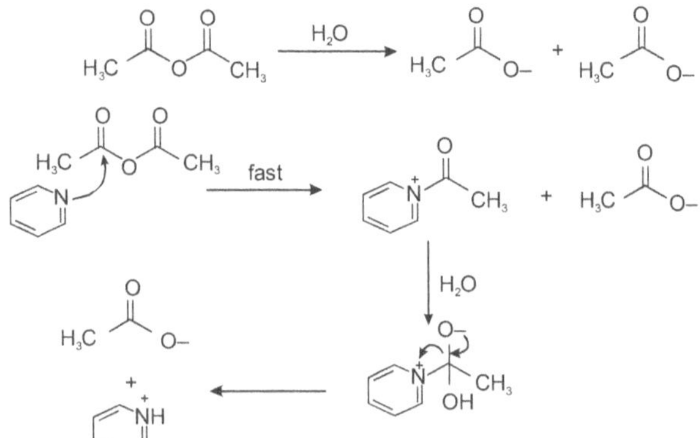

There must be some advantage in any particular enzymatic reaction that proceeds via covalent catalysis. This reaction is catalysed by pyridine, a better nucleophile than water (pKa = 5.5). Hydrolysis is accelerated because of loss of charge in the transition state.

Figure 7.3 Covalent catalysis

METAL ION CATALYSIS

Metal-activated enzymes may have an absolute requirement for the metal ion, or they may simply have enhanced activity in the presence of the metal ion. Phosphofructokinase is an example of a metal-activated enzyme which catalyses the reaction.

Fructose 6-phosphate + ATP → Fructose 1,6-diphosphate + ADP

A divalent metal ion (Mg^{2+}) is needed to coordinate the phosphate groups on the ATP molecule in order for phosphofructokinase to successfully catalyse the reaction. Mg^{2+},Mn^{2+}, Ca^{2+} and K^{+} often function as cofactors for metal-activated enzymes.

Most enzymes require the presence of metal ions for catalytic activity. There are two classes of metal ion-requiring enzymes that are distinguished by the strengths of their ion–protein interactions.

1. *Metalloenzymes* Metalloenzymes contain tightly bound metal ions. The most common transition metal ions are Fe^{2+}, Fe^{3+}, Cu^{2+}, Zn^{2+}, Mn^{2+} and Co^{3+}.

2. *Metal-activated enzymes* Metal-activated enzymes loosely bind to metal ions from solution, usually the alkali and alkaline earth metal ions Na^{+}, K^{+}, Mg^{2+} or Ca^{2+}.

Metal ions participate in the catalytic process in three major ways, which are as follows:

1. By binding to substrates so as to properly orient them for reaction.
2. By mediating oxidation–reduction reactions through reversible changes in the metal ion oxidation state.
3. By electrostatically stabilizing or shielding negative charges.

In many metal ion-catalysed reactions, the metal ion acts in much the same way as a proton, that is, as a Lewis acid. Yet, metal ions are often much more effective catalysts than protons because metal ions can be present in high concentrations at neutral pH and can have charges > +1. Metal ions have therefore been called **super acids**. The decarboxylation of dimethyloxaloacetate as catalysed by metal ions such as Cu^{2+} and Ni^{2+} is a non-enzymatic example of catalysis by metal ion.

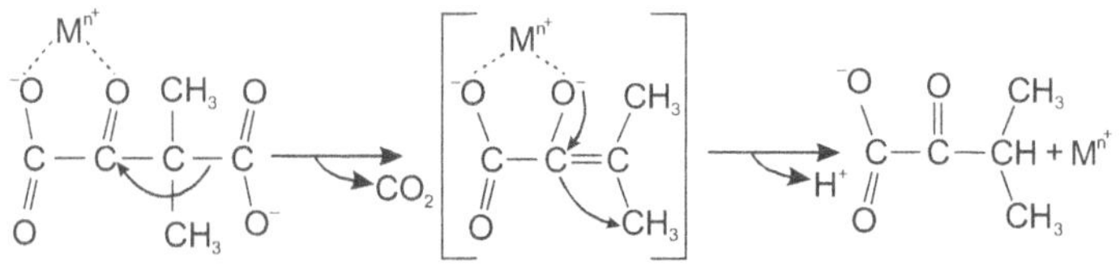

Dimethyloxaloacetate

Here the metal ion (Mn^{+}) which is chelated by the dimethyloxaloacetate, electrostatically stabilizes the developing enolate ion of the transition state. This mechanism is supported by the observation that acetoacetate, which cannot form such a chelate, is not subject to metal ion catalysed decarboxylation. Most enzymes that decarboxylate oxaloacetate require a metal ion for activity. Another important enzymatic function of metal ions is charge shielding. For example, the actual substrates of kinases are Mg^{2+}–ATP complexes such as

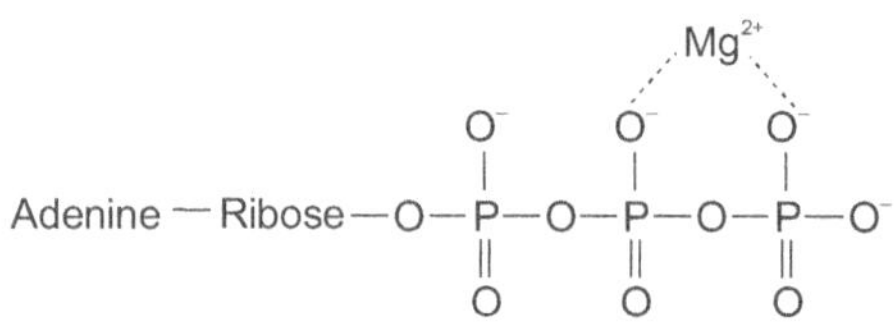

ATP

rather than just ATP. Here, the Mg^{2+} ion's role, in addition to its orienting effect, is to electrostatically shield the negative charges of the phosphate groups. Otherwise, these charges would tend to repel the electron pairs of attacking nucleophiles, especially those with anionic character (Figure 7.4).

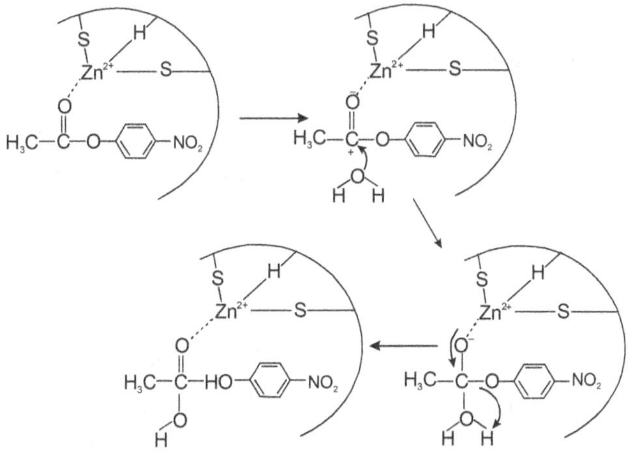

(a) Metal ions that are bound to the protein (prosthetic groups or cofactors) can also aid in catalysis. In this case, zinc is acting as a Lewis acid. It coordinates to the non-bonding electrons of the carbonyl, inducing charge separation, and making the carbon more electrophilic, or more susceptible to nucleophilic attack.

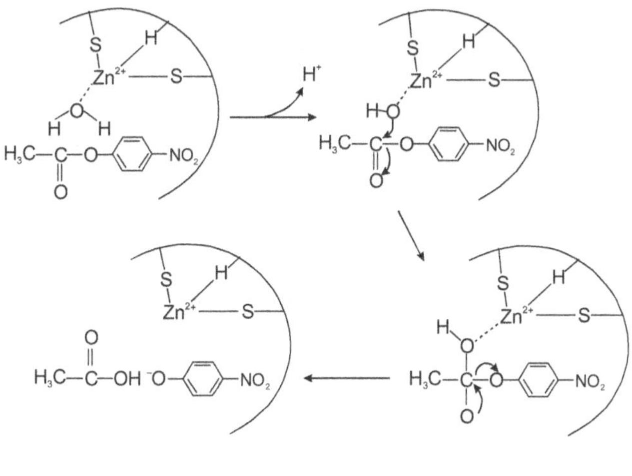

(b) Metal ions can also function to make potential nucleophiles (such as water) more nucleophilic. For example, the pKa of water drops from 15.7 to 6–7 when it is coordinated to zinc or cobalt. The hydroxide ion is 4 orders of magnitude more nucleophilic than is water.

Figure 7.4 Metal ion catalysis

Metalloenzymes

Metalloenzymes are metal-dependent enzymes, in which the metal ion is bound tightly to the enzyme and is not dissociated even after several extensive steps of purification. Metal ions can be involved in enzyme catalysis in a variety of ways:

1. They may accept or donate electrons to activate electrophiles or nucleophiles, even in neutral solution.
2. They themselves may act as electrophiles; they may mask nucleophiles to prevent unwanted side reactions.
3. They may bring together enzymes and substrate by means of co-ordination bonds, possibly causing strain to the substrate in the process.
4. They may hold reacting groups in the required three-dimensional orientation.
5. They may simply stabilize a catalytically active conformation of the enzyme (Figure 7.5).

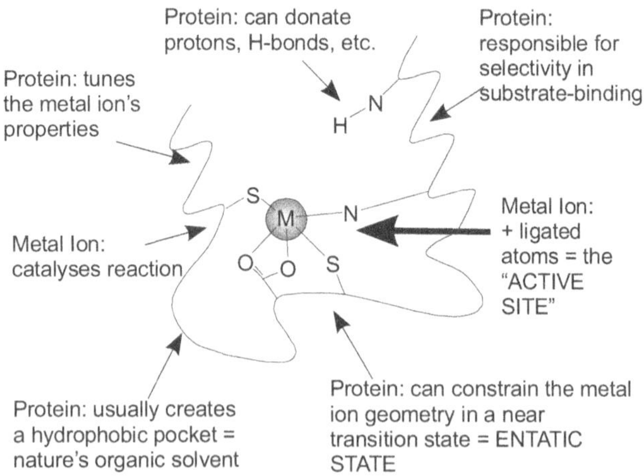

Figure 7.5 Function of active sites of metalloenzyme

The metal ion is bound to the protein with one labile coordination site. As with all enzymes, the shape of the active site is crucial. The metal ion is usually located in a pocket whose shape fits the substrate. The involvement of metal ions in enzymes may be investigated by NMR, ESR and proton relaxation rate (PRR) enhancement techniques.

Carbonic anhydrase Carbonic anhydrase catalyses the combination of CO_2 and H_2O to form carbonic acid, and also catalyses the reverse reaction.

$$CO_2 + H_2O \rightleftharpoons H_2CO_3$$

It is a zinc-containing enzyme; there are three coordinating histidine residues to which the zinc ion is coordinated. It requires the hydroxide ion, which acts as a catalyst and in its presence the reaction is quite fast.

$$CO_2 + OH \rightleftharpoons HCO_3^-$$

The structure of the active site in carbonic anhydrase is well-known from a number of crystal structures, it contains a zinc ion coordinated by three imidazole nitrogen atoms from three histidine units. The fourth coordination site is occupied by a water molecule. The coordination sphere of the zinc ion is approximately tetrahedral. The positively charged zinc ion polarizes the coordinated water molecule and the nucleophilic attack by the negatively charged hydroxide portion on the carbon dioxide (carbonic anhydride) proceeds rapidly. The catalytic cycle produces the bicarbonate ion and the hydrogen ion as the equilibrium favours dissociation of carbonic acid at biological pH values.

$$H_2CO_3 \rightleftharpoons HCO_3^- + H^+$$

Vitamin B_{12}-dependent enzyme This enzyme catalyses the transfer of methyl ($-CH_3$) groups between two molecules, which involves the breaking of C–C bonds, a process that is energetically expensive in organic reactions. It consists of a cobalt (II) ion coordinated by four nitrogen atoms of a corrin ring and a fifth nitrogen atom from an imidazole group. In the resting state there is a Co–Cσ bond with the 5′ carbon atom of adenosine. This is a naturally occurring organometallic compound, it participates in, trans-methylation reactions, such as the reaction carried out by methionine synthase.

Nitrogenase This enzyme is involved in the fixation of atmospheric nitrogen which is a very energy-intensive process. It involves the breaking up the very stable triple bond between the nitrogen atoms. The enzyme occurs in certain bacteria and there are three components to its action: a molybdenum atom at the active site, iron sulphur clusters which are involved in transporting the electrons needed to reduce the nitrogen and an abundant energy source. The energy is provided by a symbiotic relationship between the bacteria and the host plant, often a legume. The reaction may be written as:

$$N_2 + 16MgATP + 8e^- \rightarrow 2NH_3 + 16MgADP + 16P_i + H_2$$

P_i stands for inorganic phosphate. The precise structure of the active site has been difficult to determine. It appears to contain a $MoFe_7S_8$ cluster which is able to bind the dinitrogen molecule and, presumably, enables the reduction process to begin. The electrons are transported by the associated "P" cluster, which contains two cubical Fe_4S_4 clusters joined by sulphur bridges.

Superoxide dismutase Superoxide dismutase is a copper-metalloenzyme which catalyses the removal of the highly reactive O_2 (superoxide ion). The superoxide ion is generated in biological systems by the reduction of molecular oxygen.

It has an unpaired electron, so it behaves as a free radical. It is a powerful oxidizing agent therefore, must be removed from the cell before it does unwanted damage to the cell.

Superoxide dismutase catalyses the dismutation of superoxide ion into molecular oxygen and hydrogen peroxide.

$$2O_2^- + 2H^+ \rightarrow O_2 + H_2O_2$$

It involves both oxidation and reduction of superoxide ions.

Bovine erythrocyte superoxide dismutase is a dimeric protein containing two Cu^{2+} ions and two Zn^{2+} ions. The Zn^{2+} ions appear to have a structural rather than a catalytic role, while the Cu^{2+} ions are involved in the reaction sequence:

$$E - Cu^{2+} + O_2^- \rightarrow E - Cu^+ + O_2$$

$$E - Cu^{2+} + O_2^- \rightarrow E - Cu^{2+} + H_2O_2$$

The copper ions are coordinated tetrahedrally by four histidine residues. This enzyme also contains zinc ions. Other isozymes may contain iron, manganese, magnesium or nickel. Ni-SOD is particularly interesting because it contains nickel(III), an unusual oxidation state for this element. The active site Ni geometry cycles from square planar Ni(II), with thiolate (Cys2 and Cys6) and backbone nitrogen (His1 and Cys2) ligands, to square pyramidal Ni(III) with an added axial His1 side chain ligand.

Alkali metal cations (Na^+ and K^+) bind only weakly to form complexes with enzymes, but K^+ is known to activate a great many enzymes particularly those catalysing phosphoryl transfer or elimination reactions. It appears that the role of K^+ is largely to bind to negatively charged groups on an inactive form of the enzyme and thus cause a change in conformation to a more active form. Pyruvate kinase catalyses the reaction:

$$\text{PEP} + H^+ + \text{ADP} \rightarrow \text{Pyruvate} + \text{ATP}$$

This has a requirement for alkali metal cation and for Mn^{2+} (or Mg^{2+}), all of which bind to the region of the active site. Various studies have indicated that the carboxyl group or PEP binds to the enzyme-bound K^+, which brings about a conformational change that facilitates the progress of the reaction via the E–Mn^{2+} –PEP complex.

ELECTROSTATIC CATALYSIS

The binding of substrate generally excludes water from an enzyme's active site. The local dielectric constant of the active site therefore resembles that in an organic solvent where electrostatic interactions are much stronger than they are in aqueous solutions. The charge distribution in a medium of low dielectric constant can greatly influence chemical reactivity (Figure 7.6).

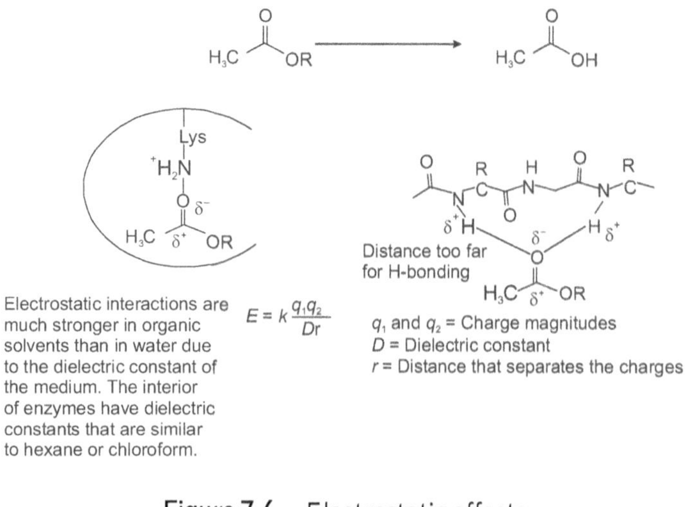

Figure 7.6 Electrostatic effects

The charge distribution in the relatively anhydrous active site may influence the chemical reactivity of the substrate. In addition, weak electrostatic interactions such as those between permanent and induced dipoles in both the reactive site and the substrate are believed to contribute to catalysis. A more efficient binding of substrate lowers the free energy of the transition state, which accelerates the reaction. Thus enzymes act by stabilizing the distribution of electrical charge in the transition states. Many inhibitors of enzymes mimic the transition state. The enzyme binds the inhibitor at a stable energy state, and these inhibitors bind with a much higher affinity than the normal substrate. Therefore, the pK's of amino acid side chains in proteins may vary by several units from their nominal values because of the proximity of charged groups.

Although experimental evidences are sparse, there is mounting evidence that the charge distribution about the active sites of enzymes are arranged so as to stabilize the transition states of the catalysed reactions. Such a mode of rate enhancement, which resembles the form of metal ion catalysis, is termed as electrostatic catalysis.

Moreover, in many enzymes these charge distributions apparently serve to guide polar substrates towards their binding sites so that the rates of these enzymatic reactions are greater than their apparent diffusion-controlled limits.

CATALYSIS THROUGH PROXIMITY AND ORIENTATION EFFECTS

Although enzymes employ catalytic mechanisms that resemble those of organic model reactions, they are far more catalytically efficient than those models. Such efficiency must arise from the specific physical conditions at enzyme catalytic sites that promote the corresponding chemical reactions. The most obvious effects are proximity and orientation.

The substrate must come into close proximity to catalytic functional groups (side chains of amino acids which are involved in catalytic reactions) within the active site. In addition, the substrate must be precisely oriented to the catalytic groups. Once the substrate is correctly positioned, a change in the enzyme's conformation may result in a strained enzyme–substrate complex. This strain helps to bring the enzyme–substrate complex into the transition state. In general, the more tightly the active site can bind the substrate while it is in the transition state, the greater the rate of the reaction.

Proximity alone can account for a fivefold increase in activity. Reactants must come together with the proper spatial relationship for a reaction to occur. This increases the rate of the reaction as enzyme–substrate interactions align reactive chemical groups and hold them close together. When two molecules collide with enough energy to cause a reaction, they don't necessarily form products. They have to be oriented so that the energy of the colliding molecules is transferred to the reactive bond. Enzymes bind substrates so that the reactive groups are steered to the direction that can lead to a reaction. This reduces the entropy of the reactants and thus makes reactions such as ligations or addition reactions more favourable. There is a reduction in the overall loss of entropy when two reactants become a single product. This effect is analogous to an effective increase in concentration of the reagents. The binding of the reagents to the enzyme gives the reaction intramolecular character, which gives a massive rate increase. For example, the biomolecular reaction of imidazole with *p*-nitrophenylacetate is as follows:

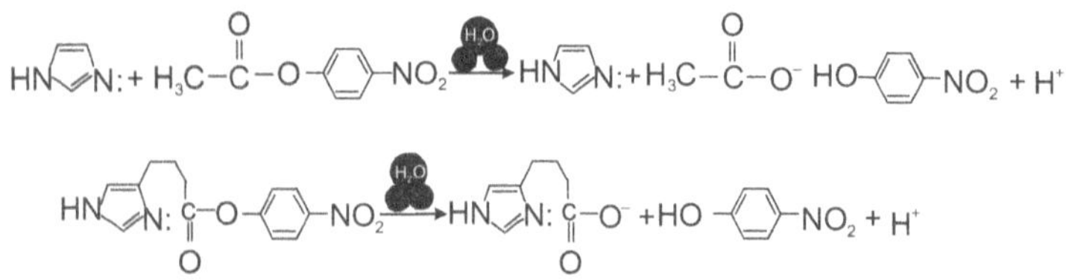

The progress of the reaction is conveniently monitored by the appearance of the intense yellow *p*-nitrophenolate ion. Here k'_1, the pseudo first-order rate constant, is 0.0018 s^{-1} when [imidazole] = 1 M (Φ = phenyl). However, for the intramolecular reaction, the first-order rate constant k_2 = 0.043 s^{-1}, that is, $k_2 = 24\,k'_1$. Thus, when the 1M imidazole catalyst is covalently attached to the reactant, it is 24-fold more effective than when it is free in solution, that is, the imidazole group in the intramolecular reaction behaves as if its concentration is 24 M.

PREFERENTIAL BINDING OF THE TRANSITION STATE COMPLEX

The concept of transition state binding proposed that enzymes mechanically strain their substrates towards the transition state geometry through binding sites into which undistorted substrates do not fit properly. This is known as **rack mechanism**. This mechanism was based on the extensive evidence for the role of strain in promoting organic reactions. For example, the rate of the reaction is 315 times faster when R is CH_3 rather than H because of the greater steric repulsions between the CH_3 groups and the reacting groups.

Similarly, ring-opening reactions are considerably more facile for strained rings such as cyclopropane than for unstrained rings such as cyclohexane. In either process, the strained reactant more closely resembles the transition state of the reaction than does the corresponding unstrained reactant.

Thus, as suggested by Linus Pauling and Gustav Lienhard, interactions that preferentially bind the transition state increase its concentration and therefore proportionally increase the reaction rate. This can be explained by considering the kinetic consequences of preferentially binding the transition state of an enzymatically catalysed reaction involving a single substrate. The substrate S may react to form product P either spontaneously or through enzymatic catalysis.

$$\mathrm{S} \xrightarrow{K_N} \mathrm{P}$$

$$\mathrm{ES} \xrightarrow{K_E} \mathrm{EP}$$

Here K_E and K_N are the first-order rate constants for the catalysed and uncatalysed reactions, respectively. The relationships between the various states of these two reaction pathways are indicated in the following scheme.

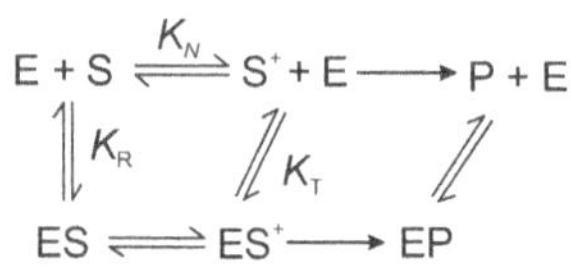

where,

$$K_R = \frac{[\mathrm{ES}]}{[\mathrm{E}][\mathrm{S}]} \qquad K_T = \frac{[\mathrm{ES}]}{[\mathrm{E}][\mathrm{S}]}$$

$$K_N^{+} = \frac{[\mathrm{E}][\mathrm{S}^{+}]}{[\mathrm{E}][\mathrm{S}]} \quad \text{and} \quad K_E^{+} = \frac{[\mathrm{ES}]}{[\mathrm{ES}]}$$

are all associations constants. Consequently,

$$\frac{K_T}{K_R} = \frac{[\mathrm{S}][\mathrm{ES}]}{[\mathrm{S}][\mathrm{ES}]} = \frac{K_E}{K_N}$$

According to transition state theory, the rate of the uncatalysed reaction can be expressed as follows:

$$v_N = K_N[\mathrm{S}] = \left(\frac{kK_{\mathrm{B}}^{\mathrm{T}}}{h}\right)[\mathrm{S}^{+}] = \left(\frac{kK_{\mathrm{B}}^{\mathrm{T}}}{h}\right)K_N[\mathrm{S}]$$

Similarly, the rate of the enzymatically catalysed reaction is

$$v_E = K_E[\mathrm{ES}] = \left(\frac{kK_B^T}{h}\right)\left[\mathrm{ES}^+\right] = \left(\frac{kK_B^T}{h}\right)K_E^+[\mathrm{ES}]$$

Therefore, combining equations

$$\frac{K_E}{K_N} = \frac{K_E^+}{K_N^+} = \frac{K_T}{K_R}$$

This equation indicates that the more tightly an enzyme binds its reaction's transition state (K_T), the greater the rate of the catalysed reaction (K_E), relative to that of the uncatalysed reaction (K_N); that is, catalysis results from the preferential binding and therefore the stabilization of the transition state relative to that of the substrate (S).

If an enzyme preferentially binds its transition state, then it can be expected that transition state analogs, stable molecules that resemble S or one of its components are potent competitive inhibitors of the enzyme. For example, the reaction catalysed by

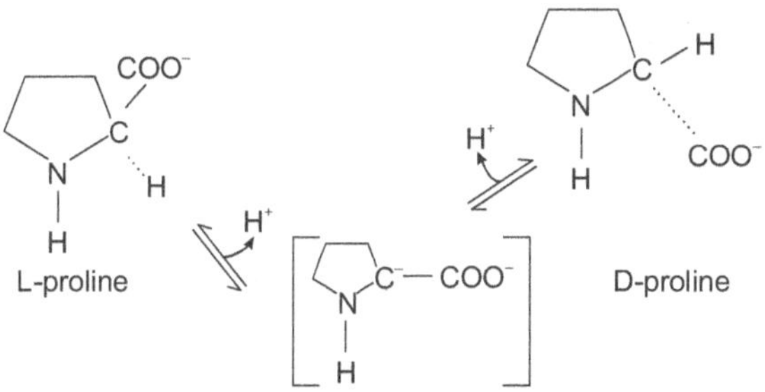

proline racemase from *Clostridium sticklandii* is thought to occur via a planar transition state. Proline racemase is competitively inhibited by the planar analogs of proline, pyrrole 2-carboxylate and 1-pyrroline 2-carboxylate, both of which bind to the enzyme

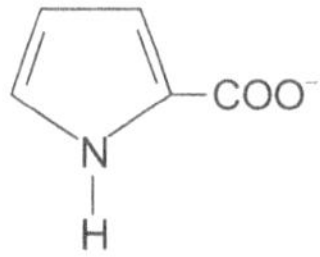

Pyrrole 2-carboxylate

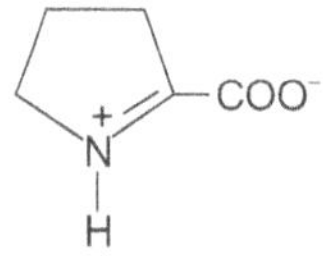

1-pyrroline 2-carboxylate

with tenfold greater affinity than does proline. These compounds are therefore thought to be analogs of the transition state in the proline racemase reaction. In contrast, tetrahydrofuran 2-carboxylate, which more closely resembles the tetrahydral structure of proline, is not nearly as good an inhibitor as these compounds.

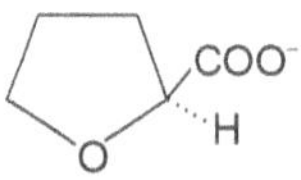

Tetrahydrofuran-2-carboxylate

Hundreds of transition state analogs for various enzymatic reactions have been reported. Some are naturally occurring antibiotics. Others were designed to investigate the mechanisms of particular enzymes and/or to act as specific enzymatic inhibitors for therapeutic or agricultural use. Indeed, the theory that enzymes bind transition states with higher affinity than substrates has led to a rational basis of drug design based on the understanding of specific enzyme reaction mechanisms.

To understand the mechanism of an enzyme-catalysed reaction, the mechanism of some enzymes—lysozyme, chymotrypsin ribonuclease and DNA polymerase— have been explored in detail in the following sections.

LYSOZYME

Lysozyme is a small stable enzyme, ideal for research into protein structure and function. Lysozyme is a single polypeptide chain of 129 amino acids cross-linked with four disulphide bridges. It hydrolyses ($\beta1 \rightarrow 4$) linkages between *N*-acetylmuramic acid and *N*-acetyl D-glucosamine residues in peptidoglycan and between N-acetyl D-glucosamine residues in chitodextrin (Figure 7.7).

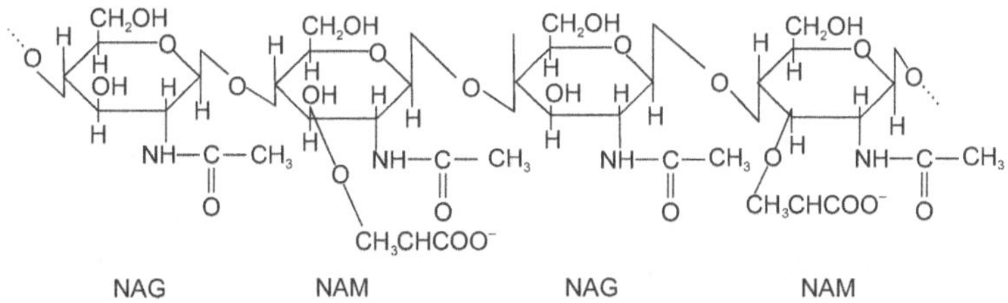

Figure 7.7 Structure of NAM and NAG

Lysozyme is an asymmetrical ellipsoidal protein having a long cleft or valley that runs the length of one side of the molecule. This cleft contains the active site of the enzyme that binds the carbohydrate substrate. It is a highly specific enzyme that accommodates only six sugar molecules of a polysaccharide and cleaves them into a disaccharide and a tetrasaccharide subunit. Common substrates include the polysaccharides, chitin and peptidoglycans (murein) found in bacterial cell walls.

The enzyme is active over a broad pH range (6.0–9.0). At pH 6.2, maximal activity is observed over a wider range of ionic strengths (0.002–0.100 M) and at pH 9.2, maximal activity is observed in the ionic strength range of 0.01–0.06 M.

Lysozyme is one of the hydrolytic enzymes of the lysosomal compartment of the phagocytic cells such as monocytes/macrophages and neutrophils. Lysozyme is secreted into the extracellular compartment and has been shown to be bactericidal. This enzyme along with other enzymes and proteins is involved in the first line of defence against any pathogenic infections. The biological importance of lysozyme lies in its ability to hydrolyse the polysaccharide portion of bacterial cell walls.

Lysozyme causes lysis of gram-positive and some gram-negative bacteria by specific cleavage of the $(\beta 1 \rightarrow 4)$ linkage between *N*-acetylglucosamine and *N*-acetylmuramate in the peptidoglycan backbone of bacterial cell wall. Lysozyme has been shown to specifically bind glucose-modified proteins bearing advanced glycation end products (AGEs). Exposure to AGE-modified protein has been shown to inhibit the enzymatic and bactericidal activity of lysozyme. It has been shown that the binding of lysozyme with lipopolysaccharide (LPS) affects the enzymatic activity of lysozyme as well as the immunostimulatory effect of LPS.

The antibacterial nature of enzymes was first witnessed by Alexandar Flemming in 1922. He treated a plated culture of bacteria with samples of his own nasal mucus, and over time the bacteria near the mucus dissolved away. The bacterial cells appeared to dissolve away because the cells were being lysed or broken apart. Thus, he named the enzyme "Lysozyme". Although lysozyme did not prove to be useful as an all-purpose antibiotic, it was subsequently found that the lysozyme is produced by a wide variety of organisms including bacteriophages, bacteria, plants, and animals.

Lysozyme serves as a non-specific innate opsonin by binding to the bacterial surface, reducing the negative charge and facilitating phagocytosis before opsonins from the acquired immune system arrived at the scene. In other words, lysozyme makes it easier for phagocytic white blood cells to engulf bacteria. The enzyme functions by attacking peptidoglycans and hydrolysing the glycosidic bond that connects

N-acetylmuramic acid with the fourth carbon atom of *N*-acetylglucosamine. It does this by binding to the peptidoglycan molecule in the binding site within the prominent cleft between its two domains (Figure 7.8). This causes the substrate molecule to adopt a strained conformation similar to that of the transition state.

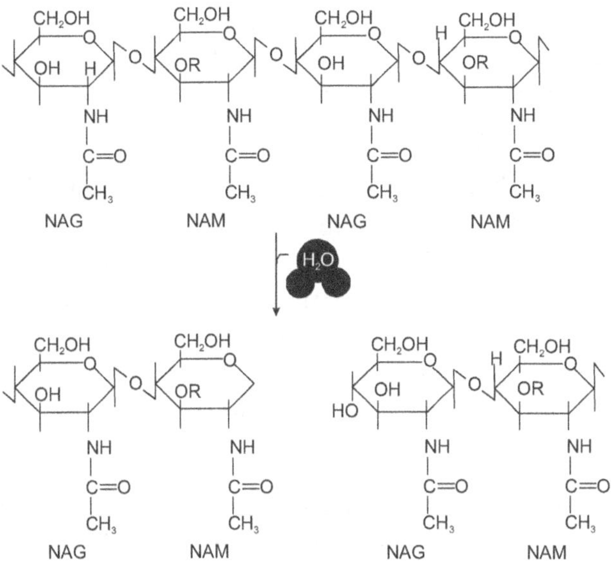

Figure 7.8 Mechanism of action of lysozyme

According to the Philips mechanism, lysozyme binds to a hexasaccharide. The lysozyme then distorts the fourth sugar in hexasaccharide (D-ring) into a half-chair conformation. In the stressed state, the glycosidic bond is easily broken. The amino acid side chains glutamic acid 35 (Glu 35) and aspartate 52 (Asp 52) have been found to be critical to the activity of this enzyme. Glu 35 acts as a proton donor to glycosidic bond, cleaving C–O bond in the substrate, whilst Asp 52 acts as a nucleophile to generate a glycosyl-enzyme intermediate. The glycosyl-enzyme intermediate then reacts with a water molecule, to give the product of hydrolysis and leaving the enzyme unchanged.

CHYMOTRYPSIN

Chymotrypsin is formed by the activation of its inactive precursor chymotrypsinogen, which is secreted and synthesized by the mammalian pancreas. Chymotrypsin catalyses the cleavage of the peptide bond at the carboxyl side of aromatic amino acid (phenylalanine, tyrosine or tryptophan) residues (Figure 7.9).

X-ray diffraction studies revealed a hydrophobic binding pocket at the active site for aromatic side chains, and showed that aspartate 102, buried in the interior of the molecule, could be closely linked to the action of histidine 57 and serine 195. A suggestive mechanism could be a charge relay system with aspartate 102 removing a proton from histidine 57 making it easier for the latter to remove a proton from serine 195 during the course of the reaction.

Histidine 57 acts as a base catalyst to enable the oxygen of serine 195 to make a nucleophilic attack on the carboxyl group of the enzyme-bound substrate. An unstable tetrahedryl intermediate is formed whose negatively charged oxygen atom may be stabilized by hydrogen-bonding with the backbone –NH of glycine 193. Similarly, the positive charge on the imidazole ring of histidine 57 is stabilized by electrostatic interaction with aspartate 102. The imidazole group of histidine 57 then acts as an acid catalyst to facilitate the liberation of the first product leaving behind the acyl enzyme. The second stage of the reaction is also initiated by a nucleophilic attack; however this time it is by water. The liberation of the second product (RCOOH) is by acid catalysis.

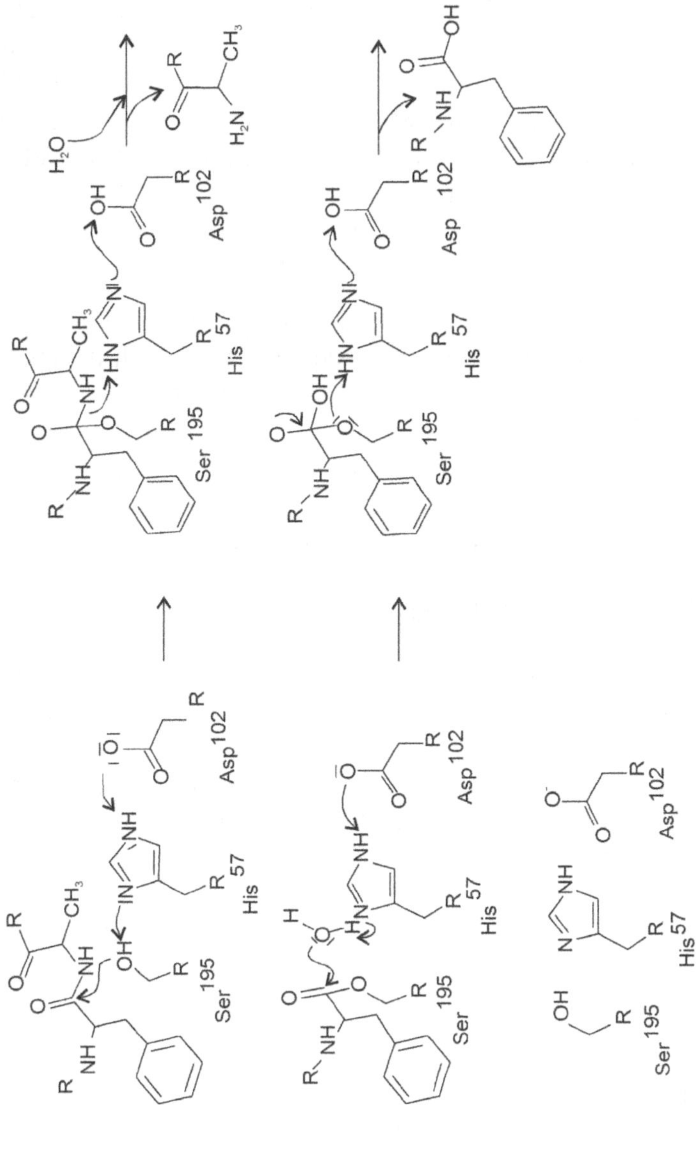

Figure 7.9 Mechanism of action of chymotrypsin

RIBONUCLEASE

Ribonuclease catalyses the cleavage of the phosphodiester backbone of ribonucleic acids by a reaction involving transfer of a phosphate group from the 5´-position of one nucleotide to the 3´-position of the next nucleotide in the chain. Histidine 12 and histidine 119 are the amino acids present at the catalytic site one possibly acting as a proton donor and the other as a proton acceptor. Lysine 41 is also reported to be involved in the catalytic action of ribonuclease.

Histidine 12 removes a proton from the substrate, producing a nucleophilic oxygen atom which attacks the phosphate. Histidine 119 donates a proton to the complex and the first product R´OH is liberated. The rest of the reaction is the reversal of the first with water replacing R´OH, while, lysine 41 helps to stabilize the complex by an electronic interaction with the negative charge of the phosphate.

DNA POLYMERASE

Several mechanisms control the fidelity of the DNA replication process. These include correct base selection, removal of base insertion errors by 3´ exonucleolytic proofreading and correction by DNA mismatch repair. This base selection and proofreading are performed by the DNA polymerase enzyme, the enzyme that replicates the bacterial chromosome. DNA polymerase is considered to be a holoenzyme since it requires a magnesium ion as a cofactor to function properly. In the absence of the magnesium ion, it is referred to as an apoenzyme.

DNA polymerase initiates DNA replication by binding free nucleotides only to the 3´end of the newly forming strand. This results in elongation of the new strand in the 5´–3´ direction. DNA polymerase can add a nucleotide only onto a pre-existing 3´OH group and therefore, needs a primer at which it can add the first nucleotide. Primers consist of RNA and DNA bases with the first two bases always being RNA and are synthesized by another enzyme called primase.

Error correction is a property of some, but not of all DNA polymerases. This process corrects mistakes in newly synthesized DNA. When an incorrect base pair is recognized, DNA polymerase reverses its direction by one base pair of DNA. The 3′–5′ exonuclease activity of the enzyme allows the incorrect base pair to be excised (proofreading). Following base excision, the polymerase can reinsert the correct base and replication can continue. Based on sequence homology, DNA polymerases can be further subdivided into seven different families: A, B, C, D, X, Y and RT.

i. Family A polymerases contain both replicative and repair polymerases, e.g. T7 DNA polymerase, mitochondrial DNA polymerase δ, *E. coli* DNA pol I.
ii. Family B polymerases mostly contain replicative polymerases and include the major eukaryotic DNA polymerases α, δ, ε and ζ.
iii. Family C polymerases are the primary bacterial chromosomal replicative enzymes.
iv. Family D polymerases are not well characterized, e.g. DNA polymerases found in the Euryarchaeota of Archaea are found to be replicative polymerases.
v. Family X contains the well-known eukaryotic polymerase pol β, pol σ, pol μ and pol λ.
vi. Family Y polymerases differ from others in having a low fidelity on undamaged templates and in their ability to replicate through damaged DNA. Members of this family are hence called translesion synthesis (TLS) polymerases.

vii. Family RT, the reverse transcriptase family, contains examples from both retroviruses and eukaryotic polymerases. The eukaryotic polymerases are usually restricted to telomerases. These polymerases use RNA template to synthesize the DNA strand.

Bacteria has 5 known DNA polymerases, they are as follows:

Pol I Implicated in DNA repair; has both 5´–3´ (polymerase) activity and 3´–5´ (proofreading) exonuclease activity.

Pol II Involved in replication of damaged DNA; has 3´–5´exonuclease activity.

Pol III The main polymerase in bacteria (elongates in DNA replication); has 3´–5´ exonuclease proofreading activity.

Pol IV A Y-family DNA polymerase

Pol V A Y-family DNA polymerase; participates in bypassing DNA damage.

Eukaryotes have at least 15 DNA polymerases including the following important ones.

Pol α Acts as a primase and then as a DNA polymerase elongating the primer with DNA nucleotides.

Pol β Implicated in repairing DNA, in base-excision repair and gap-filling synthesis.

Pol γ Replicates mitochondrial DNA

Pol δ Highly processive and has proofreading 3´–5´ exonuclease activity

Pol ε Has proofreading 3´–5´ exonuclease activity.

DNA Polymerase I

DNA polymerase I is an enzyme that participates in the process of DNA replication in prokaryotes. It is composed of 928 amino acids and can sequentially catalyse multiple polymerizations. It was discovered by Arthur Kornberg (1956). The structure of DNA polymerase I is given in Figure 7.10 Polymerase I possesses three enzymatic activities.

1. A 5´–3´ (forward) DNA polymerase activity, requiring a 3´primer site and a template strand.
2. A 3´–5´ (reverse) exonuclease activity that mediates proofreading activity.
3. A 5´–3´ (forward) exonuclease activity mediating nick translation during DNA repair.

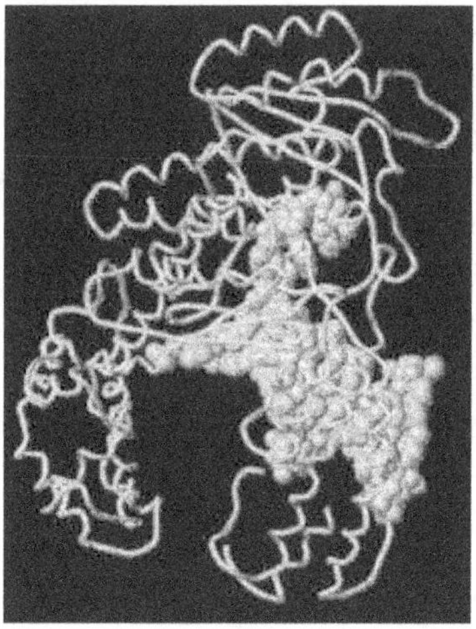

Figure 7.10 DNA polymerase

In the replication process, DNA polymerase I removes the RNA primer from the lagging strand and fills in the necessary nucleotides of the okazaki fragments in 5´–3´ direction, proofreading for mistakes as it goes. It is a template-dependent enzyme—it only adds nucleotides that correctly base-pair with an existing DNA strand acting as a template. Ligase then joins the various fragments together into a continuous strand of DNA.

DNA Polymerase II

DNA polymerase II is a prokaryotic DNA polymerase involved in DNA repair. The enzyme is 90 kDa in size and is coded by the polymerase B gene. Polymerase II differs from polymerase I in that it lacks a 5´–3´ exonuclease activity and cannot use a nicked duplex template.

DNA Polymerase III

DNA polymerase III holoenzyme is the pimary enzyme complex involved in prokaryotic DNA replication. It was discovered by Thomas Kornberg (Son of Arthur Kornberg) and Malcolm Gefter in 1970. DNA polymerase III works in conjunction with four other DNA polymerases (Pol I, Pol II, Pol IV and Pol V). DNA polymerase III holoenzyme also has proofreading capabilities that correct replication mistakes by means of exonuclease activity working in 3´–5´direction.

REVIEW QUESTIONS

1. What do you mean by active site? Bring out the importance of determining the active site by chemical methods.
2. What are TCPK and TLCK? Discuss how they are used in determining the active site.
3. Discuss how aspartate and glutamate can be ascertained at the active site.
4. How are sulphur-containing amino acids detected at the active site?

8

CONTROL OF ENZYME ACTIVITY

The control of enzyme activity is one of the remarkable properties of enzymes. It should be self-evident why enzyme activity needs to be controlled in living organisms, since if no such control existed, all metabolic processes would tend towards states in which they were at equilibrium with their surroundings. Thus, for instance, storage of glycogen as an energy reserve in muscle would be impossible if the enzyme phosphorylase (present in muscle) catalysing the breakdown of glycogen into glucose units is not perfectly controlled. Clearly, the existence of glycogen as an energy store shows that the activity of phosphorylase can be controlled in such a way as to allow the store to supply when there is demand for glucose.

Regulation of enzyme activity can be brought about in two ways:

- Firstly, the amount of an enzyme can be altered as a result of changes in the rate of enzyme synthesis or degradation (or both). This type of regulation is suitable for long-term regulation, e.g. in response to changes in nutrients.

- Secondly, the activities of enzymes already present in a cell or organism can be altered, permitting a rapid response to changes in conditions.

CONTROL OF ACTIVITIES OF SINGLE ENZYMES

The principle mechanism that exists for control of enzyme activity can be broadly classified into two categories:

1. Those mechanisms that involve a change in the covalent structure of enzymes.
2. Those mechanisms that involve conformational changes in an enzyme caused by the reversible binding of "regulator" molecules.

Control of Activity by Changes in the Covalent Structures of Enzymes

This category of control can be subdivided into

1. those cases which are essentially irreversible and
2. those cases in which the change can be reversed provided that a source of energy, such as ATP, is available.

(i) *Control of activity by essentially irreversible changes in covalent structure* Generally a number of proteases are synthesized in the form of inactive precursors or zymogens, which can then be activated by the action of other proteases.

The first four enzymes mentioned in Table 8.1 are synthesized as zymogens in the acinar cells of the pancreas and stored in zymogen granules. The release of the zymogens from these granules into the duodenum is under the control of hormones, primarily cholecystokinin-pancreazymin and secretin.

Table 8.1 Precursors and functions of enzymes

Enzyme	Precursor	Function
Trypsin	Trypsinogen	
Chymotrypsin	Chymotrypsinogen	Pancreatic secretions
Elastase	Pro-elastase	
Carboxypeptidase	Pro-carboxypeptidase	
Pepsin	Pepsinogen	Secreted into gastric juice
Thrombin	Prothrombin	Part of the blood coagulation system
Phospholipase A2	Pro-phospholipase A2	Pancreatic secretions

Digestion of proteins requires the coordinated action of these various enzymes with their different specificities; coordination is achieved by the common factor that the zymogens are all activated by the action of trypsin (Figure 8.1).

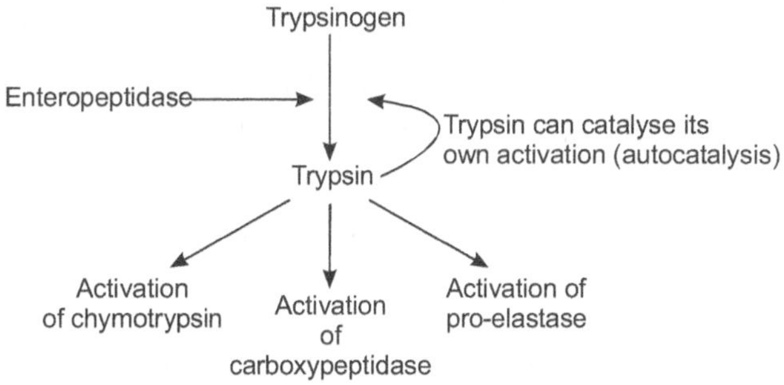

Figure 8.1 Coordinated activation of the pancreatic proteases

In each case, the peptide bond cleaved during the activation of the zymogen lies on the C-terminal side of a lysine or arginine side-chain in accordance with the known specificity of trypsin. For example, the activation of trypsinogen involves the removal of a hexapeptide Val-Asp-Asp-Asp-Asp-Lys from the N-terminus of the polypeptide chain. In order to account for the role of trypsin in initiating the activation of zymogens, we must consider the involvement of another protease, enteropeptidase. It is synthesized in the brush border of the epithelial cells of the small intestine; it is an enzyme of high specificity catalysing the activation of trypsinogen as it enters the duodenum. The trypsin produced then catalyses the activation of the zymogens, since this would result in considerable damage to the pancreas. In order to achieve this, the initial trigger provided by enteropeptidase is located at a site distinct from the production site of zymogens. The amount of active proteases produced is controlled by the amount of zymogens secreted into the duodenum, which is in turn under hormonal control. Eventually, the proteases digest themselves into amino acid and small peptides, which are reabsorbed from the intestinal tract and used for protein synthesis. The average daily production of the short-lived hydrolytic enzymes by the human pancreas is of the order of 10 g.

An additional mechanism to prevent premature activation of trypsinogen is provided by the presence of a trypsin inhibitor protein in the pancreatic secretion. This protein, which is present to the extent of about 2% of the protein content of the secretion, binds very tightly to any active trypsin that might be present in the zymogen granules, and thus ensures that the activation of trypsinogen occurs only at the physiologically appropriate time and place, i.e., when it encounters enteropeptidase in the small intestine. However, the control of the production of enteropeptidase is less well-understood.

Trypsin activates the inactive precursor chymotrypsinogen into an active form called chymotrypsin. Chymotrypsin is a digestive enzyme that hydrolyses proteins in the small intestine (proteolytic enzyme). Its inactive precursor, chymotrypsinogen is synthesized in the acinar cells of the pancreas and stored in the zymogen granules. When there is a hormonal signal, these are released from the zymogen granules into the duodenum.

Chymotrypsinogen is a single polypeptide chain consisting of 245 amino acid residues. It is cross-linked by five disulphide bonds. This precursor is virtually devoid of enzymatic activity (inactive) because the substrate-binding pocket is not properly formed, and thus a substrate cannot be positioned precisely inside that pocket. It is converted into a fully active enzyme when the peptide bond joining arginine 15 and isoleucine 16 is cleaved by trypsin.

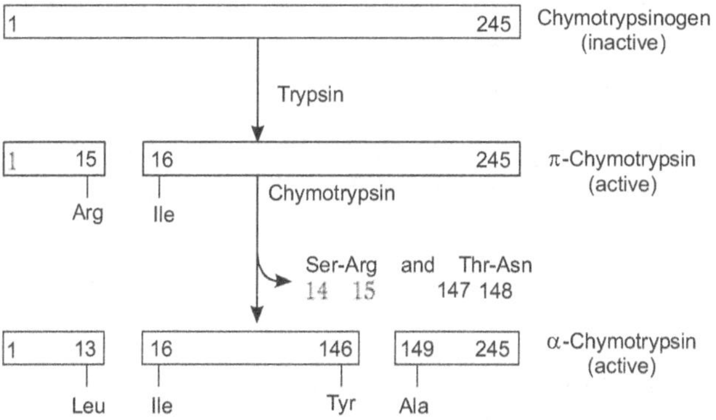

Figure 8.2 Chymotrypsin activation

The resulting active enzyme called π-chymotrypsin, then acts on other π-chymotrypsin molecules. Two dipeptides are removed to yield α-chymotrypsin, the stable form of the enzyme. The striking feature of this activation process is that cleavage of a single, specific peptide bond transforms the protein from a catalytically inactive form into one that is fully active (Figure 8.2).

The proteolytic activation of chymotrypsinogen leads to the formation of a substrate-binding site; the hydrolysis of the peptide bond between arginine 15 and leucine 16 results in the newly formed amino terminal group of isoleucine 16 which turns inward and forms an electrostatic bond with aspartate 194 in the interior. This interaction triggers a series of local conformational changes that results in the creation of a pocket for the binding of the aromatic side chain of the substrate. In addition, an oxyanion hole that stabilizes the transition state is formed. A molecule of pepsinogen activates itself in acidic media by hydrolysing a specific peptide bond.

The coordinated activation of pancreatic proteases illustrates the principle of signal amplification, i.e., a small amount of enteropeptidase can produce rapidly a larger amount of trypsin since one molecule of enzyme can act on many molecules of substrate in a short time. The trypsin, in turn, gives rise to an even larger quantity of active protease.

Thus, the pancreatic proteases synthesized as zymogens are irreversibly covalently modified to their active forms. A similar irreversible covalent modification is also viewed in the blood clotting scheme, and in complement activation.

In the blood coagulation systems, a series of enzymes and other protein factors act on each other in a sequential fashion known as cascade, eventually leading to the activation of zymogen prothrombin to yield thrombin. Thrombin then catalyses the hydrolysis of limited number of Arg–Gly bonds in fibrinogen to produce fibrin which spontaneously forms an insoluble clot, strengthened by cross-linking between fibrin molecules. There are two pathways of activation of blood coagulation: extrinsic, which is triggered by a lipoprotein released from injured tissue, and intrinsic, which is triggered by contact of one of the factors with a surface such as the fibrous protein collagen that surrounds the damaged blood vessel (Figure 8.3).

In both pathways, the sequence of the enzyme-catalysed reaction provides a large amplification of the initial signal to ensure that sufficient fibrin is formed rapidly at the site of injury to restrict the blood loss. A striking feature of this process is that the activated form of one clotting factor catalyses the activation of the next factor.

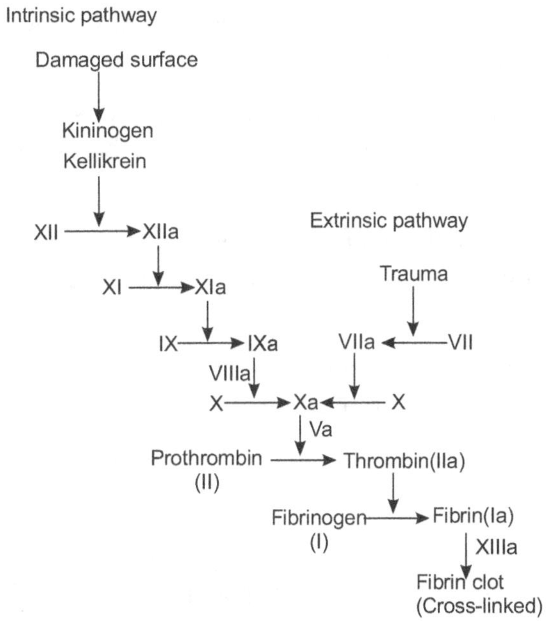

Figure 8.3 Blood clotting mechanism

The complement system consists of a series of blood plasma proteins that cause damage to, and eventually lysis of, an invading organism, such as a bacterium. The trigger for the system is usually provided by the aggregation of antibody molecules bound to the surface of the invading organism. This aggregation activates the components of the complement system by a sequence of activations of zymogens, leading to the rupture of the membrane of the invading organism.

In the various cases discussed above, the sequences of enzyme-catalysed reactions provides a rapid, amplified response to a small initial signal. Once the zymogens have been activated, and the resulting proteases have performed their particular functions, the active enzymes are degraded rather than being converted back to zymogens. It is obviously important that the systems are "switched off" at an appropriate time, so that for instance fibrin is formed only at the site of an injury and not in the entire blood system.

(ii) *Control of activity by reversible changes in covalent structure* A number of enzymes may be listed under this category. However we illustrate a few:

A large number of enzymes exist in two forms (of different catalytic properties), which can be interconverted by the action of other enzymes.

A classical example of this type of enzyme is phosphorylase, which catalyses the reaction:

$$(\text{Glycogen})_n + \text{Orthophosphate} \rightleftharpoons (\text{Glycogen})_{n-1} + \alpha\text{-D-glucose 1-phosphate}$$

This enzyme exists in two forms: **phosphorylase b,** which requires AMP for activity and **phosphorylase a,** which is active in the absence of AMP, although AMP activates it to a small extent. The only difference in the covalent structure between the **a** and **b** forms of phosphorylase is that the side chain of ser 14 is phosphorylated in the **a** form but not phosphorylated in the **b** form. Comparison of the three-dimensional structures of these two forms shows that the first 19 amino acids from the N-terminus are markedly different. This segment is flexible in the **b** form, but well-ordered on the **a** form, with the phosphorylated side chain of ser 14 interacting with the positively charged side chain of Arg 69.

In the direction, **b** → **a**, the reaction requires ATP and Mg^{2+} and is catalysed by phosphorylase kinase, and in the direction

a $\rightarrow$ **b,** the reaction is catalysed by phosphorylase phosphatase and liberates orthophosphate. If the reaction proceeds in a

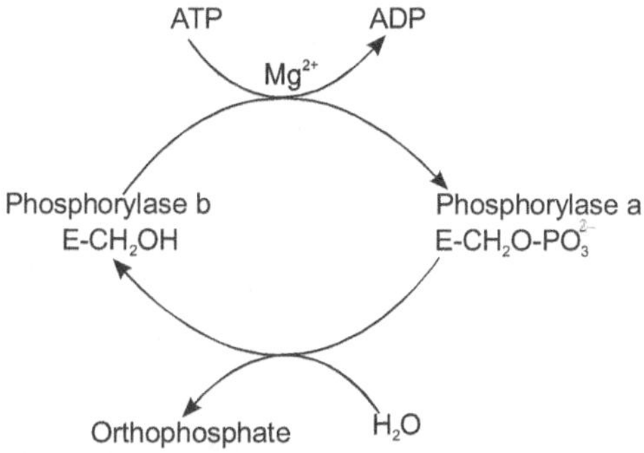

cyclic fashion, the net result would be the hydrolysis of ATP to ADP and orthophosphate. Over 140 enzymes and other proteins can be controlled by reversible covalent modification; some of them are phosphorylase, glycogen synthase, phosphorylase kinase, pyruvate dehydrogenase complex, glutamine synthetase, phosphatase-inhibitor protein, troponin-1, etc.

The majority of the reversible covalent modifications involve a phosphorylation–dephosphorylation cycle, usually at a serine side chain, but in the case of phosphatase-inhibitor protein, this cycle is at a threonine side chain. In the case of glutamine synthetase, a tyrosine side chain (tyr 397) in each of the 12 subunits of the enzyme can be adenylated.

The other types of modification include methylation, acetylation, tyrosinolation and sulphation.

The two important aspects about the covalent modification of enzymes are:

1. The conversion of an enzyme from one form to another is catalysed by an enzyme. There can be a rapid change in the amount of active enzyme present and a large amplification of an initial signal.

2. Reversible modification permits much more controlled responses to different metabolic circumstances than is the case with irreversible covalent modification. In the former case, the system can be viewed as being poised for response; in the latter case, there is a continual activation and inactivation of the enzyme.

Under the influence of an appropriate stimulus, a system can rapidly be activated to produce almost exclusively the active form of the required enzyme. When the stimulus is removed, the system can be converted back to its resting stage (almost exclusively the inactive form of the enzyme) and it is also possible to generate a whole range of intermediate levels of response. By contrast, in the cases of irreversible covalent modification, such as blood coagulation and complement activation, there is a rapid amplified response to some emergency (e.g. injury or infection) but, after use, the whole cascade systems must be replenished, since the steps in the cascades operate in one direction only.

Control of Activity by Ligand-induced Conformational Changes in Enzymes

The importance of ligands in controlling the activity of enzymes was demonstrated by Cori and her workers, who showed that AMP was required to bring about the breakdown of glycogen in muscle. Many biosynthetic pathways strengthened this fact.

Example 1 The inhibition of threonine dehydratase, by L-isoleucine, the end product of isoleucine biosynthetic pathway in *E. coli,* is a good example. Here, isoleucine is synthesized from L-threonine and the first step of this pathway is catalysed by the enzyme threonine dehydratase. Only this first enzyme in the pathway shows this inhibition. The end-product inhibitor isoleucine bears only a limited structural resemblance to the substrate or the product of the reaction.

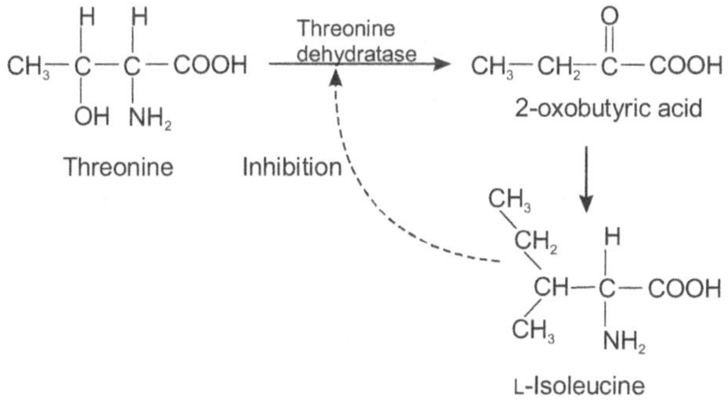

Example 2 Another striking example is provided by aspartate carbamoyl transferase. It catalyses the first step in pyrimidine biosynthesis in *E. coli* and is inhibited by the end products of the pathway especially CTP. However the purine nucleotide ATP also can activate this enzyme.

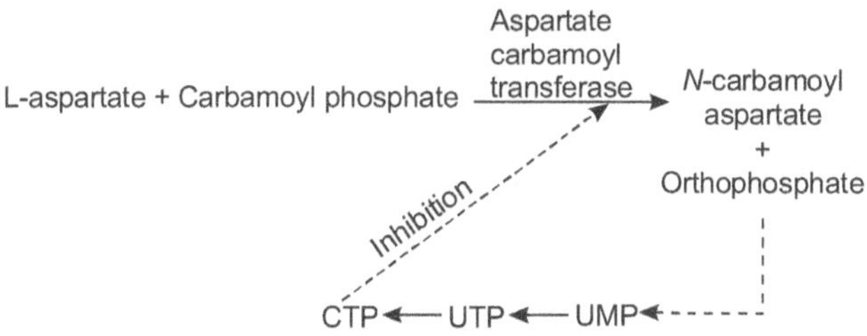

This provides a mechanism for achieving a balance between the production of pyrimidine and purine nucleotides, which is required for the synthesis of nucleic acids.

These two types of regulation are said to be feedback inhibition.

The common features of this type of control are:

- The ligands that are involved in the regulation of enzyme activity are structurally distinct from the substrates or products of the relevant enzyme-catalysed reactions. They bind to a distinct site in the enzyme. The inhibition caused by them may be of mixed type rather than competitive, non-competitive or uncompetitive types.

- Enzymes which are subject to such a control do not show the normal type of kinetic behaviours. Their Michaelis–Menten curve is sigmoidal rather than hyperbolic.
- It is often possible to distinguish between the binding of effector and the binding of substrate to a regulatory enzyme (Figure 8.4). Various physical or chemical treatments could lead to desensitization of the enzyme, i.e., loss of response to regulator molecules with no loss of catalytic activity. In the case of aspartate carbamoyl transferase, mild heat treatment yielded an active enzyme that was not inhibited by CTP. The desensitized enzyme showed normal, hyperbolic kinetic behaviour.
- In general, the regulatory enzymes are oligomeric, i.e., they have multiple subunits held together by noncovalent forces.

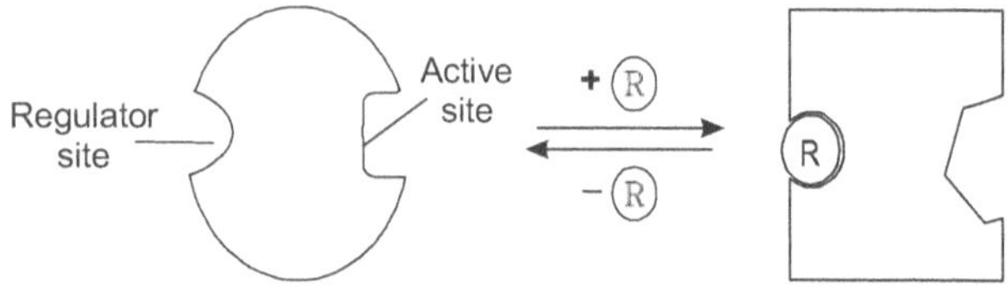

Figure 8.4 Alteration of the structure of the active site of an enzyme by reversible binding of a regulator R to a regulator site

Such regulatory enzymes are said to be allosteric. The binding of the regulator molecule to the regulator site induces a reversible conformational change in the enzyme that causes an alteration of the structure of the active site and a consequent change in the kinetic properties of the enzyme. A detailed explanation of allosteric modification is explained in chapter 10.

Feedback Inhibition

An even more subtle type of control of enzyme action is designated as feedback inhibition. This is demonstrated most easily by considering the following sequence.

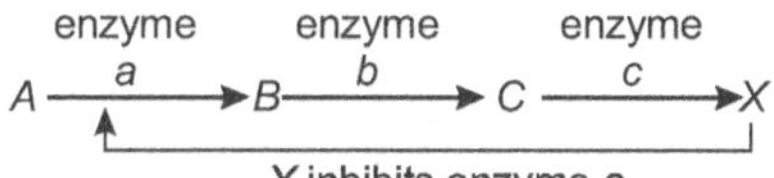

Here, X, the ultimate product of the sequence serves to prevent the formation of one of its own precursors by inhibiting the action of enzyme *a,* the first enzyme of the sequence, which is also called a monovalent regulatory or allosteric enzyme Figure 8.5. It can also be called the pacemaker since the entire sequence is effectively regulated by inhibiting it.

Let us now consider the different feedback inhibitions of metabolic sequences. The regulation of a linear sequence as referred is a straightforward end-product inhibition of the first enzyme in the sequence, a monovalent allosteric enzyme. But the regulation of a branched biosynthetic pathway of *X* and *Y* would lead to a situation in which an excess of one end product would lead not only to a decrease in the synthesis of *X*, but also the other end product, and is not a very good control system.

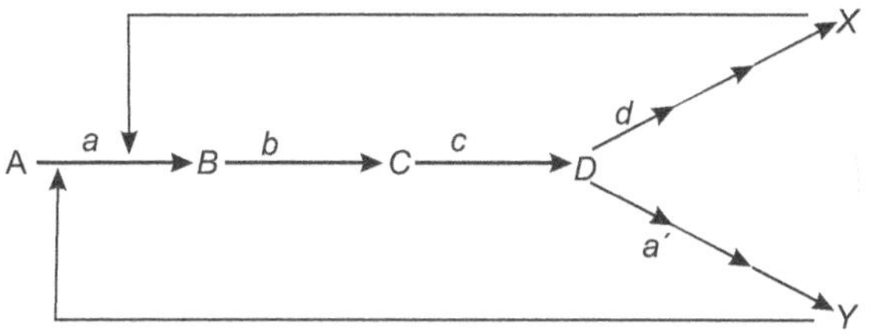

Figure 8.5 Monovalent feedback

Isofunctional enzymes In this mechanism (Figure 8.6), the first common step is catalysed by two different or isofunctional enzymes that convert the same substrate to the same product. However, enzyme *a´* is insensitive to *X* but is under the specific control of *Y*.

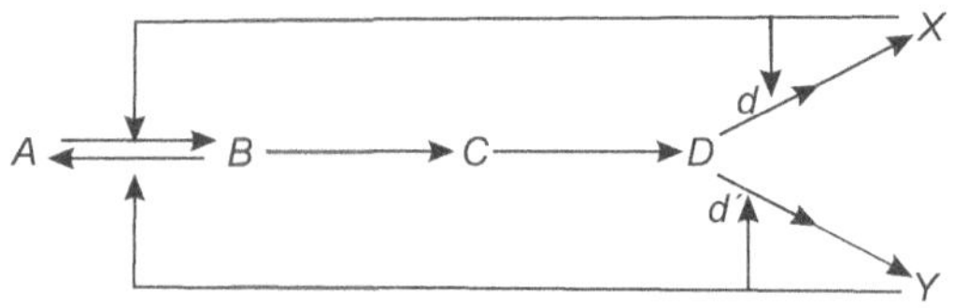

Figure 8.6 Mechanism of isofunctional enzymes

Since enzyme *a* in the latter instance would still be involved in the synthesis of *B*, *C*, and *D*, a secondary feedback control must be exerted by the two end products, namely, *X* on enzyme *d* and *Y* on enzyme *d´*. Thus, if an excess of *X* is formed, it will not only inhibit enzymes *a* and *d*, but also will not interfere with the synthesis of *Y*.

Sequential feedback control In this mechanism (Figure 8.7), enzyme *a* is not regulated by either of the end products of the branched pathway. However, *X* will inhibit the enzyme that converts the last common substrate *D* to precursors of *X*, and

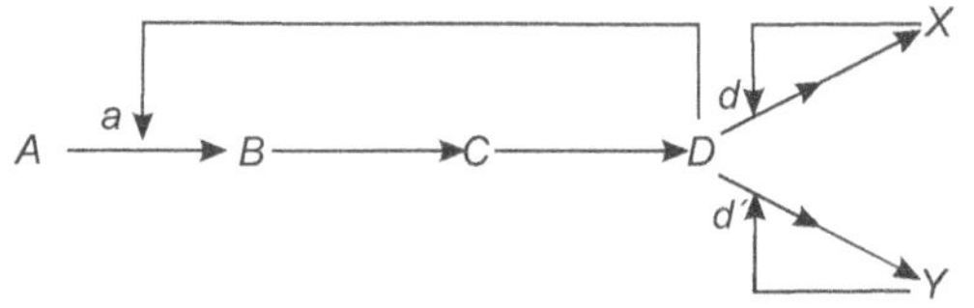

Figure 8.7 Sequential feedback control

Y will inhibit the enzyme that will convert *D* to precursors of *Y*. Thereby, *D* will accumulate and inhibit enzyme *a*, which shuts off the entire pathway. An example of this pathway is observed in the biosynthesis of aromatic amino acids in a number of bacteria and the regulation of threonine and isoleucine biosynthesis in *Rhodopseudomonas spheroids*.

Concerted feedback inhibition In this system, enzyme *a* is insensitive to *X* or *Y* alone, but when both are present, they act in concert to inhibit enzyme *a*. Again, both *X* and *Y* exert secondary controls by *X* inhibiting enzyme d and Y enzyme d′ (Figure 8.8).

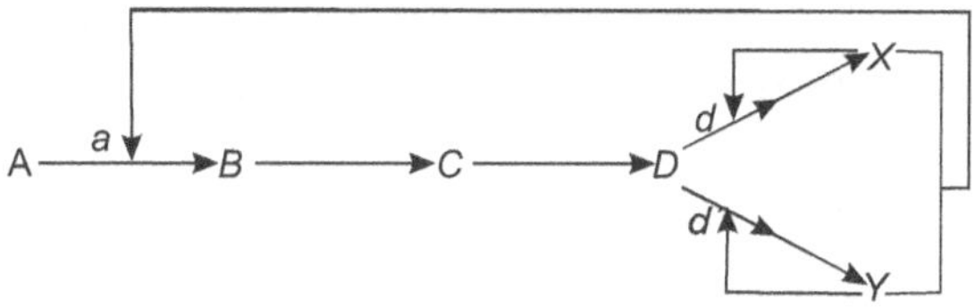

Figure 8.8 Concerted feedback inhibition

Thus, if there is an excess of *X* synthesized, it will only inhibit its own synthesis by controlling the activity of enzyme *d*, allowing *Y* to be synthesized. As *Y* accumulates, both *X* and *Y* can now, in concert, inhibit enzyme a, which is sensitive to inhibition only in the presence of both *X* and *Y*.

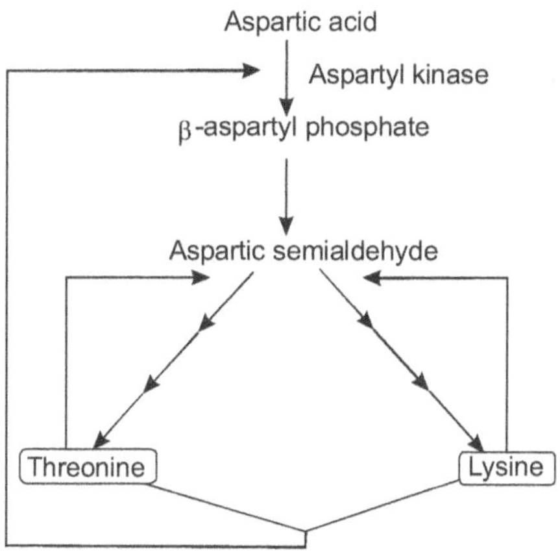

Figure 8.9 Inhibition of aspartyl kinase

A good example is the inhibition of aspartyl kinase (Figure 8.9) from *Rhodopseudomonas capsulatus* by the combination of both threonine and lysine. These amino acids are ineffective individually.

Cumulative feedback inhibition In this type of inhibition *X* and *Y*, in saturating concentrations, only cause partial inhibition of enzyme *a*, but when they both are present simultaneously, a cumulative effect is observed. Thus if X at saturating concentration inhibits 30% of the reaction, 70% by *Y* alone and 50% when both X and Y are present in saturating concentrations, the residual activity will be 35% of the total activity. Inhibition will be 65% (Figure 8.10).

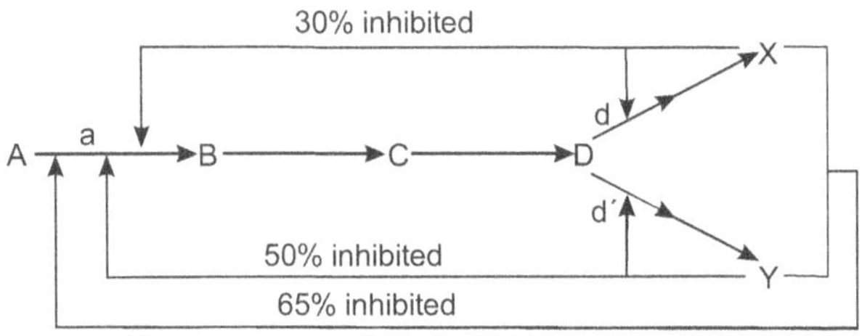

Figure 8.10 Cumulative feedback inhibition

REVIEW QUESTIONS

1. Define
   i. Transition state
   ii. Metalloenzymes
   iii. Metal-activated enzymes
2. Elaborately discuss the following with suitable examples.
   i. Acid–base catalysis
   ii. Covalent catalysis

iii. Metal ion catalysis

iv. Electrostatic catalysis

3. What are the effects of proximity and orientation in catalysing a reaction?
4. Write notes on preferential binding of the transition state complex?
5. Discuss the mechanism of action of

   i. Lysozyme

   ii. Chymotrypsin

9

KINETICS OF MULTISUBSTRATE–ENZYME REACTIONS

Kinetics is the study of the rates at which chemical reactions occur. It is through kinetic studies that the binding affinities of substrates and inhibitors to an enzyme can be determined and the maximum catalytic rate of an enzyme can be established. By observing how the rate of an enzymatic reaction varies with the reaction conditions and combining this information with that obtained from chemical and structural studies of the enzyme, the enzyme's catalytic mechanism may be elucidated. The study of the kinetics of an enzymatic reaction leads to an understanding of that enzyme's role in an overall metabolic process. Under proper conditions, the rate of an enzymatically catalysed reaction is proportional to the amount of the enzyme present and therefore most enzyme assays (measurement of the amount of enzyme present) are based on kinetic studies of the enzyme. Kinetic measurements of enzymatically catalysed reactions are among the most powerful techniques for elucidating the catalytic mechanisms of enzymes.

The study of enzyme kinetics began in 1902 when Adrian Brown reported an investigation of the rate of hydrolysis of sucrose as catalysed by the yeast enzyme, invertase (now known as β-fructofuranosidase).

$$\text{Sucrose} + H_2O \xrightarrow{\text{Invertase}} \text{Glucose} + \text{Fructose}$$

Brown demonstrated that when the sucrose concentration is much higher than that of the enzyme, the reaction rate becomes independent of the sucrose concentration, i.e., the rate is zeroth-order with respect to sucrose. It is therefore proposed that the overall reaction is composed of two elementary reactions in which the substrate forms a complex with the enzyme that subsequently decomposes to products and enzyme.

$$E + S \underset{k_{-1}}{\overset{k_1}{\rightleftharpoons}} ES \xrightarrow{k_2} P + E$$

Here, E, S, ES and P symbolize the enzyme, substrate, enzyme–substrate complex and products respectively.

According to this model, when the substrate concentration becomes high enough to entirely convert the enzyme to the ES form, the second step of the reaction becomes rate-limiting and the rate of the overall reaction becomes insensitive to further increases in substrate concentration. The general expression for the velocity (rate) of this reaction is

$$V = \frac{d[P]}{dt} = k_2[ES]$$

The overall rate of production of [ES] is the difference between the rates of the elementary reactions.

$$\frac{d[ES]}{dt} = k_1[E][S] - k_{-1}[ES] - k_2[ES]$$

This equation can be integrated by simplifying assumptions. Two possible assumptions are as follows:

1. Assumption of Equilibrium

In 1913, Leonor Michaelis and Maude Menten assumed that $k_{-1} >> k_2$ so that the first step of the reaction achieves equilibrium.

$$K_S = \frac{k_{-1}}{k_1} = \frac{[E][S]}{[ES]}$$

Here, K_s is the dissociation constant of the first step in the enzymatic reaction.

2. Assumption of Steady State

With the exception of the initial stage of the reaction the so-called transient phase which is usually over within milliseconds of mixing the enzyme and substrate [ES] remains approximately constant until the substrate is nearly exhausted. Hence the rate of synthesis of ES must equal its rate of consumption over most of the course of the reaction, that is, [ES] maintains a steady state. One can therefore assume with a degree of accuracy that [ES] is constant, i.e.,

$$\frac{d[ES]}{dt} = 0$$

This steady-state assumption was first proposed by G.E. Briggs and James B.S. Haldane (1925). In order to be of use, kinetic expressions for overall reactions must be formulated in terms of experimentally measurable quantities. The quantities [ES] and [E] are not in general, directly measurable but the total enzyme concentration is usually readily determined.

The initial velocity of the reaction can then be expressed in terms of the experimentally measurable quantities $[E_t]$ and [S].

$$V_o = \left(\frac{d[P]}{dt}\right)_{t=0} = k_2[ES] = \frac{k_2[E_t][S]}{K_m + [S]} \tag{1}$$

The use of initial velocity rather than just the velocity minimizes such complicating factors such as the effects of reversible reactions, inhibition of the enzyme by product and progressive inactivation of the enzyme.

The maximal velocity of a reaction V_{max} occurs at high substrate concentrations when the enzyme is saturated, that is when it is entirely in the ES form.

$$V_{max} = k_2[E_t] \tag{2}$$

Therefore, combining equations (1) and (2), we obtain

$$V_o = \frac{V_{max}[S]}{K_m + [S]}$$

This expression, the Michaelis–Menten equation is the basic equation of enzyme kinetics. It describes a rectangular hyperbola as plotted in Figure 9.1.

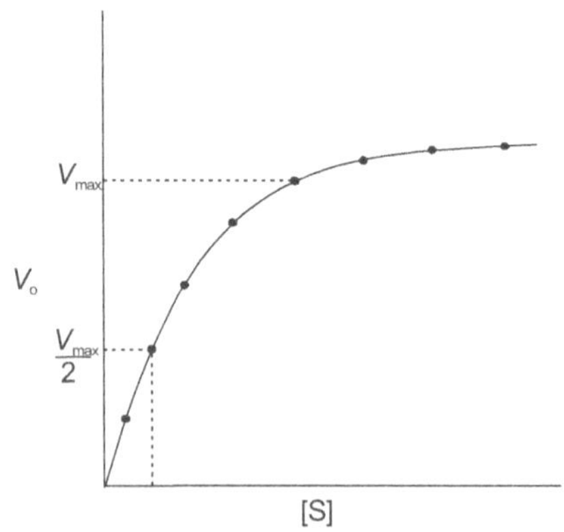

Figure 9.1 Plot of the initial velocity V_o of a simple Michaelis–Menten reaction versus the substrate concentration [S].

DERIVATION OF MICHAELIS–MENTEN EQUATION

The derivation starts with the two basic steps involved in the formation and breakdown of ES.

$$E + S \underset{k_{-1}}{\overset{k_1}{\rightleftharpoons}} ES \underset{k_{-2}}{\overset{k_2}{\rightleftharpoons}} E + P \tag{1}$$

In this reaction, the concentration of the product [P] is negligible, and we make the assumption that k_{-2} can be ignored. Therefore the overall reaction reduces to

$$E + S \underset{k_{-1}}{\overset{k_1}{\rightleftharpoons}} ES \overset{k_2}{\longrightarrow} E + P \tag{2}$$

V_0 is determined by the breakdown of ES to form product, which is determined by [ES].

$$V_0 = k_2[ES] \tag{3}$$

As [ES] is not easily measured experimentally, an alternative expression for [ES] must be found. For this, the term $[E_t]$ can be introduced, which represents the total enzyme concentration (the sum of free and substrate-bound enzyme). Free or unbound enzyme can then be represented by $[E_t]$–[ES].

Also, because [S] is greater than $[E_t]$, the amount of substrate bound by the enzyme at any given time is negligible compared with the total [S]. With these conditions in mind, the following steps lead to an expression for V_0 in terms of parameters that are essentially measured.

Step 1 The rates of formation and breakdown of ES are determined by the steps governed by the rate constants k_1 (formation) and $k_{-1} + k_2$ (breakdown) according to the expressions

$$\text{Rate of ES formation} = k_1[E_t] - [ES][S] \tag{4}$$

$$\text{Rate of ES breakdown} = k_{-1}[ES] + k_2[ES] \tag{5}$$

Step 2 An assumption is made that the initial rate of the reaction reflects a steady state in which [ES] is constant, that is, the rate of formation of ES is equal to its rate of breakdown. This is called the steady-state assumption. The expressions in equations (4) and (5) can be equated for the steady state, giving

$$k_1\left([\mathrm{E_t}]-[\mathrm{ES}]\right)[\mathrm{S}] = k_{-1}[\mathrm{ES}] + k_2[\mathrm{ES}] \tag{6}$$

Step 3 A series of algebraic steps is taken to solve the equation (6) for [ES]. First, the left side is multiplied out and the right side is simplified.

$$\left(k_1[\mathrm{E_t}][\mathrm{S}] - k_1[\mathrm{ES}][\mathrm{S}]\right) = \left((k_{-1}+k_2)[\mathrm{ES}]\right) \tag{7}$$

$$k_1[\mathrm{E_t}][\mathrm{S}] = \left((k_{-1}+k_2)[\mathrm{ES}]\right) + k_1[\mathrm{ES}][\mathrm{S}] \tag{8}$$

$$k_1[\mathrm{E_t}][\mathrm{S}] = \left(k_1[\mathrm{S}] + (k_{-1}+k_2)\right)[\mathrm{ES}] \tag{9}$$

We then solve this equation for [ES]

$$[\mathrm{ES}] = \frac{k_1[\mathrm{E_t}][\mathrm{S}]}{k_1[\mathrm{S}] + k_{-1} + k_2} \tag{10}$$

The above equation can be further simplified by combining the rate constants into one expression

$$[\mathrm{ES}] = \frac{[\mathrm{E_t}][\mathrm{S}]}{[\mathrm{S}] + \left[k_2 - k_{-1}\right]/k_{-1}} \tag{11}$$

The term $(k_2 - k_1)/k_1$ is defined as the Michaelis constant (K_m). Substituting this into equation (11) simplifies the expression to

$$[\mathrm{ES}] = \frac{[\mathrm{E_t}][\mathrm{S}]}{K_m + [\mathrm{S}]} \tag{12}$$

Step 4 V_o can now be expressed in terms of [ES]. Substituting the right side of equation (12) for [ES] in equation (3)

$$V_o = \frac{k_2[E_t][S]}{K_m + [S]} \tag{13}$$

This equation is further simplified because the maximum velocity occurs when the enzyme is saturated with $[ES] = [E_t]$.

Therefore, V_{max} can be defined as k_2 $[E_t]$. Substituting this in equation (13) gives

$$V_o = \frac{V_{max}[S]}{K_m + [S]}$$

This is the Michaelis–Menten equation, the rate equation for a one-substrate enzyme-catalysed reaction. It is a statement of the quantitative relationship between the initial velocity V_o, the maximum initial velocity V_{max} and the initial substrate concentration [S], all related through the Michaelis constant K_m.

K_m has units of concentration. This can be confirmed by considering the limiting situations where [S] is very high or very low as shown in Figure 9.2.

In a special case, when V_o is exactly one-half V_{max} then, the Michaelis–Menten equation can be written as

$$\frac{V_{max}}{2} = \frac{V_{max}[S]}{K_m + [S]} \tag{14}$$

On dividing by V_{max} on both sides, we obtain,

$$\frac{1}{2} = \frac{[S]}{K_m + [S]} \tag{15}$$

Solving for K_m, we get,

$$\begin{aligned} 2[S] &= K_m + [S] \\ K_m &= [S] \end{aligned} \tag{16}$$

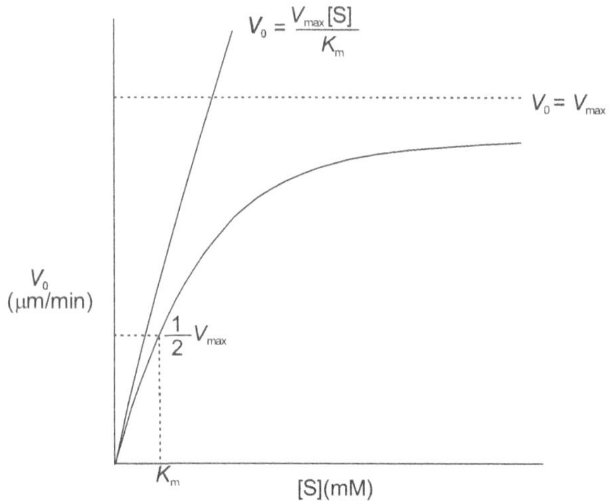

Figure 9.2 Dependence of initial velocity on substrate concentration

This equation represents K_m which is equivalent to the substrate concentration at which V_o is one half $V_{max.}$

The practical rule that K_m = [S] when $V_o = ½\ V_{max}$ holds for all enzymes that follow Michaelis–Menten kinetics. However, the Michaelis–Menten equation does not depend on the relatively simple two steps reaction proposed by Michaelis–Menten. Many enzymes that follow Michaelis–Menten kinetics have quite different reaction mechanisms and enzymes that catalyse reactions with six or eight identifiable steps exhibit the same steady-state kinetic behaviour.

K_m is the substrate concentration, at which the reaction velocity is half-maximal. Therefore, if an enzyme has a small value of K_m, it achieves maximal catalytic efficiency at low substrate concentrations. Even though equation (16) holds true for many enzymes, both the magnitude and the real meaning of V_{max} and K_m can differ from one enzyme to the next. This is an important limitation of the steady-state approach to enzyme kinetics.

The Double-Reciprocal Plot

The Michaelis–Menten equation

$$V_o = \frac{V_{max}[S]}{K_m + [S]}$$

can be transformed into equations that are more useful in plotting experimental data. One common transformation is derived by taking the reciprocal on both sides of the Michaelis–Menten equation.

$$\frac{1}{V_o} = \frac{K_m + [S]}{V_{max}[S]}$$

Separating the components of the numerator on the right side of the equation gives

$$\frac{1}{V_o} = \frac{K_m}{V_{max}[S]} + \frac{[S]}{V_{max}[S]}$$

which simplifies to

$$\frac{1}{V_o} = \frac{K_m}{V_{max}[S]} + \frac{1}{V_{max}}$$

This form of the Michaelis–Menten equation is called the Lineweaver–Burk Plot equation.

For enzymes obeying the Michaelis–Menten relationship, a plot of $1/V_o$ versus $1/[S]$ yields a straight line. This line has a slope of K_m/V_{max}, an intercept of $1/V_{max}$ on the $1/V_o$ axis, and an intercept of $-1/K_m$ on the $1/[S]$ axis (Figure 9.3).

The double-reciprocal presentation, has the great advantage of allowing a more accurate determination of V_{max} which can be approximated from a simple plot of V_o versus [S]. A disadvantage of this plot is that most experimental measurements

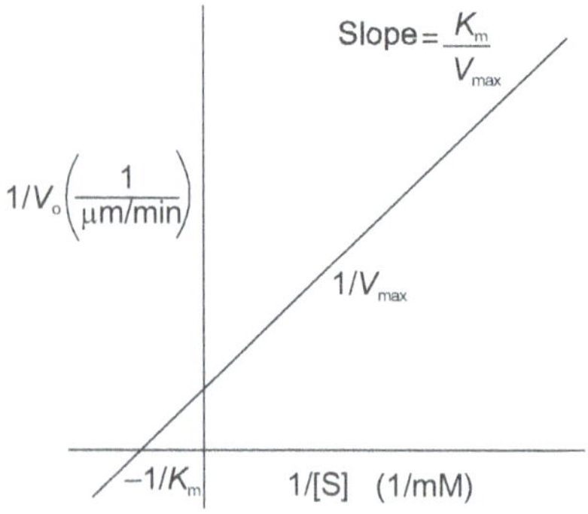

Figure 9.3 A double-reciprocal or Lineweaver–Burk plot

involve relatively high [S] and are therefore crowded onto the left side of the graph. Furthermore, for small values of [S], small errors in V_o lead to large errors in $1/V_o$ and hence to large errors in K_m and $V_{max.}$

Several other types of plots, each with its advantages and disadvantages, have been formulated for the determinations of V_{max} and K_m from kinetic data.

The Haldane Relationship

The equation derived from by Michaelis–Menten was modified by Briggs and Haldane (1925) who introduced a more generally valid assumption that of the steady state.

All reactions are to some degree reversible and many enzyme-catalysed reactions can function in either direction within the cell. It is therefore of interest to compare the kinetics of the forward and backward reactions.

Consider a single-substrate reaction, S□ P, proceeding via the formation of a single-intermediate complex: In the forward reaction this would be regarded as an ES complex, but in the reverse direction, the complex would be EP.

$$\mathrm{E + S} \underset{K_{-1}}{\overset{k_1}{\rightleftharpoons}} \mathrm{ES} \underset{K_{-2}}{\overset{k_2}{\rightleftharpoons}} \mathrm{E + P}$$

The Michaelis–Menten equation in the forward direction at fixed $[E_o]$ gives

$$V_f = \frac{V^s_{max}[S_o]}{[S_o] + K^s_m}$$

where, V_f is the initial velocity in the forward direction and V^s_{max} is the maximum initial velocity in the forward direction.

$$K^s_m = \frac{k_{-1} + k_2}{k_1}$$

The Michaelis–Menten equation for the reverse reaction (at constant $[E_o]$) gives

$$V_b = \frac{V^p_{max}[p_o]}{[p_o] + K^p_m}$$

where, V_b is the initial velocity in the backward direction and V^p_{max} is the maximum initial velocity for the backward reaction, and

$$K^p_m = \frac{k_{-1} + k_2}{k_{-2}}$$

Haldane derived a useful relationship between the kinetic constants and the equilibrium constants of the reaction. At equilibrium the rate of the forward reaction equals the rate of the backward reaction.

Under these conditions,

$$k_{-1}[ES] = k_1[ES][S]$$

Therefore

$$\frac{[ES]}{[E]} = \frac{k_1}{k_{-1}}[S]$$

Also under these conditions,

$$k_2[ES] = k_{-2}[E][P]$$

Therefore

$$\frac{[ES]}{[E]} = \frac{k_{-2}}{k_2}[P] = \frac{k_1}{k_{-1}}[S]$$

Therefore

$$K_{eq} = \frac{[P]}{[S]} = \frac{k_1}{k_{-1}}\frac{k_2}{k_{-2}}$$

But,

$$V^{s}_{max} = k_2[E_o]$$

And

$$V^{p}_{max} = k_{-1}[E_o]$$

$$\therefore \frac{V^{s}_{max}}{V^{p}_{max}} = \frac{k_2}{k_{-1}}$$

Also

$$\frac{K^{s}_{m}}{K^{p}_{m}} = \frac{(k_{-1}+k_2)}{k_1}\frac{k_{-2}}{(k_{-1}+k_2)} = \frac{k_{-2}}{k_1}$$

$$\therefore K_{eq} = \frac{k_1 k_2}{k_{-1}k_{-2}}$$

$$K_{eq} = \frac{V^{s}_{max}K^{p}_{m}}{V^{p}_{max}K^{s}_{m}}$$

This is called the Haldane relationship.

This Haldane relationship demonstrates that the kinetic parameters of a reversible, enzymatically catalysed reaction are not independent of one another. Rather, they are related by the equilibrium constant for the overall reaction which is independent of the presence of the enzyme.

The Eadie–Hofstee and Hanes Plot

Eadie–Hofstee plot is a graphical representation of enzyme kinetics in which reaction velocity is plotted as a function of the velocity vs substrate concentration ratio.

$$V = -K_m \frac{V}{[S]} + V_{max}$$

where V represents reaction velocity, K_m is the Michaelis–Menten constant, [S] is the substrate concentration, and V_{max} is the maximum reaction velocity. It can be derived from the Michaelis–Menten equation as follows.

$$V_o = \frac{V_{max}[S]}{K_m + [S]}$$

Inverting and multiplying with V_{max}

$$\frac{V_{max}}{V_o} = \frac{V_{max}\left(K_m + [S]\right)}{V_{max}[S]} = \frac{K_m + [S]}{[S]}$$

Rearranging

$$V_{max} = \frac{V_o K_m}{[S]} + \frac{V_o[S]}{[S]} = \frac{V_o K_m}{[S]} + V_o$$

Isolating V_o

$$V_o = -K_m \frac{V_o}{[S]} + V_{max}$$

A plot of V vs. $V/[S]$ will yield V_{max} at the intercept with the y-axis and the slope is $-K_m$ (Figure 9.4). Eadie–Hafstee plot was used historically for rapid identification of important kinetic terms like K_m and V_{max} but has been superseded by nonlinear regression methods that are significantly more accurate and no longer computationally inaccessible.

One drawback from the Eadie–Hofstee approval is that neither ordinate nor abscissa represent independent variables: both are dependent on reaction velocity. Thus, any experimental error will be present in both axes.

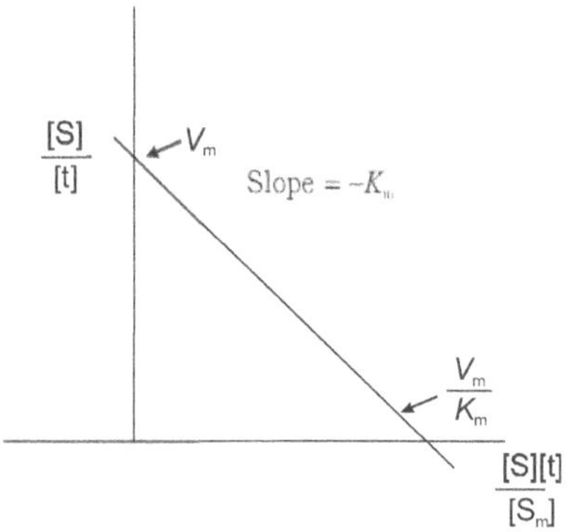

Figure 9.4 Eadie–Hofstee plot

The Hanes plot (Figure 9.5) similarly states with the Lineweaver–Burk plot equation, which in this instance is multiplied throughout by $[S_0]$.

$$\frac{1}{V_0}[S_0] = \frac{K_m}{V_{max}}\frac{1}{S_0}[S_0] + \frac{1}{V_{max}}[S_0]$$

Therefore,

$$\frac{[S_o]}{V_o} = \frac{1}{V_{max}}[S_o] + \frac{K_m}{V_{max}}$$

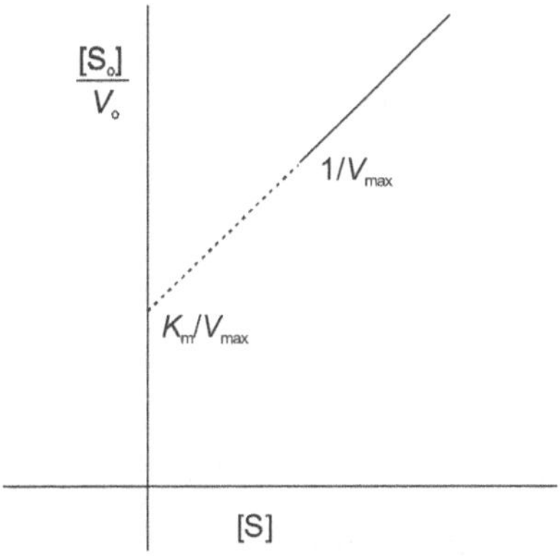

Figure 9.5 The Hanes plot

PRE-STEADY-STATE KINETICS

A complete description of the enzyme-catalysed reaction requires direct measurements of the rates of individual reaction steps. For example, the measurement of the association of enzyme and substrate to form the ES complex. It is during the pre-steady state that the rates of many reaction steps can be measured independently. The pre-steady-state phase is generally very short, which often requires specialized techniques for rapid mixing and sampling.

The chief factors which determine the initial velocity of a particular reaction are enzyme concentration, substrate concentration, pH, temperature and the presence of activators or inhibitors.

STEADY-STATE KINETICS

The two most important experimental parameters provided by steady-state kinetics are k_{cat} and k_{cat}/K_m.

k_{cat}/K_m—Measure of Catalytic Efficiency

An enzyme's kinetic parameters provide a measure of its catalytic efficiency. The catalytic constant of an enzyme is defined as

$$k_{cat} = \frac{V_{max}}{[E_t]}$$

This quantity is also known as the **turnover number** of an enzyme because it is the number of reaction processes (turnovers) that each site catalyses per unit time.

When the concentration of the substrate [S] is less than K_m the formation of ES becomes less. k_{cat}/K_m is the apparent second-order rate constant of the enzymatic reaction. The rate of the reaction varies directly with how often enzyme and substrate encounter one another in solution. The quantity k_{cat}/K_m is therefore a measure of an enzyme's catalytic efficiency.

INHIBITION KINETICS

Inhibitors are substances which tend to decrease the rate of an enzyme-catalysed reaction. The binding of an inhibitor can stop a substrate from entering the enzyme's active site and/or hinder the enzyme from catalysing its reaction. Inhibitor binding is either reversible or irreversible. Irreversible inhibitors usually react with the enzyme and change it chemically. These inhibitors modify key amino acid residues needed for enzymatic activity. In contrast, reversible inhibitors bind non-covalently and different types of inhibition are produced depending on whether these inhibitors bind the enzyme, the enzyme–substrate complex, or both.

Competitive Inhibition

Competitive inhibition occurs when substrate (S) and inhibitor (I) both binds to the same site on the enzyme. In effect, they compete for the active site and bind in a mutually exclusive fashion. There is another type of inhibition that would give the same kinetic data. If S and I bound to different sites, and S bound to E and produced a conformational change in E such that I could not bind (and vice versa), then the binding of S and I would be mutually exclusive. This is called **allosteric competitive inhibition**. Inhibition studies are usually done at several fixed and non-saturating concentrations of I and varying S concentrations. The key kinetic parameters to understand are V_{max} and K_m.

However, if at this same inhibitor and enzyme concentration, the substrate concentration is high, then the inhibitor will be much less successful in competing with the substrate for the available binding sites and the degree of inhibition will be less marked. At very high-substrate concentrations, the molecules of substrate will greatly out number the molecules of inhibitor and the effect of the inhibitor will be negligible. Hence V_{max} for the reaction is unchanged. However, the apparent K_m, the substrate concentration when $V_o = 1/2\ V_{max,}$ is clearly increased as a result of inhibition.

Let us investigate the steady-state kinetics of a simple, single-substrate, single binding-site, single-intermediate enzyme-catalysed reaction in the presence of a competitive inhibitor I.

The dissociation constant for the reaction between E and I is $K_{i,}$ where

$$K_i = \frac{[E][I]}{[EI]}$$

$$[EI] = \frac{[E][I]}{K_i} \qquad (1)$$

In this context, K_i is called the inhibitor constant. Equilibrium between enzyme and inhibitor will normally be established almost instantaneously on mixing.

If we derive the initial velocity equation using the steady-state assumption, we reach the assumption

$$\frac{[E][S]}{[ES]} = \frac{k_{-1} + k_2}{k_1} = K_m \qquad (2)$$

Here, we make a substitution for [E] in this equation and we take the inhibitor into account, since

$$[E_o] = [E] + [ES] + [EI] \qquad (3)$$

Substitute (1) in (3)

$$= [E] + [ES] + \frac{[E][I]}{K_i}$$

$$= [E]\left(1 + \frac{[I]}{K_i}\right) + [ES]$$

Therefore,

$$[E] = \frac{[E_o] - [ES]}{\left(1 + \frac{[I]}{K_i}\right)} \qquad (4)$$

Substitute [E] in equation (2)

$$\frac{([E_o]) - [ES][S]}{\left(1 + \frac{E}{K_i}\right)[ES]} = K_m$$

Therefore,

$$\frac{([E_o]) - [ES][S]}{[ES]} = K_m\left(1 + \frac{[I]}{K_i}\right)$$

This is an equation of the same form as the Michaelis–Menten equation. The only difference being that K_m has been increased by a factor $[1+([I_0]/K_i]$. Therefore, for simple competitive inhibition, V_{max} is unchanged, but K_m is altered so that $K'_m = K_m[1+[I_o]/K_i)]$, where K'_m is the apparent K_m in the presence of an initial concentration $[I_0]$ of competitive inhibitor. It can be seen that K_i is equal to the concentration of competitive inhibitor which apparently doubles the value of K_m.

The Lineweaver–Burk equation in the presence of a competitive inhibitor will be

$$\frac{1}{I_0} = \frac{K'_m}{V_{max}}\frac{1}{S_0} + \frac{1}{V_{max}}$$

Lineweaver–Burk plots showing the effect of competitive inhibition are shown in Figure 9.6a and b.

It must be pointed out that this identical expression would be obtained if the inhibitor-binding site was separate from the substrate-binding site, provided the binding of the substrate to the enzyme resulted in the blockage of the inhibitor-binding site by a conformational change or other mechanism. In this situation the inhibitor could bind to E but not to ES, exactly as discussed above. For this reason, it has become common to classify an inhibitor as competitive if in its presence, a Lineweaver–Burk plot is obtained with changed K_m but unchanged V_{max} irrespective of the actual mechanism involved.

Once the type of inhibitor has been established, it is desirable to determine the inhibitor constant K_i. It is obtained from the expression $K'_m = K_m[1+([I_0]/K_i)]$, but a graphical method is preferred to a direct substitution of numbers to allow errors in individual determinations to be averaged out. From the above expression

$$K'_m = \frac{K_m}{K_i}[I_o] + K_m$$

The simplest forms of competitive inhibition considered above are sometimes termed **as linear competitive inhibition** because both primary and secondary plots are linear. In more complicated systems the primary plots may be linear but the secondary plots non-linear.

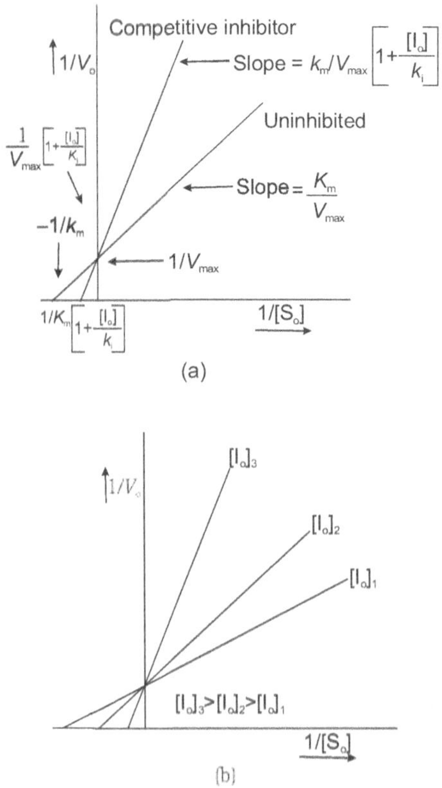

Figure 9.6 (a) Lineweaver–Burk plot showing the effect of competitive inhibition. (b) The Lineweaver–Burk plot, showing plots for several inhibitor concentrations at fixed enzyme concentrations.

Competitive inhibitors, like other types of inhibitors, may be used to elucidate metabolic pathways by causing accumulation of intermediates. There are a few instances where competitive inhibition by a product may play an important role in metabolic regulation within the cell. For example, 2,3-bisphosphoglycerate inhibits its own formation from 3-phosphoglycerol phosphate, a reaction catalysed by bisphosphoglycerate mutase.

Uncompetitive Inhibition

Uncompetitive inhibition occurs when I binds only to ES and not to a free E. One can hypothesize that on binding S, a conformational change in E occurs which presents a binding site for I. Inhibition occurs since ESI cannot form product. It is a dead-end complex which has only one fate, to return to ES.

Both K_m and V_{max} are altered, but a distinctive kinetic pattern emerges under steady-state conditions.

Let us consider,

$$E + S \rightleftharpoons ES \rightarrow E + P$$

$$-I \uparrow\downarrow +I$$

$$ESI$$

ESI is a dead-end complex; the inhibitor constant K_i = ([ES][I]/[ESI]. Under steady state conditions.

$$\frac{[E][S]}{[ES]} = K_m$$

For this system,

$$[E_o]=[E]+[ES]+[ESI]$$

$$=[E]+[ES]+\frac{[ES][I]}{K_i}$$

$$=[E]+[ES]\left(1+\frac{[I]}{K_i}\right)$$

Therefore

$$[E]=[E_o]-[ES]\left(1+\frac{[I]}{K_i}\right)$$

Substituting [E]

$$V_o=\frac{V_{max}[S_o]}{[S_o]\left(1+\frac{[S_o]}{K_i}\right)+K_m}$$

Dividing throughout by $(1+([I_o]/K_i)$ gives:

$$V_o=\frac{\frac{V_{max}}{\left(1+\frac{[I_o]}{K_i}\right)}[S_o]}{[S_o]+\frac{K_m}{\left(1+\frac{[I_o]}{K_i}\right)}}$$

This is an equation of the same form as the Michaelis–Menten equation, the constants K_m and V_{max} both being derived by a factor $(1+([I_o]/K_i))$. Thus for competitive inhibition,

$$V'_{max}=\frac{V_{max}}{\left(1+\frac{[I_o]}{K_i}\right)} \quad \text{and} \quad K'_m=\frac{K_m}{\left(1+\frac{[I_o]}{K_i}\right)}$$

where V'_{max} is the value of V_{max} in the presence of an initial concentration $[I_0]$ of uncompetitive inhibitor and K_m is the apparent value of K_m under the same conditions. An inhibitor concentration equal to K_i will halve the values of both V_{max} and K_m.

The Lineweaver–Burk equation in the presence of uncompetitive inhibitor is

$$\frac{1}{V_0} = \frac{K'_m}{V'_{max}} \frac{1}{[S_0]} + \frac{1}{V_{max}}$$

and the slope of a Lineweaver–Burk plot is equal to

$$\frac{K'_m}{V'_{max}} = \frac{K_{max}}{V_{max}} \frac{\left(1+\frac{[I_0]}{K_i}\right)}{\left(1+\frac{[I_0]}{K_i}\right)} = \frac{1}{V_{max}} \frac{K_m}{V_{max}}$$

In other words, the slope of a Lineweaver–Burk plot is not altered by the presence of an uncompetitive inhibitor, but both intercepts change (Figure 9.7a and b).

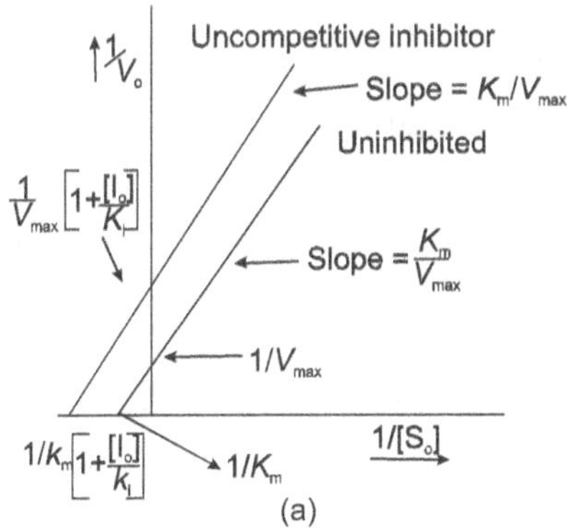

Figure 9.7 (*continues*)

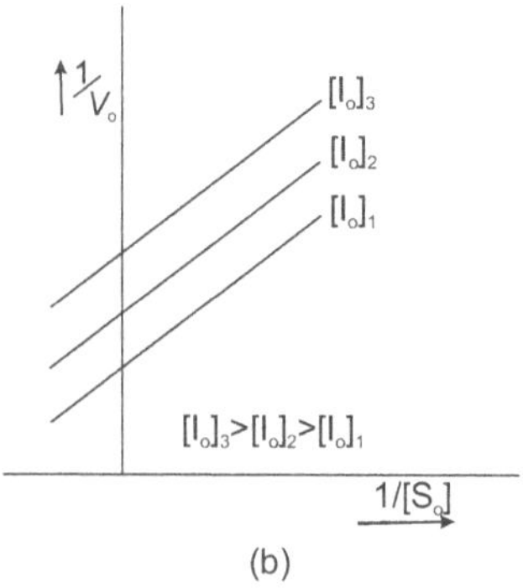

Figure 9.7 (a) Lineweaver–Burk plot showing the effect of uncompetitive inhibition. (b) The Lineweaver–Burk plot showing plots for several inhibitor concentrations at fixed enzyme concentrations.

Therefore, uncompetitive inhibition of single-substrate enzyme-catalysed reaction is a rare phenomenon, one of the few possible examples known being the inhibition of arylsulphatase by hydrazine. However, uncompetitive inhibition patterns are seen with two substrate reactions and this may help in the elucidation of the reaction mechanism.

Noncompetitive Inhibition

Noncompetitive inhibition occurs when I binds to both E and ES. We will look at only the special case in which the dissociation constants of I for E and ES are the same. This is called noncompetitive inhibition. In the more general case, the Kd's are different, and the inhibition is called **mixed inhibition**. Since inhibition occurs, we will hypothesize that ESI cannot form product. It is a dead-end complex which has only one fate, to return to ES or EI. Therefore the inhibitor must bind to the site different from that of the active site, bringing about a conformational change affecting the catalytic site.

Even this is a complex situation, for ES can be arrived at by alternative routes, making it impossible for an expression of the same form as the Michaelis–Menten equation to be derived using the general steady-state assumption.

However, noncompetitive inhibition is consistent with Michealis–Menten type equation and a linear Lineweaver–Burk plot can occur with the inhibitor at equilibrium.

Consider the reaction

$E + I \rightleftharpoons EI$ and $ES + I \rightleftharpoons ESI$ have an identical dissociation constant K_i, again called the inhibitor constant. The total enzyme concentration is effectively reduced by the inhibitor, decreasing the value of V_{max} but not altering K_m, since neither the inhibitor nor the substrate affects the binding of the other.

In the presence of a noncompetitive inhibitor which will bind equally well to E or to ES, i.e., where

$$K_i = \frac{[E][I]}{[EI]} = \frac{[ES][I]}{[ESI]}$$

$$[E_0] = [E] + [ES] + [EI] + [ES]$$

$$= [E] + [ES] + \frac{[E][I]}{K_i} + \frac{[ES][I]}{K_i}$$

$$= ([E] + [ES]) = \left(1 + \frac{[I]}{K_i}\right)$$

$$([E] + [ES]) = \frac{[E_0]}{\left(1 + \frac{[I]}{K_i}\right)}$$

$$[E] = \frac{[E_0]}{\left(1 + \frac{[I]}{K_i}\right)} - [ES]$$

$$V_0 = \frac{V_{max}}{\left(1+\frac{[I_0]}{K_i}\right)} \frac{[S_0]}{([S_0]+K_m)}$$

This is the form of the Michealis–Menten equation with V_{max} being divided by a factor $[1+([I_0]/K_i)]$. Thus, for simple, linear, noncompetitive inhibition, K_m is unchanged and V_{max} is altered so that,

$$V'_{max} = \frac{V_{max}}{\left(1+\frac{[I_0]}{K_i}\right)} \text{ or } \frac{1}{V'_{max}} = \frac{1}{V_{max}}\left(1+\frac{[I_0]}{K_i}\right)$$

where V'_{max} is the value of V_{max} in the presence of a concentration $[I_0]$ of noncompetitive inhibitor. It follows that K_i for such a system is the inhibitor concentration which halves the value of V_{max}. The Lineweaver–Burk equation for simple linear noncompetitive inhibition is

$$\frac{1}{V_0} = \frac{K_m}{V'_{max}} \frac{1}{[S_0]} + \frac{1}{V_{max}}$$

And Lineweaver–Burk plots showing the effects of such inhibition are shown in Figure 9.8a and b.

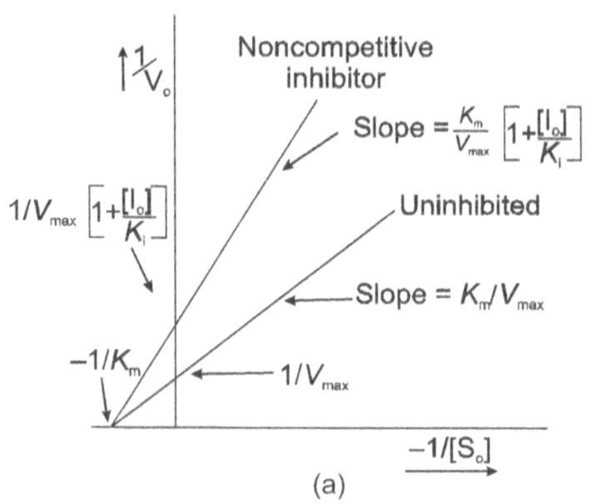

(a)

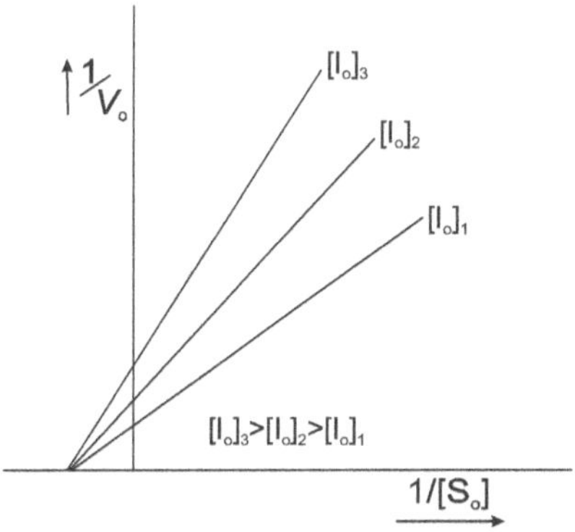

Figure 9.8 (a) Lineweaver–Burk plot showing the effect of simple linear noncompetitive inhibition. (b) The Lineweaver–Burk plot showing several inhibitor concentrations at fixed enzyme concentrations.

REACTIONS INVOLVING TWO SUBSTRATES

Enzymatic reactions involving two substrates and yielding two products $A + B \overset{E}{\rightleftharpoons} P + Q$ account for ~ 60% of known biochemical reactions. Almost all of these bisubstrate reactions are either transferase reactions in which the enzyme catalyses the transfer of a specific functional group, X from one of the substrate to the other or oxidation–reduction reactions in which reducing equivalents are transferred between the two substrates.

$$P-X+B \overset{E}{\rightleftharpoons} P+B-X$$

In trypsin, hydrolysis of peptide bond takes place, involving transfer of the peptide carbonyl group from the peptide nitrogen atom to water.

$$\underset{\text{Polypeptide}}{R_1-\overset{\overset{O}{\|}}{C}-NH-R_2} + H_2O \xrightarrow{\text{Trypsin}} R_1-\overset{\overset{O}{\|}}{C}-O^- + H_3N^+-R_2$$

Similarly in alcohol dehydrogenase reaction, a hydride ion is formally transferred from ethanol to NAD^+.

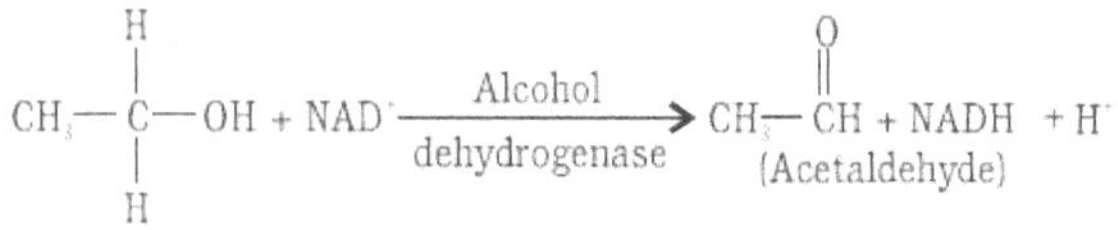

Types of Bi-Bi Reactions

Enzyme-catalysed group transfer reactions fall under two major groups:

1. *Sequential reactions* Reactions in which all substrates must combine with the enzyme before a reaction can occur and products be released are known as the sequential reactions.

In such reactions, the group being transferred X is directly passed from A(= P – X) to B yielding P and Q (= B – X). Hence such reactions are called single-displacement reactions.

Ordered mechanism Sequential reactions can be subclassified into those with a compulsory order of substrate addition to the enzyme, which are said to have an ordered mechanism. In the ordered mechanism, the binding of the first substrate is apparently required for the enzyme to form the binding site for the second substrate.

Figure 9.9 Ordered bi-bi reaction

The Figure 9.9 describes the ordered bi-bi enzymatic reactions using Cleland's shorthand notation. The enzyme is represented by horizontal line and successive addition of substrates and release of products are denoted by vertical arrows.

Random mechanism Sequential reactions that have no preference for the order of the substrate addition are described as having random mechanism (Figure 9.10). In this random mechanism both the binding sites are present on the free enzyme.

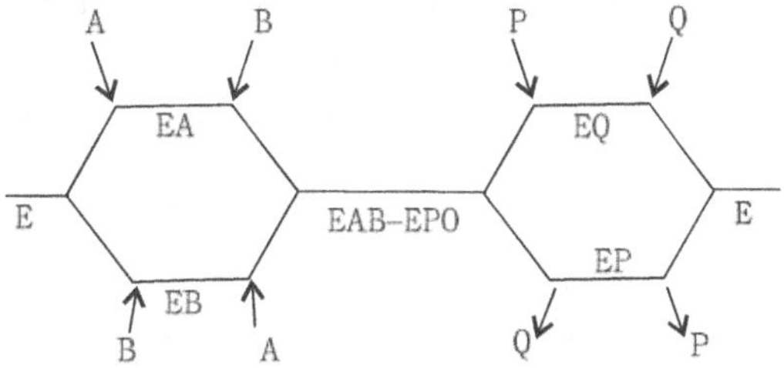

Figure 9.10 A random bi-bi reaction using Cleland's shorthand notation

Some dehydrogenases and kinases operate through random bi-bi mechanisms.

2. *Ping-pong reactions* Mechanisms in which one or more products are released before all the substrates have been added are known as ping-pong reactions. The ping-pong Bi-Bi reaction is represented in Figure 9.11.

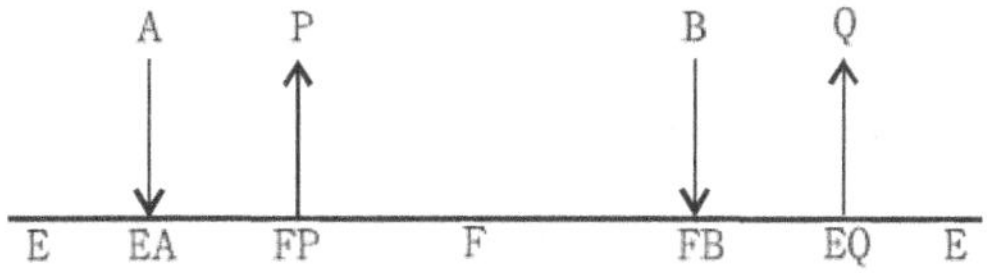

Figure 9.11 A ping-pong bi-bi reaction

In it a functional group X of the first substrate A is displaced from the substrate by the enzyme E to yield the first product P and a stable enzyme form F in which X is tightly bound to the enzyme (ping). In the second stage of the reaction, X is displaced from the enzyme by the second substrate B to yield the second product Q thereby regenerating the original form of the enzyme E (pong). Such reactions are therefore known as double-displacement reactions.

Many enzymes including chymotrypsin, transaminases and some flavo-enzymes react with ping-pong mechanism.

Isotope exchange—evidence favouring a ping-pong Sequential (single-displacement) and ping-pong (double-displacement) bisubstrate mechanisms may be differentiated through the use of isotope exchange studies. Double-displacement reactions are capable of exchanging an isotope from the first product P back to the first substrate A in the absence of the second substrate. An overall ping-pong reaction catalysed by the bisubstrate enzyme E is as follows.

$$P - X + B \overset{E}{\rightleftharpoons} P + B - X$$

In which A = P—X, Q = B—X and X is the group that is transferred from one substrate to the other in the course of reaction. Only the first step of the reaction can take place in the absence of B if a small amount of isotopically labelled P, denoted P^* is added to this reaction mixture then, in the reverse reaction P^*—X will form,

$$\text{Forward Reaction} \quad E + P - X \rightarrow E - X + P$$

$$\text{Reverse Reaction} \quad E - X + P^* \rightarrow E + P^* - X$$

That is, isotopic exchange will occur.

In contrast, let us consider the first step of a sequential reaction. Hence a non-covalent enzyme–substrate complex forms.

$$E + P - X \rightleftharpoons EP - X$$

Addition of P^* cannot result in an exchange reaction because no covalent bonds are broken in the formation of EP–X, that is, there is no P released from the enzyme to exchange with P*. The demonstration of isotopic exchange for a bisubstrate enzyme is therefore convincing evidence favouring a ping-pong mechanism, e.g. the enzyme sucrose phosphorylase is best example of enzymatically catalysed, isotopic, exchange reactions. Sucrose phosphorylase catalyses the overall reaction.

$$\text{Glucose-Fructose} + \text{Phosphate} \underset{\text{Sucrose}}{\overset{E}{\rightleftharpoons}} \text{Glucose 1-phosphate} + \text{Fructose}$$

If the enzyme is incubated with sucrose and isotopically labelled fructose in the absence of phosphate, it is observed that the label passes into the sucrose.

$$\text{Glucose-Fructose} + \text{Fructose}^* \underset{\text{Sucrose}}{\overset{E}{\rightleftharpoons}} \text{Glucose-Fructose}^* + \text{Fructose}$$

For the reverse reaction, if the enzyme is incubated with glucose 1-phosphate and ^{32}P labelled phosphate. This label exchanges into the glucose 1-phosphate.

$$\text{Glucose 1-phosphate} + \text{Phosphate}^* \overset{E}{\rightleftharpoons} \text{Glucose 1-phosphate}^* + \text{Phosphate}$$

These observations indicate that a tight glucosyl-enzyme complex is formed with the release of fructose thereby establishing that the sucrose phosphorylase reaction occurs through a ping-pong mechanism.

ENZYMES CATALYSING TWO REACTIONS SIMULTANEOUSLY

When an enzyme is not absolutely specific for one substrate only, the possibilities arises that it may act on two different substrates present at the same time. This cannot be considered simply as a case of two independent parallel reactions, because the same active centre acts on both the substrates and therefore there will be competition between them.

For an enzyme acting on two substrates A and B

$$E + A \rightleftharpoons EA \qquad (1)$$

$$EA \longrightarrow E + \text{Products} \qquad (2)$$

$$E + B \rightleftharpoons EB \qquad (3)$$

$$EB \longrightarrow E + \text{Products} \qquad (4)$$

When A is considered in a reaction, B behaves simply as a competitive inhibitor and vice versa. The rate of breakdown of A will therefore be given by

$$V_a = \frac{V_a}{1 + \frac{K_a}{a}\left(1 + \frac{b}{K_b}\right)} \qquad (5)$$

And that of B by

$$V_b = \frac{V_b}{1 + \frac{K_b}{b}\left(1 + \frac{a}{K_a}\right)} \qquad (6)$$

If we write the sum of reactions (5) and (6)

$$V_1 t = V_a + V_b = \frac{V_a \frac{a}{K_a} + V_b \frac{b}{K_b}}{1 + \frac{a}{K_a} + \frac{b}{K_b}}$$

In any mixture containing concentrations a and b of A and B respectively, the total velocity V_t must lie between the two velocities which would be obtained with these concentrations of A and B separately.

Thus the ratio of the affinities of an enzyme for two substrates can be determined from these measurements of velocity namely the maximum velocities with the two substrates separately and with an equimolar mixture of both.

TWO ENZYMES ACTING ON ONE SUBSTRATE

The disappearance of a substrate may be due to the simultaneous action of two or more enzymes. If we consider two enzymes, distinguished by the use of two subscripts 1 and 2, the total rate of disappearance will be the sum of two separate reactions, so that we may write

$$V_t = \frac{V_1}{1+\frac{k_1}{S}} + \frac{V_2}{1+\frac{k_2}{S}}$$

In reciprocal form this equation reduces to

$$\frac{1}{V_t} = \frac{1+\frac{k_1+k_2}{S}+\frac{k_1+k_2}{S^2}}{V_1+V_2+\frac{V_1 k_2+V_2 k_1}{S}}$$

The shape of the curve obtained will depend on the relative values of k_1 and k_2. If $k_1 = k_2$, a straight-line plot is obtained (in case of a single enzyme). If the constants differ, a straight line will not be obtained and the results will not be easy to interpret.

REVIEW QUESTIONS

1. Define steady state, K_m, turnover number.
2. Write notes on Michaelis–Menten equation.
3. Derive and write the significance of Lineweaver–Burk plot.
4. Write notes on Haldane–Briggs modification of Michaelis–Menten equation?
5. Discuss Eadie–Hoftsee plot.
6. Write notes on the kinetics of:
   i. Competitive inhibition
   ii. Uncompetitive inhibition
   iii. Noncompetitive inhibition
7. Describe the types of reactions involving two substrates explaining
   i. Sequential reactions
   ii. Ping-pong reactions

10

ALLOSTERIC MODIFICATION OF ENZYMES

Physiologically, the rates of some enzyme-catalysed reactions are regulated by reversibly binding compounds, termed effectors, at specific sites other than their substrate-binding sites, which accordingly are called as allosteric site(s). The term allosteric is derived from Greek *allo* meaning "other" and *steros* meaning "space" or "site". Allosteric enzymes are those having "other sites". Hence the allosteric site is a unique region of the enzyme that differs from the substrate-binding site or catalytic site. The modulating ligand of the allosteric enzyme that binds at the allosteric site are known as allosteric effectors or modulators. Just as the catalytic site of an enzyme is specific for its substrate, the allosteric site is specific for its modulator. Noncovalent binding of allosteric effector causes an allosteric transition in the enzyme that is, the conformation of the enzyme changes, so that the affinity for the substrate or other ligand changes. Positive (+) allosteric effectors increase the enzyme affinity for the substrate or other ligand. The reverse is true for negative effectors (allosteric inhibitors). The allosteric site at which the positive effector binds is referred to as the activator site. The negative effector binds at an inhibitory site.

Allosteric inhibition may be achieved either by reducing the binding affinity of the enzyme for its substrate, evident from an increase in K_m, or by increasing the time required for each catalytic turnover evident from a decrease in V_{max} or both. Conversely, allosteric activation may occur either by reduction in K_m or by an increase in V_{max}, or both.

Most commonly, various metabolites, as well as hormones, metal ions and coenzymes act as allosteric effectors for enzymes. Occasionally, the role of an allosteric effector can be taken up by the substrate molecule itself. In such enzymes, the active centre is similar to the allosteric centre configurationally, but the latter possesses no catalytic site, which is its major distinction from the active centre of the enzyme.

If the allosteric effector is the substrate itself, the effect is said to be homotropic, and if the effector is other than the substrate it is said to be heterotropic. Some enzymes have multiple allosteric sites, some specific for positive effectors and others for negative effectors. Allosteric sites exhibit a range of binding specificities analogous to those of substrate-binding at the catalytic site, i.e., either absolute specificity for a single compound or variable affinity for a class of related compounds. Most importantly, *in vivo* the effectors are all compounds that occur normally in the life of the cell and by their presence can modulate the rate of a given reaction. The magnitude of the effect is the function of their own concentration in the cell at any instant.

All known allosteric enzymes have two or more subunits, which may have the same or different amino acid sequences and contain more than one substrate-binding site per molecule. The interaction of an effector or substrate with an allosteric enzyme produces a change in conformation of the subunit that alters the catalytic properties of the enzyme. When the conformation of one subunit changes on binding, the effector may also influence the conformation of a second subunit that has no bound effector and alter the catalytic activity of the second subunit.

Enzymes with single active sites display normal Michaelis–Menten kinetics. Allosteric enzymes have multiple active sites and show cooperative binding. As a result allosteric enzymes display a sigmoidal dependence on the concentration of their substrates, allowing them to greatly vary in catalytic output in response to small changes in effector concentration.

The fact that an enzyme molecule is built of several subunits does not mean that it necessarily is subject to allosteric modification. Many multisubunit enzymes with more than one active site and obeying Michealis–Menten kinetics, have no effectors and have active sites unaffected by partial saturation of active sites on neighbouring subunits.

Allosteric enzymes, however, may have more complex kinetics. They often give sigmoidal curves in *V* vs [S] plot and double reciprocal plots are non-linear.

Their kinetic behaviour is described by an equation

$$K = [S]^n H \frac{V_{max} - V}{V}$$

$$-\log \frac{V_{max} - V}{V} = n_H \log[S] - \log K$$

where n_H is the Hill coefficient that differs from MM equation.

The allosteric effects can be explained by the "concerted MWC model" put forth by Monod, Wyman and Changuex or the "Sequential model" by Koshland, Nemethy and Filmer. Both postulate that enzyme subunits exist in one of two conformations, tensed (T) or relaxed (R), and that relaxed subunits bind substrate more readily than those in the tensed state. The two models differ most in their assumptions about subunit interaction and the pre-existence of both states.

THE MONOD–WYMAN–CHANGUEX (MWC) MODEL

This mode is sometimes called as a symmetrical model because it is based on the assumption that, in an enzyme (protein) all the promoters must be in the same conformational state. All of them must be either in the R-form or all in the T-form, no hybrids found at any point of time. The two conformational forms are in equilibrium in the absence of ligand, and the equilibrium is disturbed by the binding of the ligand. And the R-state has higher affinity for the ligand than the T-state. Because of that, although the ligand may bind to the subunit when it is in either state, the binding of a ligand will increase the equilibrium in favour of the R-state.

The concerted model is so named because during the transition between the two conformations, all subunits of the enzyme exist either in the T-form or the R-form. That is all subunits undergo simultaneous all-or-none concerted changes in their conformation.

If we were to consider a dimeric protein having two identical binding sites for a substrate or ligand (S), in the absence of the ligand, there will be equilibrium between the two conformational states R2 □ T2, the equilibrium constant termed the allosteric constant and given the symbol L. The hybrid RT is held to be unstable and ignored.

The ligand binds to either of the sites on the R2 molecule as shown in Figure 10.1.

This model explains sigmoidal binding properties as change in concentration of ligand over a small range will lead to a vast increase in the proportion of molecules in the R-state, and thus will lead to a high association of the ligand to the protein. A characteristic positive homotropic cooperativity will be apparent from the above MWC model, however it cannot explain negative cooperativity.

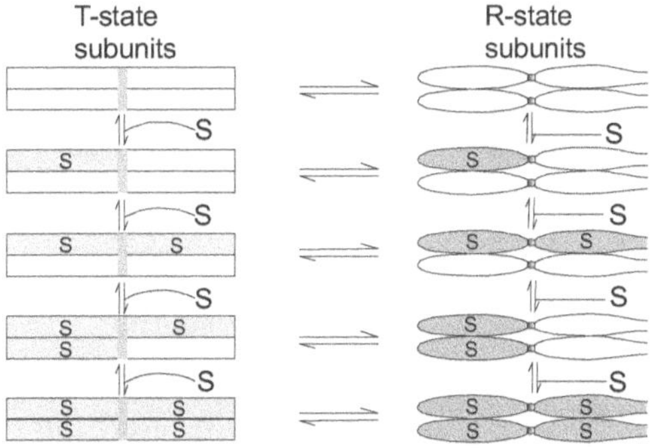

Figure 10.1 Concerted model

Cooperative Binding

When the substrate binds to one enzymatic subunit, the rest of the subunits are stimulated and become active. Ligands can either have non-cooperativity, positive cooperativity, or negative cooperativity.

An example of positive cooperativity is the binding of oxygen to haemoglobin. One oxygen molecule can bind to the iron (II) in the porphyrin ring of a haem molecule in each of the four chains of the haemoglobin molecule. Deoxy-haemoglobin has a relatively low affinity for oxygen, but when one molecule binds to a single haem, the oxygen affinity increases, allowing the second molecule to bind more easily, and the third and the fourth even more easily. The oxygen affinity of 3-oxy-haemoglobin is ~300 times greater than that of deoxy-haemoglobin. This behaviour leads the affinity curve of haemoglobin to be sigmoidal, rather than hyperbolic as with monomeric myoglobin (Figure 10.2). By the same process, the ability for haemoglobin to lose oxygen increases as fewer oxygen molecules are bound.

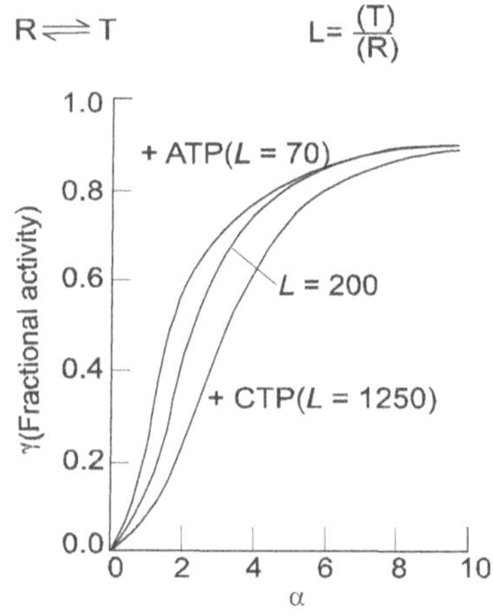

Figure 10.2 Sigmoidal curves of the Concerted Model

In the concerted model, sigmoidal curves can be described by 3 parameters

L—the allosteric constant equal to the ratio of [T]/[R]

K_T—the dissociation constant for each site in the T-state

K_R—the dissociation constant for each site in the R-state

where *n* is the number of active sites

THE KOSHLAND–NEMETHY–FILMER (KNF) MODEL

The KNF or "sequential model" of allosteric regulation holds that subunits are not connected in such a way that a conformational change in one induces a similar change in the others (Figure 10.3). Thus, all enzyme subunits do not necessitate the same conformation. Moreover, the sequential model dictates that molecules of substrate bind via an induced-fit protocol. In general, when a subunit randomly collides with a molecule of substrate, the active site, in essence, forms a glove around its substrate. While such an induced fit converts a subunit from the tensed state to the relaxed state, it does not propagate the conformational change to adjacent subunits. Instead, substrate-binding at one subunit only slightly alters the structure of the other subunits so that their binding sites are more receptive to the substrate.

To summarize:

1. Subunits need not exist in the same conformation.
2. Molecules of substrate bind via induced-fit protocol.
3. Conformational changes are not propagated to all subunits.
4. Substrate-binding causes increased substrate affinity in adjacent subunits.

Therefore, for a dimeric protein where each protomer can exist in R- and T- forms, the species R2, T2, R2S, R2S2, R.TS, RS.TS, T2S and T2S2 can all exist.

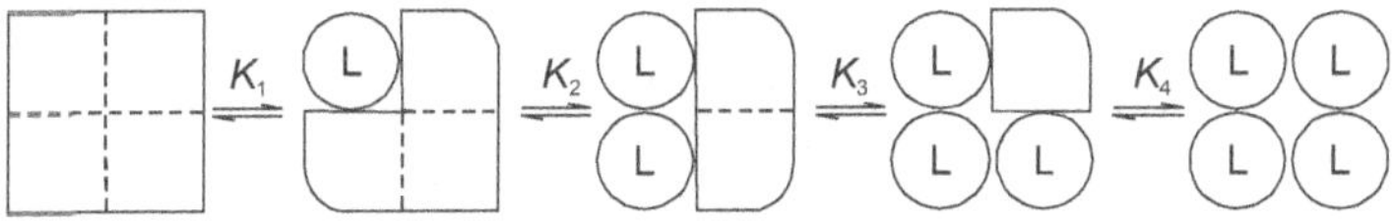

Figure 10.3 Sequential model (KNF)

ALLOSTERIC ACTIVATION AND INHIBITION

Activation

Allosteric activation, such as the binding of oxygen molecules to haemoglobin, occurs when the binding of one ligand enhances the attraction between the substrate molecules and other binding sites, e.g. oxygen is effectively bound to both the substrate and the effector. The binding of oxygen to one subunit induces a conformational change in that subunit which interacts with the remaining active sites to enhance their oxygen affinity.

Inhibition

Allosteric inhibition occurs when the binding of one ligand decreases the affinity for the substrate at other active sites. For example, when 2,3-BPG binds to an allosteric site on haemoglobin, the affinity for oxygen of all subunits decreases.

ASPARTATE TRANSCARBAMOYLASE (ATCASE)

A classical example for allosteric regulation is aspartate transcarbamyolase (ATCase) (Figure 10.4).

ATCase's c_6r_6 structure contains two catalytic trimers and 3 regulatory dimers. The catalytic trimers are stacked on top of each other. The regulatory dimers have significant contacts with the trimers through structural domains stabilized by 4 Cys residues interacting with a zinc ion (Figure 10.4a and b).

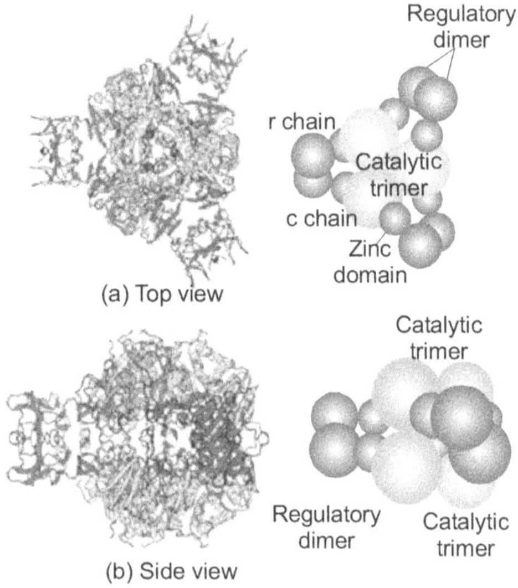

Figure 10.4 ATCase structure

It catalyses the reaction:

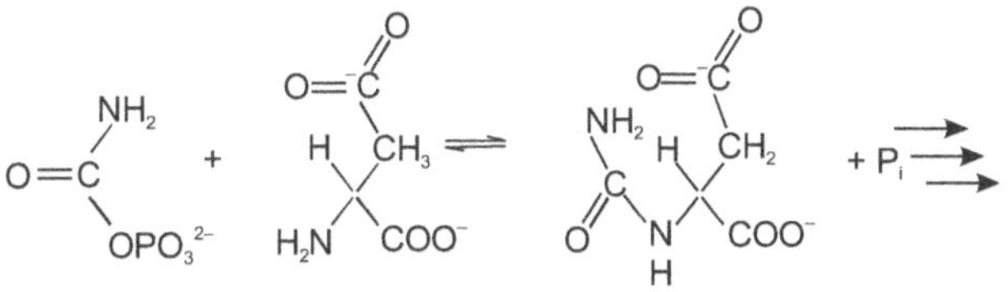

Carbamoyl phosphate   Aspartate   *N*-Carbamoylaspartate

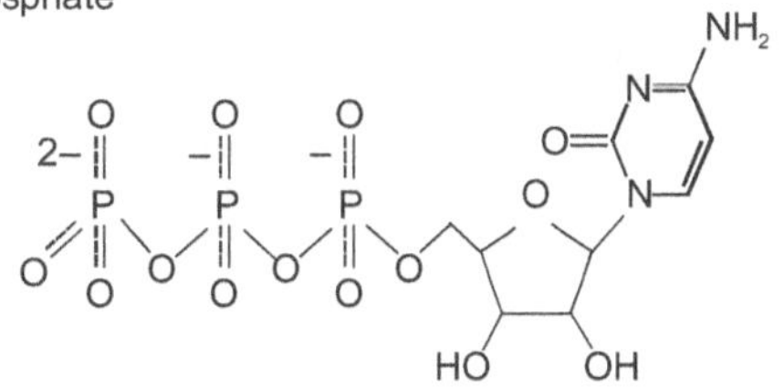

Cytidine triphosphate (CTP)

This is first in the sequence of reactions leading to the biosynthesis of the pyrimidine nucleotides UMP, UDP, UTP and finally CTP. Two important control mechanisms regulate pyrimidine formation in *E. coli*; the synthesis of ATCase is repressed in the presence of uracil, and the metabolic end product, CTP, controls its own synthesis by feedback inhibition of ATCase (Figure 10.5).

Figure 10.6 shows the crystallization of ATCase in the presence of CTP resulting in binding of the CTP on the regulatory subunit. CTP binds to regulatory subunits in the T-state at least 50 Å away from the active sites of the catalytic subunits but exerts its effect by preferentially stabilizing the T-state.

High concentrations of aspartate, the substrate of the enzyme, ATCase reduces the inhibitory effect of CTP. Aspartate transcarbamoylase activity is increased by an activator molecule, ATP, which also binds the same allosteric site. Here, ATP functions as a positive modulator.

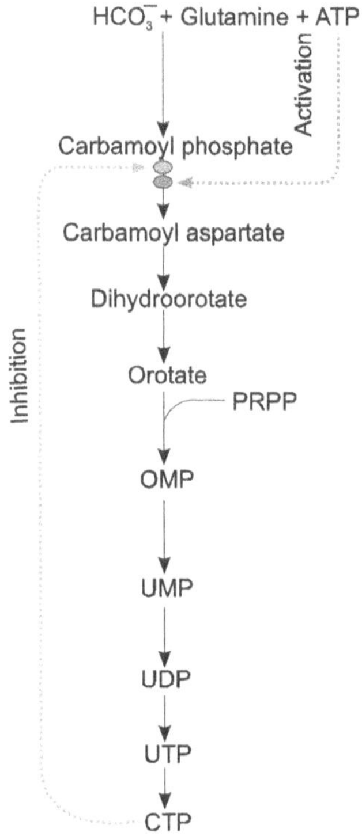

Figure 10.5 *E. coli* pyrimidine biosynthesis

Figure 10.7 shows ATCase's allosteric behaviour that can be explained by transitions between a less active T-state and a more active R-state. T-state is stabilized by CTP binding to regulatory subunit, but not as active. R-state is more active and induced by substrate (or analogue) binding to catalytic subunits (also induced by ATP binding). Each regulatory dimer can bind two molecules of CTP, but although the two subunits are apparently identical, the affinity for the first molecule of CTP is considerably greater than that for the second; the precise explanation for this negative homotropic cooperativity has yet to be found. Also it is known that ATP and CTP compete for the same site. Both CTP and ATP are allosteric efforts of

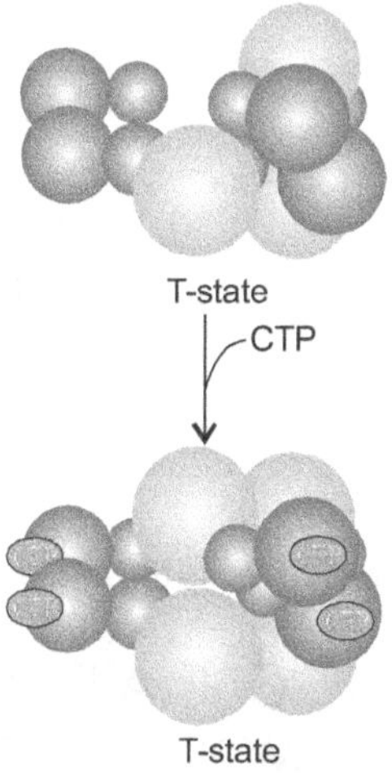

Figure 10.6 CTP binding to ATCase

ATCase. CTP is an allosteric inactivator, stabilizing the T-state and making the enzyme less active. ATP is an allosteric activator, stabilizing the R-state and making it easier for the substrate to bind. When ATP concentration is high, the cell has energy to expend for making nucleotides, and making pyrimidines will balance the nucleotide pools (especially if CTP concentration is low) (Figure 10.8).

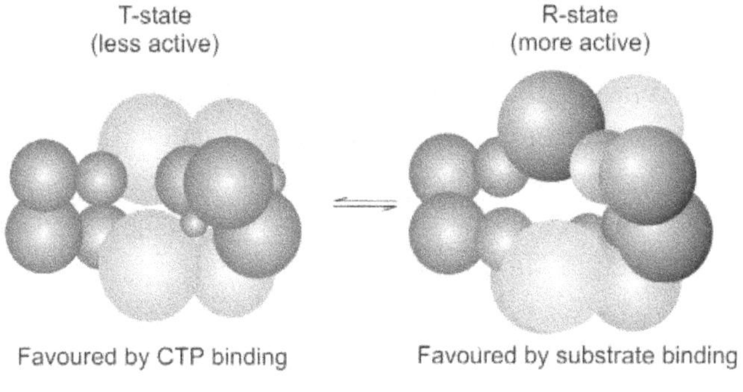

Figure 10.7 ATCase allosteric behaviour

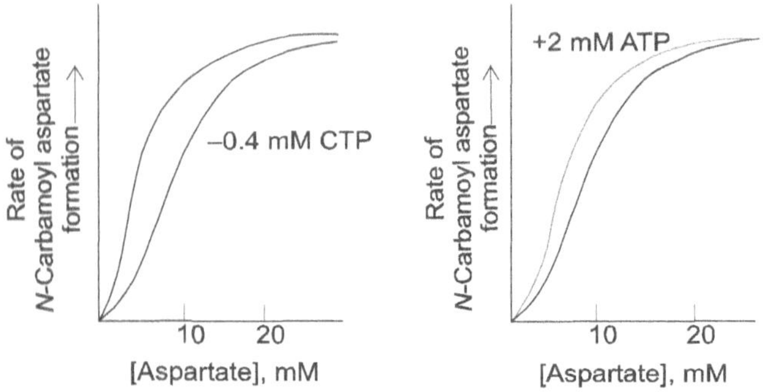

Figure 10.8 Allosteric regulators modulate the T or R transitions

PHARMACOLOGICAL APPLICATIONS OF ALLOSTERIC MODULATION

Allosteric modulation of a receptor results from the binding of allosteric modulators at a different site (regulatory site) other than that of the endogenous ligand and enhances or inhibits the effect of the ligand. It normally acts by causing a conformational change in a receptor molecule, which results in a change in the binding affinity of the ligand. In this way, an allosteric ligand "modulates" its activation by a primary "ligand" and can be thought to act like a dimer switch in an electrical circuit, adjusting the intensity of the receptor's activation.

For example, the anti-anxiety drugs Valium, Xanax, Librium and Ativan, "potentiate" or turn up the activity of the neurotransmitter gamma-aminobutyric acid (GABA) when it binds to its benzodiazepine receptor. More recent examples of drugs that allosterically modulate their drug targets include Cinacalcet and Maraviroc.

Allosteric Sites as Drug Targets

Allosteric sites may represent a novel drug target. There are a number of advantages in using allosteric modulators as preferred therapeutic agents over classic orthosteric ligands. For example, GPCR allosteric binding sites have not faced the same evolutionary pressure as orthosteric sites to accommodate an endogenous ligand, and so are more diverse. Therefore, greater GPCR selectivity may be obtained by targeting allosteric sites. This is particularly useful for GPCRs where selective orthosteric therapy has been difficult because of sequence conservation of the orthosteric site across receptor subtypes. Also, these modulators have a decreased potential for toxic effects, since modulators with limited cooperativity will have a ceiling level to their effect, irrespective of the administered dose. Another type of pharmacological selectivity that is unique to allosteric modulators is based on cooperativity. An allosteric modulator may display neutral cooperativity with an orthosteric ligand at all subtypes of a given receptor except the subtype of interest, which is termed as absolute subtype selectivity. If an allosteric modulator does not possess appreciable efficacy, it can provide another powerful therapeutic advantage over orthosteric ligands, namely, the ability to selectively tune up or down tissue responses only when the endogenous agonist is present.

REVIEW QUESTIONS

1. Differentiate allosteric site from catalytic site.
2. What is allosteric modification?
3. Discuss the Monod–Wyman–Changeux model of allosteric modification.
4. What do you mean by cooperative binding and heterotropic effects?

5. Explain the Koshland–Nemethy–Filmer (KNF) model of allosteric modification.
6. Aspartate transcarbamoylase (ATCase) is modified allosterically—justify.
7. Discuss the pharmacological applications of allosteric modulation.
8. Describe how allosteric sites can be used as drug targets.

11

ENZYME ISOLATION, FRACTIONATION AND PURIFICATION

NEED FOR ENZYME PURIFICATION

Enzymes are found in complex mixtures in cells, which contain hundred or more different enzymes. In order to study a given enzyme properly, it must be purified. It is true that this is possible by the use of sufficiently specific test methods to study enzymes while they are still in an impure state and pure enzymes had been obtained. But in most cases, some of the other enzymes present will interfere by attacking the substrate so as to give side reactions, by transforming the product into some other substance or by attacking the coenzyme or even the enzyme itself.

It is sometimes difficult, until the enzyme in question has been purified, to say exactly what reaction it catalyses. Apparently simple metabolic reactions are often found, when the enzymes are purified, to take place in a series of steps, each catalysed by a separate enzyme. Again and again lines of metabolism have been put forward, which have remained obscure and controversial until isolation of the enzymes concerned cleared up the mechanism, e.g. the mechanism of citrate formation became clear only when the citrate-forming enzyme (EC 4.1.3.7) was isolated.

For the study of the properties and behaviour of an enzyme as a chemical catalyst, it is necessary to have it as pure as possible. For example, in the investigation of its specificity, if an impure enzyme is found to catalyse some reaction, in addition to its normal reaction, there will always be some doubt whether this is not due to some other enzymes present in the preparation. In the test tube, an enzyme's environment is very different from its natural environment in the cell, and it should not be overlooked that this difference may affect its behaviour.

Purification may affect the activity of an enzyme in various ways. It may remove any cofactors that may be essential for the reaction, but these can be restored in testing. In few cases, the properties of the enzyme molecules themselves have been found to be altered by the purification procedures employed, especially warming which is sometimes used as one of the steps. For example, some purification procedures appear to facilitate partial proteolytic degradation and alteration of the properties of fructose bisphosphatase (EC 3.1.3.11).

In general, it is true that the study of isolated enzymes has thrown a vast amount of light on their catalytic potentialities within the cell and that any observed differences are more likely to be due to permeability factors or to the effects of other components in the system, than to any modification of the enzymes as a result of the process of purification.

The serious purification of enzymes dates from about 1922; the first crystalline enzyme urease (EC 3.5.1.5) was obtained by Sumner in 1926. By 1940, twenty highly purified enzymes had been obtained. At present, dozens of enzyme purifications are announced each year and the total number purified runs into many hundreds.

GENERAL METHODS IN ENZYME ISOLATION

ESTIMATION OF ENZYME ACTIVITY

The first requirement is a convenient quantitative test of the activity by which the enzyme can be estimated. Since a large part of the time spent in enzyme isolation is used up in testing the activity of the different fractions, it is most desirable that the test should be a rapid one. Speed of the test is more important than extreme accuracy. The nature of the test method adopted will naturally depend on the type of reaction catalysed by the enzyme and a suitable method must be worked out for each case.

If the substrate A is converted by the enzymatic reaction into the product B, and the test is based on the estimation of B, it may be that in reality the conversion goes by way of an unsuspected intermediate form C, so that two enzymes are involved, one to convert A into C and one to convert C into B. In this case, any purification step which separates two enzymes will cause a total loss of activity as measured by the test. The test may then be altered to a test for C so that the first enzyme can be purified without reference to the second, or if this is inconvenient a preparation of the second enzyme may be added in excess to all the tests as one of the reagents. Similarly, when an unsuspected coenzyme is involved in a reaction catalysed by a single enzyme, again activity will be lost on purification. If the factor proves to be one of the known coenzymes, it may be simply added to the test, if it is of unknown nature, it may be possible to restore it by adding an extract of the tissue which has been boiled to remove enzymes and proteins (since coenzymes are heat-stable), or a trichloroacetic acid extract (neutralized), or possibly a dialysate (since coenzymes pass through membranes).

UNIT CONCENTRATION OF ENZYME

Since the enzyme in question has not previously been purified, its specific activity and therefore the purity of the initial preparation will be unknown. It is therefore usual to define an arbitrary unit of enzyme in terms of which the purity and activity of the various fractions may be quantitatively expressed. In most cases, this unit will be defined with reference to the estimation method used.

Units of Enzyme Activity

The amount of enzyme measured cannot be expressed in classical terms such as mg/ml, since only a very small amount of catalytically active enzyme may be present in a given weight of this substance. The result of enzyme measurements are hence reported in enzyme activity units. There is considerable confusion over enzyme units, enzyme activity and specific activity. The standard definition of the enzyme unit is given below:

1 unit (U) is the amount of enzyme that catalyses the reaction of 1 μmol of substrate per minute.

However, suppliers of enzymes seldom use this definition as it frequently requires labels to be written in fractions of units or converted into milli-units. The following definition is more commonly used:

1 unit (U) is the amount of enzyme that catalyses the reaction of 1 nmol of substrate per minute.

The assay temperature and the concentrations of substrates, buffers, cofactors and other additives may all impact the measured number of enzyme units. In practice, it is extremely rare to calculate the number of units required for an assay, either because it is difficult to replicate the assay conditions used or because different assay conditions are required.

Moreover, some activity may have been lost in storage or during transportation so it is sensible to confirm the actual number of units per ml prior to setting up large-scale experiments.

Specific Activity

Specific activity is the number of enzyme units per ml derived by the concentration of protein in mg/ml. Specific activity is an important measure of enzyme purity and quality.

Activity values (units/ml) are far more important for assay since the amount of substrate converted is determined by the number of enzyme units added. Also, it is possible to calculate the volume of enzyme required for an assay from the specific activity value alone, since the specific activity values for an undiluted enzyme and a 1/1000 dilution are identical. The enzyme unit was adopted by the International Union of Biochemistry in 1964. Since the minute is not an SI unit, the enzyme unit is discouraged in favour of the katal, the unit recommended by the General Conference on Weights and Measures in 1978 and officially adopted in 1999.

Catalytic Activity

The catalytic activity of an enzyme is the property measured by the increase in the rate of a reaction of a specified chemical reaction that the enzyme produces in a specific assay system.

Katal

The katal (Symbol: kat) is the catalytic activity that will raise the rate of the reaction by one mole per second in a specified assay system.

The nomenclature committee finds no objection to the practice of specifying catalytic activity in $mols^{-1}$. However, a reason why use of the katal may prove convenient is that 1 kat implies more than 1 mols^{-1} by showing that the quantity referred to is catalytic activity, just as 'Hz' implies frequency whereas s^{-1} does not. The committee realizes that a catalytic activity of one katal, corresponding to a rate of reaction of one mole per second, will usually be too large for practical use. In most cases, catalytic activities will be expressed in microkatals (μ kat), nanokatals (nkat) or picokatals (pkat) corresponding to the reaction rates of micromoles, nanomoles or picomoles per second respectively.

The old enzyme unit (U) may be related to the katal by the following: 1 U catalyses a rate of 1 μ mol/min = 1/60 μ mol/s □ 16.67 nmol/s, therefore 1 U corresponds to 16.67 nkat.

Quantities Derived from Catalytic Activity

One derived quantity is the specific catalytic activity of an enzyme, which may be expressed in kat.kg^{-1}. Another derived quantity is the molar catalytic activity, which may be expressed in katmol^{-1}. If the conditions specified are such that the enzyme is saturated with substrate, the molar catalytic activity will be numerically equal to the rate constant, in s^{-1}, for decomposition of the enzyme–substrate complex into enzyme and products. Its interpretation may nevertheless be complex because of phenomena such as non-productive binding and the existence of more than one catalytic site per enzyme molecule.

Thus, when a measuring catalytic reaction is sufficiently well known, it is now possible to express values for the various types of amount, concentration, etc. of inorganic or organic catalysts in the original sampled system in a convenient manner with metrological traceability to the SI. It is therefore hoped that the katal, after long and tortuous development, will supplant current off-system units of catalytic activity.

ENZYME ASSAY

The activity of an enzyme may be determined based on the rate of utilization of substrate or formation of product under controlled conditions. There are a number of methods available for estimation of enzymes. They are as follows:

Spectrophotometric Methods

In the spectrophotometric method, it is the amount of enzyme that will be estimated, which provides a certain change of optical density in a certain time under the test conditions is estimated if some product is being estimated, it may be the amount of enzyme which produces a certain amount of the substance per minute. The concentration of the enzyme in solution is expressed as units per ml.

A large number of assays are based on the interconversion of NAD^+ and NADH. Both the oxidized and reduced forms of there two nucleotides absorb at 260 nm and 340 nm, respectively. Enzymes that do not directly involve in this interconversion may be assayed by using the concept of coupled reactions. In such reactions, the enzyme to be assayed is linked to one utilizing the NAD^+/NADH system by means of common intermediates, e.g. phosphofructokinase, glyceraldehyde 3-phosphate dehydrogenase.

The visible spectrophotometric assay can be extended to the use of artificial substrates and by the production of coloured derivatives of the substrate or product. Many enzymes, especially the hydrolases, will act on synthetic analogues of their natural substrate to release a coloured product such as *p*-nitrophenol and phenolphthalein.

Spectrofluorimetric Method

Fluorimetric methods are fast and sensitive but they have the practical limitation that trace impurities in the enzyme preparation can quench the emitted radiation. Additionally many fluorescent compounds are unstable especially in the presence of ultraviolet light. Nevertheless, the technique is widely applied to enzyme assays. NAD(P)H is fluorescent, so many enzymes can be assayed by coupling to an appropriate reaction.

Luminescence Method

This method is used to study the intensity of emitted light, and the reactions involved are said to be bioluminescent reactions. One of the difficulties is the lack of reproducibility. For example, firefly luciferase catalyses the oxidation of luciferin in an ATP-dependent reaction.

$$\text{Luciferin} + \text{ATP} + O_2 \longrightarrow \text{Oxyluciferin} + \text{AMP} + \text{PPi} + CO_2 + \text{Light}$$

Manometric Method

These are convenient and accurate methods for the following reactions in which one of the components is a gas. They are therefore well adapted for the study of oxidase (O_2 uptake) or decarboxylases (CO_2 output), as well as for enzymes such as hydrogenase, urease or carbonic anhydrase. Their use however is by no means restricted to such reactions. Any of the numerous reactions which involve the production or consumption of acid or alkaline substance can be followed, if they are carried out in the presence of bicarbonate in equilibrium with a gas mixture containing a definite percentage of CO_2. Any acid produced will liberate a corresponding amount of CO_2 from the bicarbonate and this can be measured on the manometer.

An obvious case is the oxidation of an aldehyde group to a carboxyl group, e.g. in the glycolysis system by phosphoglyceraldehyde dehydrogenase. As with the spectrophotometric methods, it is possible to extend the method to other enzymes by linking them with this dehydrogenase, and many coenzyme-linked dehydrogenase reactions have been studied manometrically. Carboxyl groups may also arise by hydrolysis of a peptide bond or an ester, and protein hydrolysis has been followed manometrically. The action of esterases, especially cholinesterase, have also been studied. Ferricyanide is reduced by many oxidizing enzymes and as its reduction is accompanied by the production of a hydrogen ion, it can be measured manometrically in bicarbonate.

The Thunberg Method

A considerable amount of the work on dehydrogenases in the past has been carried out by this method, which depend on visual observation of the time taken to reduce (decolorize) a definite amount of methylene blue or some similar dye. To prevent re-oxidation of the dye by atmospheric O_2, it is essential to carry out the test in special test tubes which can be evacuated (Thunberg tubes). The amount of dye added is extremely small, so that the substrate concentration remains practically constant during the reaction. Normally only the reduction time is measured and the course of the reaction is not followed. In a few cases in which the time curve has been determined, it was found that the reaction was fairly linear, slowing down only at the end so that the reciprocal of the reduction time gives the reaction velocity. The usefulness of the method, however, does not depend on this, for it is more used for the determination of relative than for absolute velocities and the reciprocal of the reduction time will be proportional to the velocity even when the curve is not linear.

The method is simple and requires no expensive apparatus; when due precautions are taken it is fairly accurate. In the case of the coenzyme-dependent dehydrogenases, however, it is necessary to add an excess of diaphorase to catalyse the reaction between the coenzyme and the dye, and this somewhat limits its usefulness with such enzymes. A modification of the method is to follow the decolorization of 2,6-dichlorophenol indophenol in the spectrophotometer. The reduced form of this dye reacts with O_2 more slowly than that of methylene blue and it is possible to carry out the tests in ordinary open spectrophotometer cells, though Thunberg tubes with optical faces have been manufactured.

Electrode Method

The use of the glass electrode gives another method of following reactions which involve the production of acid. The simplest method is merely to follow the pH changes as the reaction proceeds, but there are two objections to this. The first is that the change of pH during the reaction will probably cause complications by altering the activity of the enzyme. The second objection is that the rate of change of pH depends not only on the rate of the reaction but also on the buffering power of the solution. This is a serious objection because proteins have a high buffering power and a crude enzyme preparation will contain a large amount of inactive protein which is removed during the purification. Thus, it is difficult to keep the buffering power constant.

A much better method is that of continuous titration, in which the pH is kept approximately constant by frequent additions of alkali; the rate of addition of alkali then gives the reaction velocity and is independent of the amount of buffer. The amount of buffer, however does affect the accuracy of the method; if too little is used, it is difficult to keep the pH sufficiently constant, while too much makes the method insensitive. A very convenient automatic apparatus is available commercially which keeps the pH constant by continuous additions of acid or alkali, and at the same time plots a curve of amount added against time. With this apparatus progress curves of many enzyme reactions can be obtained without the attention of an operator.

Electrode methods can also be used for oxidizing enzymes, by measuring the redox potential. A plain platinum electrode in a solution containing a mixture of the reduced and oxidized forms of a dye or similar hydrogen acceptor, gives a potential which depends on the ratio of the concentrations of the two forms. The reduction of the dye can therefore be followed by the change of potential of the electrode. Ferricyanide is usually more convenient than a dye, partly because it reacts directly with the coenzymes and does not require the addition of diaphorase, and partly because O_2 does not interfere and the reaction can be done in open test tubes. In the interests of linearity, it is advisable to start with a mixture of equal amounts of ferri- and ferrocyanide, then over a certain range, the change of potential is approximately proportional to the amount of ferricyanide reduced. The method has been found to be very convenient.

Polarimetric Method

Many enzymes act only on one optical isomer of their substrate. If, as often happens, the product is optically inactive, the reaction can be followed by the change of rotation. This is also the case where the substrate is optically inactive, but an optically active product is formed, or where the substrate and product are both optically active but differ sufficiently in their specific rotations. This method, while not quite so convenient as some of the other methods, can be used for a number of enzymes to which they are inapplicable, especially for enzymes acting on carbohydrates. "Sucrase" is a classical example.

In some cases the substrate has too low a rotation to measure directly, but the rotation may be greatly increased by forming a complex. This is the case with the hydroxy acids, e.g. malic acid, which form complexes with molybdate and these complexes have high specific rotations. Fumarase was first studied in this way, but since the reaction will not proceed in the presence of the molybdate reagent, a sampling method is necessary and the method has been superseded by the much more rapid spectrophotometric method.

Chromatographic Methods

Where other methods fail, it is sometimes necessary to detect the formation of a product by chromatography. This is essentially a sampling method; samples are withdrawn from the reacting mixture at intervals, placed on the chromatographic paper, dried and run as chromatograms in the usual way. The gradual appearance of a new spot gives a qualitative idea of the progress of the enzyme reaction. If quantitative results are desired, it is necessary to cut the spots, extract the material from them and estimate it by some chemical or spectrophotometric method. This method has been applied, for example, to transglycosylation reactions with flavins. The great objection to the method is the time required, especially for running the chromatograms, which takes several hours. To purify an enzyme, using such a test method would be an extremely slow and laborious procedure. For this reason, chromatographic methods have not found much application in enzyme work, except where no other method is available.

Chemical Estimations

Many enzyme reactions are followed by withdrawing samples at intervals and estimating the substrate or product by chemical methods. There is little of a general nature to be said about these methods, since the different enzymes require different estimation methods, but there are a few methods of more general application which can be applied to whole classes of enzymes. The convenient colorimetric method of Fiske and Subbarow for inorganic phosphate has wide application, since so many enzyme reactions are concerned with phosphate compounds of various kinds. Phosphatases, phosphorylases and nucleotidases can be followed directly by such estimations. Since the breaking of a glycosidic link produces a reducing group, copper reduction methods have been much used for the study of enzymes acting on carbohydrates. Formol titration methods have been much used in the study of peptidases.

As Lipman has shown, the addition of hydroxylamine to acyl phosphates yields a hydroxamic acid which gives a purple colour with ferric salts. Acetyl phosphate can be estimated in this way, and the addition of coenzyme A and excess of phosphotransacetylase gives a means of following enzyme reactions in which acetyl-CoA is involved. For instance, the oxaloacetate transacetase forms citrate from oxaloacetate and acetyl-CoA, and if acetyl phosphate, CoA and phosphotransacetylase are added instead of acetyl-CoA, the reaction can be followed by the disappearance of acetyl phosphate.

PROTEIN ESTIMATION

Protein may be estimated by a number of methods.

Kjeldahl Method

The classical method is the determination of total nitrogen by the Kjeldahl method. Material which is all protein generally has a nitrogen content of 15–16% but nitrogenous substances other than protein will be included in the estimations, e.g. ammonium sulphate. The necessity for removal of non-protein contaminants makes the method a slow one. Particularly, the

Kjeldahl incineration should be continued for at least eight hours to give complete conversion of all the amino acids.

Colorimetric Methods

Quantitative methods based on the biuret reaction have been developed. Since they depend on the peptide chain rather than the side chains, they should be more independent of the composition of the protein being assayed. The method of Gornall *et al.* (1618) involving absorption measurements at 550–650 nm is suitable for 1–10 mg of protein. Microbiuret methods depending on ultraviolet absorption of copper–protein complexes suitable for 0.1–2 mg of protein have also been described.

Folin–Ciocalteau Method

The Folin–Ciocalteau method for protein estimation depends on the chromogenic nature of some amino acid side chains, as well as on the peptide bond. Thus, there will be wide variations with different proteins. The method is sensitive (only 0.01–0.1 mg of protein is needed for the assay) but subject to interference by several non-protein materials.

Spectrophotometric Measurements

It is now quite common to rely on spectrophotometric measurements of absorption at 280 nm due to aromatic amino acids. Different proteins contain different amounts of these amino acids and therefore absorb to different extent. The importance of speed in determining the activity applies to the estimation of proteins. For this reason, a convenient rapid method such as the measurement of absorption at 280 nm is to be preferred for use during purification.

The time factor is more important than possible errors, due to the fact that the specific absorption of the protein being purified may be very different from the average absorption of the mixed proteins forming the starting material. But if this or any other method depending on one particular constituent of the protein is used, it is necessary to check that no changes in the mean content of this constituent in the protein material have been brought about by the purification. This can be done by estimation of proteins at key points using one of the reliable methods such as the Kjeldahl method or biuret method.

SOURCE OF ENZYMES

The amount of a given enzyme may be very different in different tissues. By choosing a tissue which is rich in the enzyme, the degree of purification required to obtain the pure enzyme is much less, perhaps by a factor of 10 or 100. In choosing a source, due regard must be paid to its ready availability in quantity and also to its reproducibility. In the case of bacteria, it is possible to make use of the phenomenon of induction in order to obtain cultures which are greatly enriched with respect to the enzymes by growing them in the presence of the substrate. A number of enzymes are normally situated within the intracellular particles, e.g. mitochondrial enzymes, therefore it is worth to prepare the mitochondria first by high speed centrifugation of tissue extracts and to use this preparation as the starting material. The enzymes and the other constituents of the cytoplasm are removed and a considerable degree of purification is achieved before the fractionation methods are applied.

EXTRACTION

When a suitable source has been found it is necessary to bring the enzyme into solution. For this, rupture of cell membrane is usually necessary. It is easier to extract enzymes from animal tissues than from microorganisms. In many cases, simple extraction of minced muscle with water is sufficient. With a strongly glycolysing tissue, care must be taken that it does not become sufficiently acid to damage the enzyme and it is sometimes advisable to extract with a buffer solution. It may be necessary to bring the tissue into a finely divided state by the use of a high-speed homogenizer.

In many cases, it is necessary to make more energetic measures to disrupt the cell membrane. Few such methods are mechanical breaking by grinding with sand or shaking at high speed with fine glass beads, ultrasonic or sonic oscillations, freezing and thawing, treatment with solvents such as acetone, acetolysis (with toluene or sodium sulphide) or lysis with added enzymes. Autolysis or treatment with proteinases should be avoided, as it will be necessary only to isolate the enzyme from an extremely complex mixture of protein breakdown products produced by digestion.

However, especially with yeast, autolysis is the only method which can be applied on a sufficiently large scale. Drying with acetone is not only a useful method of attacking the cell membrane, but, it gives an active dry powder which can be stored in bulk and forms a convenient starting material from which the enzyme can be extracted with water or buffer solution. The acetone treatment must always be carried out at a low temperature, and even then some enzymes are destroyed by it, e.g. enzymatic lysis of *Micrococcus lysodeikticus* by lysozyme (EC 3.2.1.17).

The extraction of enzymes from subcellular organelles is a special problem. Mitochondrial enzymes were once regarded as insoluble and inseparable from the mitochondrial particle, but now it is clear that many pass freely into solution once the mitochondrial membrane has been ruptured or made permeable. However, some enzymes are more firmly bound to the structural framework of the mitochondria, probably as a lipoprotein complex, and require special methods for their extraction.

These methods include drying with acetone, treatment with butanol–water mixtures, extraction of dried mitochondria with various organic solvents, extraction with aqueous solutions of detergents such as cholate, deoxycholate, tween, triton, emasol, etc. and treatment with "chaotropic" agents such as perchlorate, and exposure to the action of hydrolytic enzymes such as lipases, nucleases or proteolytic enzymes. The butanol procedure, with some enzymes appears to involve the disruption of the lipoprotein association, so that the enzyme passes into the aqueous phase. The detergent procedure, used for the isolation of the components of the respiratory chain sometimes results in the solubilization of the enzyme which remains in clear solution after removal of the detergent.

FRACTIONATION METHODS

The extract containing the enzyme will also contain other substances of both large and small molecular weight. The small molecules can be removed by dialysis or gel filtration leaving the large molecules which are predominantly proteins though some polysaccharides may be present. In the past, a number of different types of fractionation methods have been used, but now the majority of purifications are carried out with a relatively small number of standard procedures which have proved practically effective and convenient.

The order in which these are used must be determined by trial, but it is generally true to say that more rapid purification is achieved by alternating the different fractionation methods, than by repetition of fractionations of one type. Fractionations must always be controlled by activity tests. For each stage, a pilot fractionation on a small scale is carried out, before testing the main bulk of the solution. In the small-scale tests, it is necessary to carry out the same conditions as on a large scale. In all fractionations, attention must be given to two factors which affect the results considerably, namely, the pH and the electrolyte concentration particularly the former.

Each fractionation step consists of the separation of the total protein into a series of fractions by gradually increasing the salt concentration or the amount of adsorbent added or the concentration of organic solvent or the acidity, etc. in such a way that the greater part of the enzyme is found in one fraction if possible. Fractional denaturation by heating or by changing the pH is also sometimes used to remove unwanted proteins. The correct method of fractionation is to take only one portion of the enzyme solution and to remove each fraction before the next is precipitated. In this way, each fraction contains only those proteins which are brought down over a small range of salt concentration and a true fractionation is effected.

In large-scale fractionation, it is not usual to separate the proteins into as many fractions as the pilot tests. Usually only three fractions are obtained, two of which are discarded. This may not be correct without further tests, because the conditions of precipitation are not quite the same as in the pilot fractionation. For example the precipitation of the enzyme may be influenced by the presence of proteins which in the pilot tests, had been removed in the earliest fractions, but which in the main fractionation may cause the enzyme to precipitate at saturation. For this reason, it is advisable to do a second pilot separation into the three fractions only before carrying out the main fractionation in order to make sure that the concentration limits have been correctly determined.

Fractional Precipitation by Change of pH

Occasionally an enzyme may be precipitated in this way. The removal of nucleoprotein and particulate material from animal tissues during fractionation becomes advantageous and easier as they can be adjusted to desired pH and centrifuged. This forms the first step before the main purification begins. Also, this method results in conversion of a turbid extract to clear solution.

Fractional Denaturation by Heating

This is another treatment with which a fairly stable enzyme may be valuable as a preliminary step. By heating the solution for a definite time to a temperature just below that at which the enzyme is destroyed, it is sometimes possible to coagulate much unwanted protein, which can then be centrifuged off and discarded. Also, the presence of the substrate frequently has a specific stabilizing effect on the enzyme. It may be possible to heat the solution to a temperature 10° higher in the presence of substrate than in its absence without loss of enzyme and thus get rid of much more of the other protein. In carrying out the heating, it is best to have three water baths, one at the desired temperature, one a few degrees hotter and one cold.

The solution is placed in a round flask, which should not be more than half full. Then the flask is immersed in the hottest bath and the contents of the flask are kept rapidly swirling round so that no part becomes overheated. The rise of temperature is followed by a thermometer immersed in the solution and as soon as the desired temperature is reached the flask is transferred to the bath at that temperature, where it is left for a definite time while the next batch is heated. After, the desired heating period, the flask is transferred to the cold bath and the contents rapidly cooled to a few degrees by swirling. It can then be left until cold. The time of heating is commonly 10 or 15 minutes.

Fractional Precipitation with Organic Solvents

This is one of the standard fractionation methods. Ethanol was used extensively for separating blood proteins. The enzymes of muscle extract were separated using of various solvents of which acetone gave the sharpest separations and the lowest losses. To get good separations, a low electrolyte concentration is required, i.e., a preliminary dialysis is desirable though the solution must not be completely free from electrolytes or the precipitation will not centrifuge down. pH 6.5 was found to give the best result with the muscle extracts. Since most enzymes are inactivated by solvents at room temperature, special precautions have to be taken to keep the temperature low throughout the procedure.

The enzyme solution is kept in a metal (aluminium) vessel immersed in a bath at about -5° and provided with an efficient stirrer. Acetone is contained in a graduated cylinder also in the bath and is driven by gentle air pressure through a capillary jet touching the wall of the metal vessel. It then flows down the cold metal surface in a thin layer before mixing with the solution. The addition of the solvent begins as soon as the temperature of the solution has fallen to 0° (to prevent freezing) and is continued until the desired amount of precipitate for the first fraction is obtained. The solution is then centrifuged at the same temperature and poured back into the metal vessel. The addition of acetone is then continued. The precipitate is either freeze-dried or immediately dissolved in a sufficient quantity of cold water or buffer solution to dilute the acetone to a harmless concentration, or it may be dialysed at the freezing point. The enzyme activity tests are done on the dissolved fractions and not on the main solution in which the increasing concentrations of acetone may considerably affect the results.

Fractional Precipitation by Salts

This is a widely used method. The salt most employed is ammonium sulphate, because of its larger solubility in water and absence of harmful effects on most enzymes. It has in fact a stabilizing action on many enzymes and it is usually not necessary to carry out the fractionations at a low temperature. It is advisable to use good-quality ammonium sulphate in order to avoid toxic impurities. Even pure preparations are slightly acidic and due attention must be paid to the control of pH. It is necessary to add the solid and not a solution in order to keep the volume down. The pilot fractionation is conveniently done by having the enzyme solution in a centrifuge tube and the solid ammonium sulphate in a boiling tube. The tube of ammonium sulphate is weighed, and enough of the salt is added to the enzyme solution to bring down an amount of precipitate suitable for one fraction. The precipitates do not always centrifuge very rapidly and it may be desirable to use a high-speed centrifuge (10,000 rpm). After centrifuging the first fraction, the solution is poured off into another centrifuge tube and the addition of salt continued. The precipitate is dissolved in a small volume of water or buffer solution. The activity tests are carried out on the redissolved fractions rather than the solution, on account of the high salt concentrations in the latter. The theory of enzyme fractionation by salting out is expressed as

$$\log s = B - KI$$

Where s is the solubility, I is the ionic strength and B and K are constants. As the salt concentration is increased, a point will be reached at which the solubility of the enzyme is equal to its concentration, and it will begin to precipitate. The greater part of the enzyme will then precipitate within a range of salt concentration above this point, this range being usually (in the case of ammonium sulphate) about 5–10% of saturation. The actual position of the precipitation range is determined not only by the pH, temperature, and nature of salts, but also by the concentration of the enzyme.

Fractional Adsorption

This is one of the most important and successful of the enzyme fractionation methods. It can be used in two distinct ways: by stirring successive amounts of the adsorbent into the enzyme solution and removing each portion by centrifugation or by passing the enzyme solution through a column of adsorbent into a fraction collector, as described under chromatography in the first method. If the enzyme becomes adsorbed, it may thus be removed and separated from other components of the solution and then extracted or eluted from the adsorbent. If the enzyme is not adsorbed, treatment with the adsorbent may be used to remove much unwanted material from the enzyme.

The chief adsorbents now used are calcium phosphate gel and alumina Cγ gel which have very wide application. Charcoal has been used for adsorbing unwanted components and other adsorbents, e.g. $Zn(OH)_2$ gel is occasionally used. It seems, however, that calcium phosphate gel will answer most requirements. Adsorption usually occurs best in slightly acidic solution, about pH 5 or 6 and at low electrolyte concentration.

The procedure is to add successive portions of the gel to the enzyme solution, mixing, spinning down and removing each portion before adding the next. In contrast to the other fractionation methods, the activity tests are carried out on small samples from the main solution, and the course of the enzyme removal is done. Particularly with calcium phosphate, it is not uncommon to find that the first portions of adsorbent do not remove any of the enzyme while other proteins are being removed; then the enzyme is removed rather suddenly, perhaps in one or two fractions only, leaving an inactive solution which still contains much proteins.

It is now necessary to recover the adsorbed enzyme from the active fractions by elution from the adsorbent. Elution of the enzyme can often be accomplished by slightly alkaline buffer solutions, e.g. phosphate buffer pH 7.6, if this fails to elute the enzyme it is a very favourable circumstance, for by repeated elution with phosphate until no further protein is removed a considerable degree of further purification can be achieved. The elution of the greater part of the material remaining adsorbed can nearly always be brought about by phosphate containing 10% of ammonium sulphate. This method of elution has been successfully used with several different enzymes. The volume of the eluent used should not be too large and is commonly not greater than the volume of the centrifuged gel. It is better to elute several times in succession with small volumes than to elute once with a large volume.

Chromatography

In recent years the purification of enzymes by chromatography on columns has become the most effective of all the separation methods. The mechanism of the separation will depend in different cases on adsorption, ion exchange, specific affinity to immobilized ligands or molecular sieving effects. The procedure is to add the enzyme preparation as a suitable solution to the pre-equilibrated column, and then to develop the column by eluting with a single buffer, with a series of steps of different solute concentrations, with a gradient of solute or with a specific ligand for the desired enzyme. In each case, the effluent emerging from the column is collected as a series of fractions by a fraction collector; these are tested for enzyme activity and total protein.

The various chromatographic techniques are described below.

i. Gels used for adsorption in fractionation methods are not suitable for use in columns, but special methods have been developed for this purpose. Certain forms of calcium phosphate may be used by mixing with Super Cel, Cellulose or Celite. Calcium phosphate can be converted into the microcrystalline hydroxyapatite which can be used in columns without the addition of any granular material.

ii. Ion-exchange chromatography (Figure 11.1) usually employs resins. The celluloses may be fibrous with the native polysaccharide. The cationic exchangers bear negatively charged groups (e.g. $-O-CH_2-COO^-$ in carboxymethyl cellulose, $-O-CH_2-CH_2-SO_3^-$ in sulphoethyl cellulose, $-COO^-$ in weakly acidic acrylic resins) attached to the matrix material; the pK values of these groups are in the range 2–7. The anion exchangers most commonly used are diethylaminoethyl cellulose ($-O-CH_2-CH_2-N(C_2H_5O_2)$, triethylaminoethyl cellulose ($-O-CH_2-CH_2-N-(C_2H_5O_3)$. Carboxymethyl cellulose and diethylaminoethyl (DEAE) cellulose are the ion exchangers used most frequently for fractionations. The cellulosic ion exchangers have been reviewed by Peterson. For gradient systems, it is advisable to restrict the counterion (i.e., the ion of opposite charge to the effective ion) to one species, to avoid zwitterions in the eluting buffer, to use short columns and to load protein samples at relatively low concentrations. Improved selectivity in ion-exchange chromatography may sometimes be achieved by the technique of "substrate elution". In its simplest form this would involve the application of the enzyme to a column under conditions of weak attachment, then by adding a specific substrate, substrate analogue or inhibitor. Sufficient neutralization of the charged groups of the enzyme might allow its selective release and elution from the column.

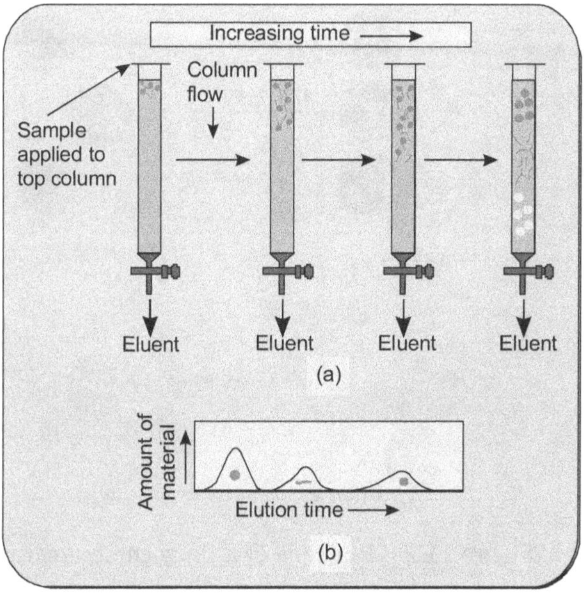

Figure 11.1 Ion-exchange chromatography

iii. Affinity chromatography (Figure 11.2) involves the immobilization of ligand on an insoluble supporting medium, which can then be used selectively to absorb enzyme from a solution passed over it. Subsequent elution can then be achieved by treatments which result in dissociation of the enzyme–ligand complex. Choice of the supporting medium, ligand and eluting conditions will depend on the enzyme under study. Agarose is the most common support, but polyacrylamide gels and glass beads are also used. The ligand must be attached to the support in such a way that its enzyme interaction function is not impaired. Elution is often achieved by a change in pH of the medium, or by including in the eluent a soluble ligand which will competitively displace the enzyme from the matrix-bound ligand.

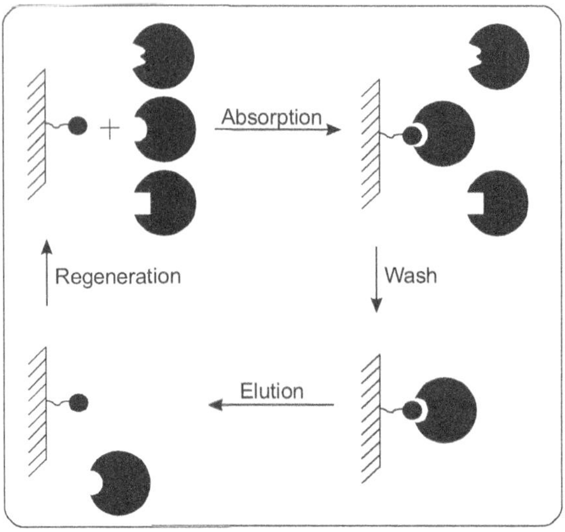

Figure 11.2 Principle of affinity chromatography

iv. Mixtures of proteins can be separated according to their molecular size by the method of gel filtration chromatography (Figure 11.3). This employs columns of cross-linked dextran gels, granulated polyacrylamide gels or bead-formed agarose gels which

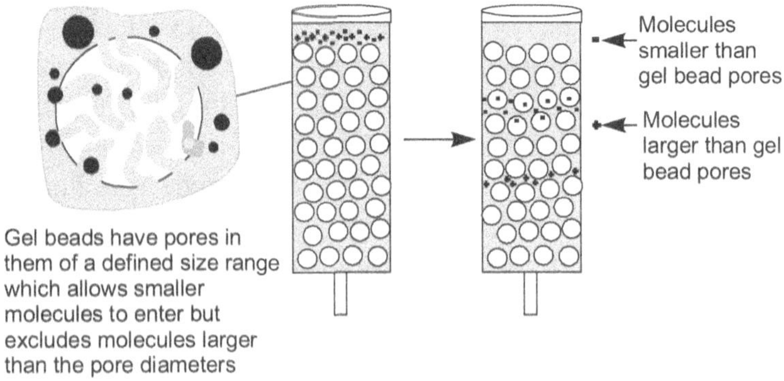

Figure 11.3 Gel filtration chromatography

act as "molecular sieves". The gel filtration mechanism is a form of partition chromatography of a solute (the enzyme) between the more mobile (external) solvent phase and the less mobile (internal) phase, the retardation of the solute depending on the extent to which it can enter the gel pores. Macromolecules thus emerge from the column in decreasing order of size.

Electrophoresis

The characteristic mobility of proteins (as electrolytes) when placed in an electric field provides the basis for the electrophoretic purification techniques for enzymes (Figure 11.4).

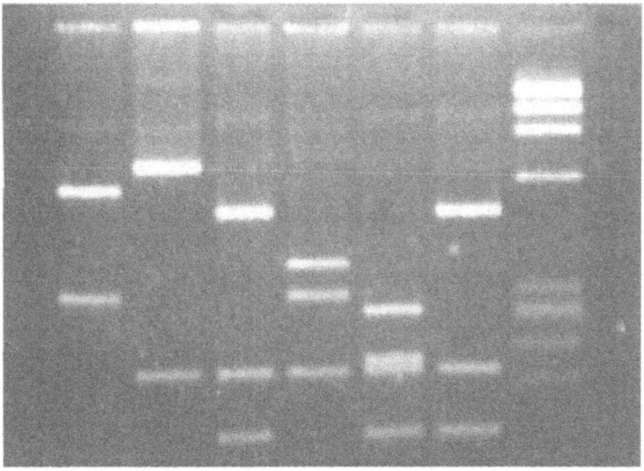

Figure 11.4 Electrophoresis

i. Free-boundary electrophoresis with the standard Tiselius apparatus 'U'-tube is more useful as a test of homogeneity. By its nature, the technique is unlikely to give complete separation of one species from another. However, it may be possible to obtain small samples of relatively pure enzyme for specific studies. Using the two-section cell, current is passed until the peaks corresponding to the various components are fairly well-separated. The current is then switched off and the liquid is very carefully driven back until only the fastest moving component remains in the upper half of the cell on the ascending limb. In the other limb it will frequently be found that by cutting off the upper halves of both limbs it is possible to obtain these two components separately in a state approximating to purity.

ii. Zone electrophoresis on inert supporting media such as paper, starch or cellulose powder can be used for separation which provides several advantages over free-boundary electrophoresis. It is mainly used for the analytical separation of enzymes for subsequent identification by an appropriate staining method. Components having different mobilities in the horizontal electric field arrive at different points along the bottom edge. In this way much larger amounts of material could be separated, although the support remains essentially two-dimensional. Packed powders of starch or cellulose provide three-dimensional supports for zone electrophoresis.

The electro-osmotic flow of solutes often complicates the electrophoretic separation. "Isotachophoresis" is the name given to a method for electrophoretically separating ionic species by maintaining them in a zone between a "leading electrolyte" of mobility higher than that of any of the ions to be separated, and a "terminating electrolyte", whose mobility should ideally be lower than the slowest ion to be separated. At equilibrium, all ions move at the same velocity; carrier ampholytes are used as "spacers" to intervene between the various ions that are to be separated. For preparative purposes, a polyacrylamide gel is used as a supporting medium and the separated zones of protein collected via an elution compartment as in conventional gel electrophoresis.

iii. Isoelectric focusing in stabilized pH gradients allow separation of proteins differing in isoelectric points. The technique employs a column of low molecular weight carrier ampholytes, with which the enzyme sample is mixed. In an electric field the ampholytes migrate appropriately to set up a pH gradient, and the individual protein components concentrate or "focus" at their isoelectric points. In this density gradient technique the resolved proteins may readily be collected sequentially after the separation by allowing the contents of the column to drain through a fraction collector.

iv. In electrodecantation, a series of connecting chambers separated by semipermeable membranes are arranged between two electrodes. In the electric field, proteins carrying a charge will collect against the semipermeable membranes and due to the increased density will decant downwards and can be collected from exit ports. Protein isoelectric under the conditions of the run will remain in the centre of the chamber, and can be collected by upward displacement and passed on to the adjacent chamber for further purification.

Concentration

There may be occasions during which a particular concentration of an enzyme is required in smaller volume. A number of methods have been described. Removal of the solvent can be achieved by evaporation, by partial freezing with removal of ice and by freeze-drying. Evaporation by suspending the enzyme solution in dialysis tubing in a current of air can be successful with relatively small volumes although salt deposition is likely to occur. Two membrane-moderated procedures are probably the most generally acceptable. Dialysis against non-permeating water, gathering macromolecules such as polyethylene glycol or dextrans can rapidly reduce the volume of enzyme solutions.

Dextrans, particularly in the form of sephadex, and some polyacrylamides may be added directly to enzyme solutions without an intervening membrane; the enzyme can be recovered in more concentrated form when the dextran has settled or been centrifuged off, although adsorption losses are inevitable.

In ultrafiltration, the enzyme solution is fed into a cell fitted with a membrane which will retain the chosen protein while being permeable to solvent and small molecules. Quite larger volumes can be reduced to a few ml in the course of one or two hours.

For retaining and concentrating proteins of low molecular weight, low-porosity ultrafilters must be used, but for larger proteins higher porosity filters will be more convenient because the solvent flow rate is likely to be faster. The successive use of ultrafilters of high porosity, which would allow a given enzyme to pass through and then of low porosity, which would retain the enzyme, provides a technique for purification as well as concentration.

Crystallization

Crystallization is a specific method of fractionation, used in the final stages when the enzyme solution is concentrated and is relatively in a small volume. When an enzyme has been brought to a reasonable state of purity it may be possible to crystallize it. Crystals of pure enzyme are required for structural studies (Figure 11.5). The most usual method of crystallization is by ammonium sulphate solution. A frequently used technique is to add the salt to a concentrated enzyme solution until a slight turbidity appears. It is then allowed to stand while the salt concentration is very slowly increased. This may be done by adding a strong solution of the salt drop by drop at long intervals, or through a fine capillary or through a dialysing membrane or the solution may simply be allowed to evaporate slowly.

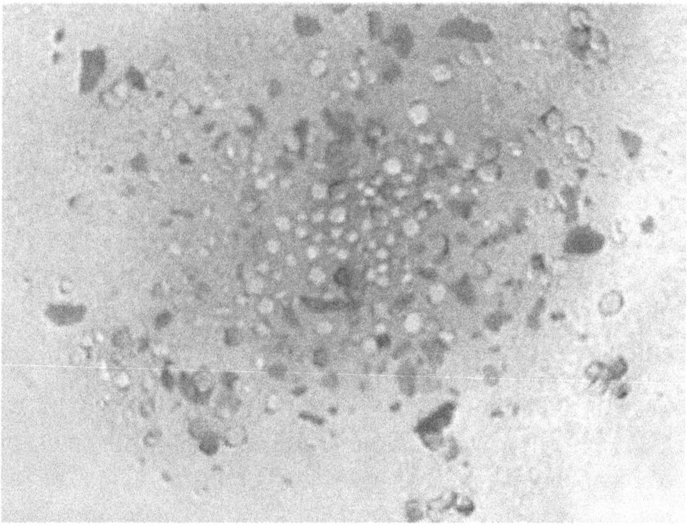

Figure 11.5 Crystallization trial of the glucose isomerase enzyme pseudo-coloured by the phase of light from a slurry/crystal mixture and captured in 3-D.

Crystallization may also be brought about at constant salt concentration by gradually changing the pH values or temperature. Crystallization seems to be facilitated if one of the previous stages has been a fractionation with organic solvent, owing to the removal of some interfering lipid material. The mere appearance of active crystals provides no proof that the enzyme itself has been crystallized; there have been many disappointments when the crystals have been found to be either inorganic or of some other protein containing some of the enzyme. The best test is recrystallization; a fall in activity indicates that the enzyme is present as an impurity in the crystals, while a rise to a constant value indicates that they are crystals of the enzyme itself.

Other Methods

A large part of the work on enzyme purification which has been carried out up to the present has been accomplished by the use of four of the main methods described above, namely, organic solvent precipitation, adsorption on gels, ammonium sulphate fractionation and column chromatography. It is probable that most enzymes could be purified by a combination of these methods alone. However, several methods other than those already quoted, have been used in particular circumstances, and may be tried if the more general techniques fail.

For example, unwanted protein can be removed without loss of enzyme by the cautious addition of salts of heavy metals, e.g. lead, but in such cases removal of the metal may be required before activity is recovered. In several cases, enzymes have been precipitated by adding nucleic acid and subsequent removal of the latter is carried out using protamine. Much inactive protein can often be removed initially by the use of Tsuchihashi method which involves denaturation by shaking with a mixture of alcohol and chloroform. It is especially useful for removing all haemoglobin from extracts of animal tissues or erythrocytes.

CRITERIA OF PURITY OF ENZYMES

Probably no single test by itself is sufficient to prove that the final product consists of only one protein, the enzyme. Each test establishes a certain degree of probability, but if the product appears homogeneous by several different tests, the probability becomes very high indeed and the enzyme may be accepted as pure. Ultracentrifugation and gel filtration provide discrimination by size, while electrophoresis and isoelectric focusing discriminate by charge; solubility tests are also possible if the enzyme is available in large amounts. Because enzymes can be measured by their catalytic behaviour, activity tests can also be applied and these may powerfully confirm the physical tests.

Sedimentation behaviour depends on molecular size and shape, but this may not be taken as proof of homogeneity. This is because sedimentation constants of many proteins fall into close proximity and it is therefore not improbable that in a mixture of two proteins, they might both have a similar sedimentation rate.

Electrophoretic separation methods are more likely to achieve clear discrimination between components. It is almost invariably true that a complex mixture of proteins, e.g. a muscle extract, can be resolved into far more elements by free boundary or zone electrophoresis than by ultracentrifugation or gel filtration.

Zone electrophoresis is more commonly used to examine homogeneity. Proteins differing from one another by only a few net charges per molecule can be successfully resolved into separate bands, particularly if the separation is carried out at a pH close to the isoelectric point of the enzyme under consideration.

Isoelectric focusing, with its ability to discriminate between proteins differing in isoelectric points by as little as about 0.01 pH units, is a very powerful analytical technique. It is also one of the most sensitive methods for testing homogeneity; at the same time it is capable of demonstrating the existence of multiple forms of a single enzyme.

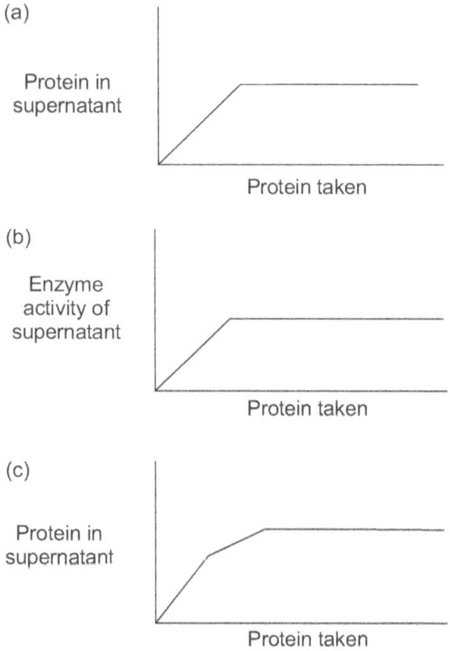

Figure 11.6 Solubility criteria of purity

Solubility test may also be applied to test protein purity (Figure 11.6). In this a series of different amounts of the enzyme are shaken, each with a given constant volume of solvent, then filtered or centrifuged and the protein in the resulting liquid estimated. This amount is then plotted against the amount taken. Therefore, solubility tests demand that a fairly large amount of protein be available in suitable form; they are used less frequently, since sensitive electrophoretic techniques are available.

The methods so far described, although sensitive, may not in fact be able to detect levels of contaminating components which are present in only very small amounts. Because the amount of material applied to the analysis may be relatively small, components which form less than 1% of the total protein, and thus comprise a very small absolute quantity, may be overlooked together.

REVIEW QUESTIONS

1. Define
   i. Unit of enzyme activity
   ii. Specific activity
   iii. Katal
2. Discuss the methods available for assay of enzymes.
3. Compare the methods available for estimation of protein during purification of enzymes.
4. Write a detailed account on the different fractionation methods employed during enzyme purification.
5. Discuss how crystallization can be used as an index of purity.
6. What do you mean by criteria of purity? Explain.

12

ENZYME IMMOBILIZATION

HISTORICAL ASPECTS OF ENZYME IMMOBILIZATION

The immobilized enzyme is defined as "the enzyme physically confined to or localized in a certain defined region of space with retention of its catalytic activity, which can be used repeatedly and continuously". In 1916, Nelson and Griffin found accidentally that yeast invertase adsorbed on activated charcoal catalyses the hydrolysis of sucrose. This experiment was the point of initiation of the enzyme immobilization process. However, enzyme immobilization came into commercial practice only when Grubhofer and Schleith immobilized various enzymes like carboxypeptidase, diastase, pepsin and ribonuclease. The enzymes were immobilized on diazotized polyaminostyrene resin by covalent binding. The first industrial application of immobilized enzyme was performed by Chibata and co-workers in 1969 when they used fungal aminoacylase immobilized on DEAE sephadex through ionic binding. The immobilized enzyme was used for the hydrolysis of *N*-acyl, L-amino acids and *N*-acyl D-amino acids.

NEED OF IMMOBILIZING ENZYME

Enzymes that are more expensive to produce, including many intracellular microbial enzymes and those employed in biosensors and analytical test systems are often used in an immobilized form. The process of enzyme immobilization is physically confining or localizing the biocatalyst in a certain defined region of space with retention of its catalytic activity, which can be used repeatedly and continuously. The greatest return from immobilization is achieved with expensive enzymes since there is a definitive cost associated with immobilizing processes. Other benefits such as ease of control and uniformity of conversions may be derived from immobilization techniques.

IMMOBILIZED ENZYME CHARACTERISTICS

It is important to understand the changes in physical and chemical properties, which an enzyme would be expected to undergo upon insolubilization if the best use is to be made of the various immobilization techniques available. Changes have been observed in the stability of enzymes and in their enzymic and kinetic properties because of the micro-environment imposed upon them by the supporting matrix and by the products of their own action.

Enzymic Action

Enzyme activities for immobilized enzymes are defined in the same way as for free enzymes, i.e., katal is the recommended unit. The percentage of the enzyme immobilized is usually calculated by measuring the amount of enzyme remaining in the supernatant after immobilization and subtracting this from the total amount of enzyme present originally. The absolute enzyme activity remaining on the support after immobilization is more difficult to determine and an apparent activity is usually measured which takes into account the mass transfer and the diffusional restrictions in the experimental procedure.

Stability

The stability of the immobilized enzymes might be expected to either increase or decrease on insolubilization depending on whether the carrier provides a micro-environment capable of denaturing the enzymic protein or stabilizing it. Inactivation due to autodigestion of proteolytic enzymes should be reduced by isolating the enzyme molecules from mutual attack by immobilizing them on a matrix. It has been found that enzymes coupled to inorganic carriers were generally more stable than those attached to the organic polymers when stored at 4° or 23°C. Stability to denaturing agents may also be changed upon insolubilization. Therefore, the recommended procedure is to store the enzyme under normal operating conditions (e.g. ambient temperature (20°C) in an appropriate buffer) and monitor its activity after fixed periods of time using the same procedure as that used for determining the activity remaining after immobilization.

Kinetic Properties

There is usually a decrease in the specific activity of an enzyme upon insolubilization and this can be attributed to denaturation of the enzymic protein caused by the coupling process. Once an enzyme has been insolubilized, however, it finds itself in a microenvironment that may be drastically different from that existing in free solution. The new microenvironment may be the result of the physical and chemical character of the support matrix alone, or it may result from interactions of the matrix with substrates or products involved in the enzymic reaction.

The effect of storage conditions, e.g. pH, temperature and ionic strength and of impurities incorporated during the immobilization steps must also be considered. Other quantifiable parameters for an immobilized enzyme are oxidation–reduction potential, working age and the apparent Michaelis constants (K_m) for appropriate substrates which also gives an indication of immobilized enzyme selectivity. These parameters can be affected by immobilization and experimental conditions and therefore comprehensive experimental details must be provided.

ASPECTS OF IMMOBILIZATION PROCEDURE

Whatever the physical form of an immobilized enzyme, i.e., solid-phase reactor membrane or solid-phase film, there are two important aspects of the immobilization procedure that must be specified in detail:

1. The type of support used
2. The methods of support activation and enzyme attachment

Enzyme Support

The support material can have a critical effect on the stability of the enzyme and the efficiency of enzyme immobilization, although it is difficult to predict in advance which support will be most suitable for a particular enzyme. The most important requirement for a support material is that it must be insoluble in water, have a high capacity to bind enzyme, be chemically inert and be mechanically stable. The enzyme-binding capacity is determined by the available surface area, both internal (pore size) and external (bead size or tube diameter), the ease with which the support can be activated and the resultant density of enzyme-binding sites. The inertness refers to the degree of non-specific adsorption and the pH, pressure and temperature stability. In addition, the surface charge and hydrophilicity must be considered. The activity of the immobilized enzyme will also depend upon the bulk mass transfer and the local diffusion properties of the system. The type of support can however be conveniently classified into one of the three following categories:

1. Hydrophilic biopolymers based on natural polysaccharides such as agarose, dextran and cellulose.
2. Lipophilic synthetic organic polymers such as polyacrylamide, polystyrene and nylon.
3. Inorganic materials such as controlled pore glass and iron oxide.

Clearly the chemical structure of the support must be defined and any quantitative physical and chemical data relevant to the above factors must also be provided.

Support Activation and Enzyme Attachment

Historically, entrapment procedures were widely used for analytical applications but currently covalent binding to a suitably attached support is now the most used and by far the most common approach. An activated support is defined as a material having an enzyme-reactive functional group covalently attached to an otherwise inert surface. The stability of the resulting bond between the enzyme and the support, the local environment of the enzyme, and the potential loss of activity on immobilization, all must be considered. For polysaccharides, activation is most commonly via derivatization of available hydroxyl functions using reagents such as cyanogen bromide, triazine derivatives, e.g. cyanuric chloride, sodium periodate, epoxides or benzoquinone. Polyacrylamide and its derivatives can be activated by reaction with diamines, and inorganic supports are most commonly silianized and activated by reaction with carbodiimides, thiophosgene, thionyl chloride, N-hydroxysuccinimides or transition metal salts such as titanium chloride. The activated support is then covalently linked to the enzyme, most commonly via direct reaction with available amino functions, e.g. on lysine residues, but also via thiol and phenol functions.

METHODS OF IMMOBILIZATION

There are five markedly different methods of immobilizing enzymes. They are as follows:

1. Adsorption of the enzyme on a carrier surface
2. Ionic binding or covalent coupling of an enzyme to a carrier surface, i.e., inert inorganic materials such as nylon, bentonite, cellulose and dextran
3. Cross-linking between enzyme molecules (co-polymerization)
4. Entrapment of an enzyme in a matrix, i.e., within gels or fibres, and intracellular enzymes may be immobilized within their producer cells
5. Encapsulation or confinement of an enzyme solution in a membrane structure

Although each immobilization technique is unique, they are by no means pure processes and are in combination with other methods. Each immobilization process has both advantages and disadvantages. The choice of technique, therefore, should be decided by the specific conditions of the application which would selectively employ the positive attribute of the specified immobilization.

Methods

- I. Carrier-binding method
  1. Covalent binding method
     - i. Cyanogen bromide method
     - ii. Acid azide derivative method
     - iii. Condensing reagent method
     - iv. Diazo coupling method
     - v. Alkylation method
     - vi. Carrier cross-linking method
  2. Ionic binding method
  3. Physical adsorption method
  4. Biospecific binding method
- II. Cross-linking method
- III. Entrapment method
  1. Polyacrylamide gel method
  2. Alginate gel method
  3. κ-carrageenan gel method
  4. Synthetic resin prepolymer method
     - i. Photo-crosslinkable resin prepolymer method
     - ii. Urethrane prepolymer method

CARRIER-BINDING METHOD

The carrier-binding method is based on binding of the biocatalyst to a water-insoluble carrier through a covalent linkage, ionic bonds, physical adsorption, or biospecific binding. Various types of insoluble materials such as water-insoluble polysaccharides (e.g. cellulose, dextran and agarose derivative polymers), proteins (e.g. gelatin and albumin), synthetic polymers (e.g. polyene derivatives, ion-exchange resins and polyurethane) and inorganic materials (e.g. brick, sand, glass, ceramics and magnetite) can be utilized directly or after proper modification or activation.

Carrier

A carrier is defined as the support material utilized to immobilize the enzyme. This support may be a matrix system, a membrane or a solid surface. The desired properties of a carrier are as follows:

1. Possessing adequate functional groups for immobilization
2. Mechanical strength
3. Physical, chemical and biological stability
4. Non-toxic

Carrier Disability Under Application Conditions

1. When a carrier is employed for the immobilization of an enzyme, the durability of that carrier during use is second in importance only to the enzyme activity.
2. If the carrier, whether it be a matrix or a membrane, is not stable under pH, ionic strengths and solvent conditions of the application, then the carrier will be disrupted or disturbed and the enzyme will be released into the solution.
3. This does not necessarily mean that the carrier must be durable under pH, ionic strengths and solvent conditions, but only those of the specific application.
4. In addition to the chemical environment, the most important consideration is that of the physical

environment of the application. A rigid structure, such as inorganic material, may be readily adsorbed in a continuous stirred reactor. This will not only lead to dissolution of the enzyme but also to the formation of non-uniform particles which may disrupt the reactor performance. Under these conditions, it would be far more advantageous to choose an elastic-type carrier of this application.

Covalent Binding Method

The covalent binding method (Figure 12.1) is one of the most frequently used techniques for the immobilization of enzymes, but this method is not so widely used for cells because of the toxicity of the agents and the difficulty of finding the proper conditions of immobilization. Organic carriers can also be used for immobilization of cells by the covalent binding method. Navarro and Durand immobilized *Saccharomyces carlsbergensis* by covalent binding to porous silica beads that had been treated with aminopropyltriethoxysilane and activated with gluteraldehyde. *Serratia marcescens*, *Saccharomyces cerevisiae* and *Saccharomyces amurcea* were also bound covalently to borosilicate glass or Zirconia ceramics by using an isocyanate coupling agent. In the case of enzymes, the amino acid residues that are not involved in the active site or substrate-binding site can be used for covalent binding with carriers. The ε-amino group of lysine, the β-carboxyl group of aspartic acid, the γ-carboxyl group of glutamic acid, the hydroxyl groups of serine and threonine, the mercapto group of cysteine, the phenolic hydroxyl group of tyrosine and the imidazole group of histidine can react with carriers having a reactive functional group such as diazonium salt, acid azide, isocyanate, activated alkyl halide or aldehyde.

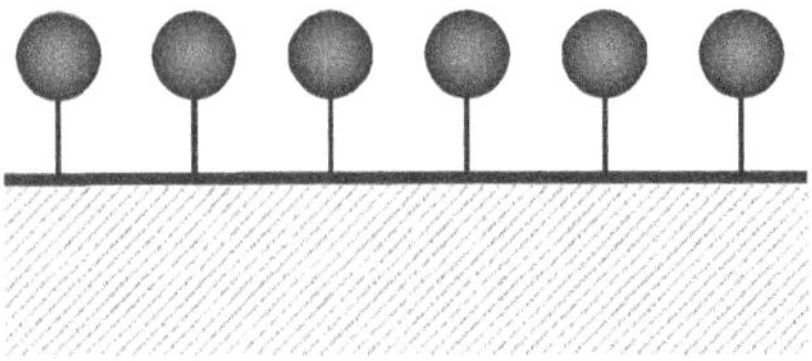

Figure 12.1 Covalent attachment

Advantages

i. The biocatalyst does not leak or detach from the carrier.
ii. The biocatalyst can easily interact with the substrate because it is on the surface of the carrier.
iii. The biocatalyst stability is often increased because of the strong interaction between it and the carrier.

Disadvantages

i. The activity yield is likely to be low owing to the exposure of the biocatalyst to toxic reagents or severe reaction conditions.
ii. The optimal conditions for the immobilization procedure are difficult to find.
iii. Renewal of the carrier and recovery of the biocatalyst from the carrier are, in general, impossible.

Hence, this method is better suited to expensive enzymes whose stability is significantly improved by covalent binding to the carrier.

Several typical covalent binding methods are described below:

i. *Cyanogen bromide method* The cyanogen bromide method was first used by Axen and co-workers in 1967. This method involves the activation of a carrier having vicinal hydroxyl groups, such as polysaccharides, glass beads or ceramics, with cyanogen bromide to yield reactive imidocarbonate derivatives. This method has been widely used for the immobilization of various enzymes, and the CNBr-activated carriers are available commercially. An aminated carrier can also be used. It is also possible to insert a spacer such as hexamethylenediamine to reduce the hindrance to the movement of the enzyme molecule, caused by the interaction between the enzymes and the carriers.

ii. *Acid azide derivative method* The acid azide method, which is also known as a method for peptide synthesis, has been used for the immobilization of various enzymes. The carboxyl group of the carrier is converted first to the methyl ester and then to the hydrazide with hydrazine. The hydrazide reacts with nitrous acid to form the azide derivative, which can then react with the amino groups of the enzyme molecules at low temperature to yield the immobilized enzyme.

iii. *Condensing reagent method* The carboxyl group or amino group of the carriers and the biocatalysts can be condensed through the formation of peptide linkages by the action of condensing reagents such as carbodiimides (dicyclohexylcarbodiimide, 1-ethyl 3-(3-dimethylaminopropyl) carbodiimide, 1-cyclohexyl 3-(2-morpholinoethyl), carbodiimide metho-*p*-toluene sulphonate, etc.) or Woodward's reagent K (N-ethyl 5-phenylisoxazolium 3´-sulphonate).

iv. *Diazo coupling method* Carriers having aromatic amino groups can be diazotized with nitrous acid to form the diazonium derivatives, which react with the enzyme molecules to yield immobilized enzymes. The functional groups of an enzyme molecule that may participate in this reaction include free amino groups, imidazole groups, phenolic hydroxyl groups and so on.

v. *Alkylation method* Carriers having an alkyl group can react with the phenolic hydroxyl groups or sulphydryl groups of the enzyme molecules. Halogenated acetyl derivatives, triazinyl derivatives, and so on can be used as carriers for this method.

vi. *Carrier cross-linking method* In the carrier cross-linking method (Figure 12.2), both the carriers and the biocatalyst having amino groups are cross-linked with a bi- or multifunctional reagent that reacts with the free amino groups. Glutaraldehyde is the most popular reagent that cross-links amino groups through a Schiff's base linkage. Diisocyanate is the other type of cross link reagent.

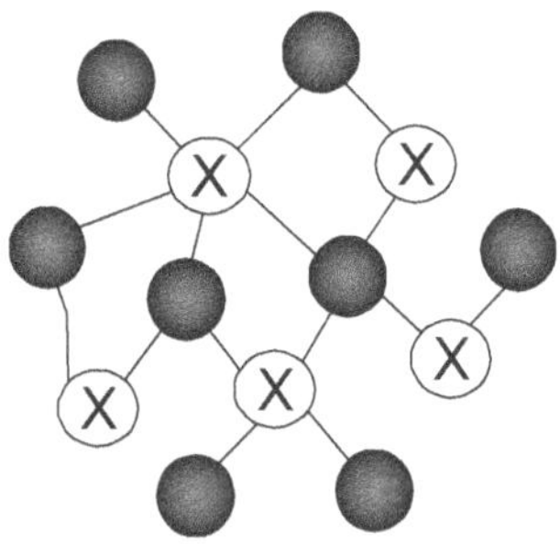

Figure 12.2 Crosslinking

Ionic Binding Method

Since catalase was found to be able to bind to the ion-exchange cellulose, the ionic binding method has been applied for the immobilization of many biocatalysts. The procedure is very simple, renewal of the carrier and recovery of the biocatalyst from the carrier are easy and the conditions of immobilization are mild. In fact, the first industrial application of immobilized enzymes employed this method for the immobilization of aminoacylase on DEAE-Sephadex to produce L-amino acids. In ionic binding, binding of the biocatalyst on the carrier is affected by the buffer used, pH, ionic strength and temperature. Although renewal of the carrier and recovery of the biocatalyst from the carrier are easy, the biocatalyst is likely to detach from the carrier. The polysaccharide derivatives having ion exchange groups, as well as various ion exchange resins, can be utilized for this purpose.

Physical Adsorption Method

The physical adsorption method (Figure 12.3) is based on the physical interactions between the biocatalyst and the carrier, such as hydrogen bonding, hydrophobic interaction, van der Waals force and their combinations. Although the biocatalyst may be immobilized without any modification, the physical interaction in general, is weaker than the ionic binding and more sensitive to environmental conditions such as temperature and concentration of solutes. Cellular organelles and various types of cells can also be immobilized with ease by this method. Renewal of the carrier can be accomplished under appropriate conditions. Physical adsorption followed by cross linking with gluteraldehyde sometimes helps to stabilize the immobilized biocatalysts. Although many carriers that efficiently adsorb biocatalysts have been developed, the one with immobilized tannin is among the most useful. Immobilization of tannin can be carried out in a reaction of aminohexylcellulose with CNBr-activated tannin.

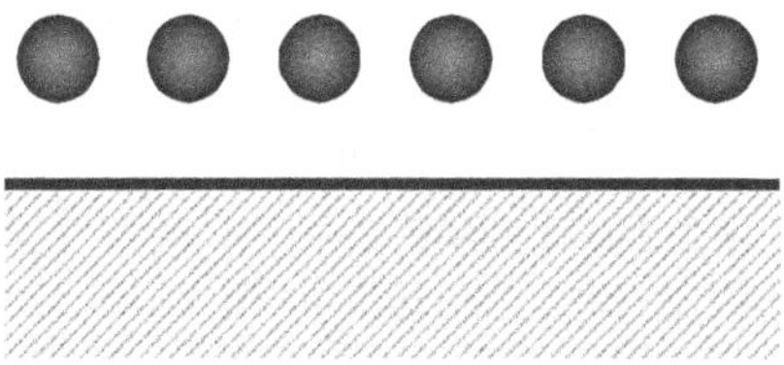

Figure 12.3 Physical adsorption

The factors affecting immobilization processes are listed in Table 12.1.

Table 12.1 Factors affecting immobilization processes

Factor	Effect
Organic compounds	Interface with bonding of waste materials
Oil and grease	Interface with the hydration of cement, reduce product strength, and weaken bonds between waste particles
Cyanides	Affect bonding of contaminants
Inorganic salts (e.g. nitrates, sulphates, chlorides)	Reduce product strength and affect curing rates
Halides (e.g. chlorides)	Retard setting and leach easily
Particle size	Small particles can coat larger particles and weaken bonds
Volatile organic compounds	May produce air emissions due to heat generation of the reaction
Solids content	Low solids content requires large amounts of reagent

Source: USEPA (1988b)

Biospecific Binding Method

The biospecific binding method is based on the biospecific interaction between the enzymes and other substances, such as, coenzymes, inhibitors, effectors, lectins, and antibodies, which are often utilized for the separation processes based on affinity. If the interaction is strong, the enzyme can be immobilized on the carrier conjugated with one of these substances. However, antibodies and inhibitors are not good choices because the enzyme is usually inactivated by binding to them. Interaction between a lectin and the carbohydrate moiety of an enzyme is useful for this application. Lectins that are glycoproteins bind tightly to specific carbohydrate residues. One most useful lectin is concanavalin A, which is obtained from the jack bean. Since many enzymes are glycoproteins, lectins can be widely used.

CROSS-LINKING METHOD

The cross-linking method utilizes a bi- or multifunctional compound as in the carrier cross-linking method; however, a carrier is not used in this method. The bi- or multifunctional compound serves as the reagent for intermolecular crosslinking of the biocatalyst. The cross-linked biocatalyst becomes water-insoluble. In addition to glutaraldehyde, which is the most popular cross-linking reagent, several other bi- or multifunctional compounds, such as, toluene diisocyanate and hexamethylene diisocyanate can be used. The activity of the biocatalyst immobilized by this method is in general, reduced.

ENTRAPMENT METHOD

The entrapment method is classified into five major types:

1. lattice,
2. microcapsule,
3. liposome,
4. membrane and
5. reversed micelle.

In the lattice type, the biocatalyst is entrapped in the matrix of one of the various polymers. The microcapsule type involves entrapment within microcapsules of a semipermeable synthetic polymer. The liposome type employs entrapment within an amphipatic liquid surfactant membrane prepared from lipid. In the membrane type, the biocatalyst is separated from the reaction solution by an ultrafiltration membrane, a microfiltration membrane or a hollow fibre. In the reversed micelle type, the biocatalyst is entrapped within the reversed micelles, which are formed by mixing a surfactant with an organic solvent.

Advantages

i. Not only single enzymes but also multiple enzymes, cellular organelles and intact or treated cells can be immobilized.

Disadvantages

i. The high molecular weight substrate may not be able to access the entrapped biocatalyst.
ii. Renewal of the carrier is difficult.

Among the entrapment methods, the lattice type is the most widely used. Several representative techniques for the lattice type method are described under the following:

Polyacrylamide Gel Method

Since Bernfeld and Wan reported the entrapment of several enzymes in polyacrylamide gel in 1963, various types of biocatalysts, including cellular organelles, microbial cells, plant cells and animal cells, have been immobilized by this method. This method (Figure 12.4) has also been applied to the industrial production of useful compounds such as L-aspartate, L-malate and acrylamide. The procedure for the preparation of the gel is identical to that employed for electrophoresis. For the immobilization of the biocatalyst by this method, acrylamide and *N′*, *N′*- methylenebisacrylamide (BIS), as the cross-linking reagents are mixed with the biocatalyst and polymerized in the presence of an initiator, such as potassium persulphate and a stimulator such as 3-dimethylaminopropionitrile (DMAPN) or *N,N,N′N′*-tetramethylethylenediamine (TEMED).

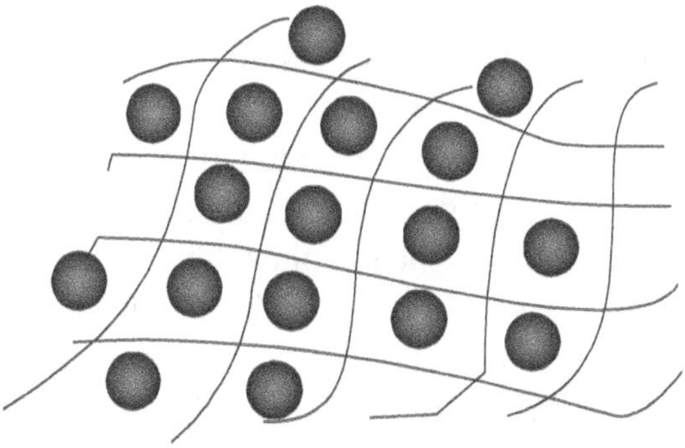

Figure 12.4 Gel entrapment

Although this technique is used for various purposes, a major disadvantage is the toxicity of the acrylamide monomer, the cross-linking reagent, the initiator and the stimulator. In some cases, free-radical polymerization results in a decrease in the activity of the biocatalyst. Several analogs or derivatives of acrylamide are also useful in this method.

Alginate Gel Method

Several natural polysaccharides such as alginate, κ-carrageenan, agar and agarose are excellent gel materials and are widely used for entrapment of biocatalysts. Alginate is a linear copolymer of D-mannuronic and L-guluronic acids and can be gelled by multivalent ions. For immobilization of the biocatalyst, sodium alginate, which is insoluble in water, is mixed with a solution or suspension of the biocatalyst and dropped into a calcium chloride solution to form the water-insoluble calcium alginate gel droplets. However, calcium alginate gels are gradually solubilized in the presence of a calcium-ion-trapping reagent such as phosphate ion.

In addition, leaking of the biocatalyst is likely to occur. The calcium ion-trapping agent frees the cells trapped in the gel, enabling the examination of the characteristics, e.g. the cell viability of the trapped cells. Aluminium ion, strontium ion or several other divalent metal ions can also be used instead of calcium ion. Treatment of calcium alginate gel with a cationic polymer such as polyethyleneamine can improve the stability of the gel in the presence of phosphate. It is also known that the addition of polyacrylacid to the sodium alginate increases the physical strength of the gel and prevents the gel from being dissolved by bridging the polyacrylate chains and alginate chains with calcium ions. The alginate method is widely applicable to the immobilization of various biocatalysts, including very fragile ones such as plant cells and protoplasts because of the simplicity of the procedure, availability of sodium alginate and mild immobilization conditions.

κ-Carrageenan Gel Method

κ-carrageenan is a readily available, non-toxic polysaccharide, which is obtained from seaweed. It is widely used in the food and cosmetic industries as a gelling, thickening and stabilizing agent. κ-carrageenan forms a gel upon cooling or in the presence of a gel-inducing reagent such as potassium chloride. Various cations such as ammonium, calcium, aluminium and magnesium also serve as good gel-inducing reagents. The conditions of immobilization by this method are mild. Another advantage of this method is that various shapes of the immobilized biocatalyst can be made. κ-carrageenan gel can be easily solubilized in saline or water to free the biocatalyst from the gel for the investigation of its properties.

On the other hand, leaking of the biocatalyst can occur easily owing to the dissolution of the gel, when a gel-inducing reagent is not present in the reaction mixture. To increase the stability of the κ-carrageenan-entrapped biocatalyst, treatment with glutaraldehyde or hexamethylenediamine after entrapment is often effective.

Synthetic Resin Prepolymer Method

With the application of the immobilized biocatalyst in a variety of bioreactions, including synthesis, transformation, degradation or analysis, each having a different desirable chemical environment, finding a suitable gel among the natural polymers may be difficult.

Synthetic resin polymers such as photo cross-linkable resin prepolymers and urethane prepolymers extend the list of polymers used for entrapment. In the synthetic resin prepolymer method (Figure 12.5), the size of the gel matrix, which affects the substrate and product diffusion in the gel, the mechanical strength of the gel formed, the biocatalyst-holding capacity and the growth capability of the cells inside the gel, can be regulated by the chain length of the prepolymer and the content of the reactive functional group. Furthermore, the hydrophilicity or hydrophobicity balance of the gel can be controlled by selecting a suitable prepolymer synthesized in advance in the absence of biocatalyst. The hydrophilicity–hydrophobicity balance of the gel is often critical in affecting the partition of the reactant to the gel in a bioconversion reaction. The ionic properties of the gel, which are an important factor for a bioconversion reaction, can be introduced to the prepolymers beforehand.

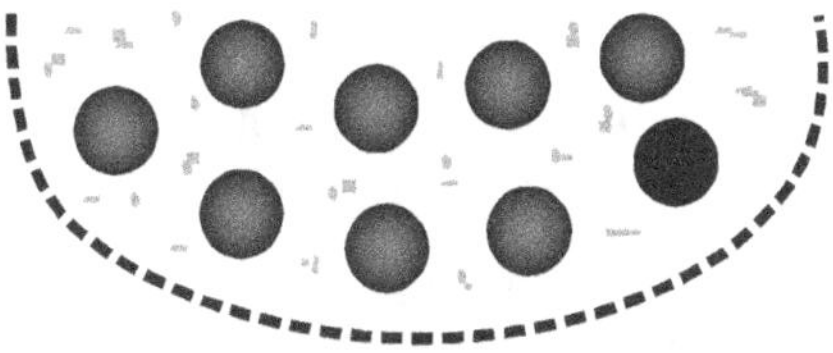

Figure 12.5 Resin prepolymer method

Advantages

i. The entrapment procedures are very simple and proceed under very mild conditions.

ii. The prepolymers do not contain monomers that may have an unfavourable effect on the biocatalyst.

Therefore, the synthetic resin prepolymer method may be one of the most widely applicable immobilization methods at present.

Photo-crosslinkable Resin Prepolymer Method

The use of photo-crosslinkable resin prepolymer that is hydrophilic or hydrophobic, cationic or anionic and of different chain lengths has been developed by Fukui and co-workers. A mixture of the prepolymer and the biocatalyst is gelled by irradiation with near UV light for several minutes in the presence of a proper photosensitizer (initiator) such as benzoin ethyl ether. This method has been applied for the pilot-scale continuous production of ethanol by immobilized, growing yeast cells. Photo-crosslinkable resin prepolymers can be autoclaved at 120°C.

Urethane Prepolymer Method

The entrapment by urethane prepolymer is quite simple. There are many types of urethane prepolymers having different hydrophilicity or hydrophobicity and chain length. The urethane prepolymers are synthesized by the Toyo Tire and Rubber Industry Co., Osaka, Japan, but are not commercially available.

The urethane prepolymers react with each other in the presence of water to form urea bonds and liberate carbon dioxide. Therefore, the entrapment can be carried out simply by mixing the prepolymer with an aqueous solution or suspension of the biocatalyst. The urethane prepolymers can be sterilized at 120°C by avoiding moisture.

BENEFICIAL FEATURES OF IMMOBILIZED ENZYMES

i. Reuse
ii. Suitability for application within continuous operations
iii. Their product is enzyme-free therefore further processing to remove or inactivate the enzyme is not required.
iv. Improved enzyme stability
v. Reduced effluent disposal
vi. The ability to stop the reaction rapidly by removing the enzyme from the reaction solution (or vice versa)
vii. Product is not contaminated with the enzyme (especially useful in the food and pharmaceutical industries)
viii. Analytical purposes
    1. Long half-life
    2. Predictable decay rates
    3. Elimination of reagent preparation

APPLICATIONS OF IMMOBILIZED ENZYMES

The first industrial use of an immobilized enzyme is the amino acid acylase by Tababe Seiyaku Company, Japan, for the resolution of racemic mixtures of chemically synthesized amino acids. Amino acid acylase catalyses the deacetylation of the L-form of the *N*-acetyl amino acids leaving unaltered the *N*-acetyl-D-amino acid that can be easily separated, racemized and recycled. Some of the immobilized preparations used for this purpose include enzyme immobilized by ionic binding to DEAE-sephadex and the enzyme entrapped as microdroplets of its aqueous solution into fibres of cellulose triacetate by means of fibre wet spinning developed by Snamprogetti. Rohm Gambh have immobilized this enzyme on macroporous beads made of flexiglass-like material (Table 12.2).

Table 12.2 Important applications of immobilized enzymes

Enzyme	Substrate	Enzyme reactor used	Product
Aminoacylase (immobilized on anion exchange resins)	*N*-acyl- DL-amino acids	PBR	L-amino acids
Aspartate ammonialyase	Fumaric acid + NH_4^+	PBR	L-aspartic acid
Cyanidase	Cyanide present in industrial waste, food or feed		Formic acid
Glucoamylase	Dextrins produced by α-amylase		D-glucose
Glucose isomerase (immobilized with glutaraldehyde by cross-linking)	D-glucose in glucose syrup	PBR	High-fructose corn syrup
Invertase	Sucrose	PBR	Invert sugar
Lactase (immobilized in cellulose triacetate fibres)	Milk and whey	STR	Lactose-free milk and whey
Lipase	Vegetable oils, e.g. palm oil		Cocoa butter substitute
Nitrile hydratase (immobilized cells)	Acrylnitrile		Acrylamide
Penicillin amylases	Penicillin G and penicillin V	STR, PBR	Penicillins
Raffinase (immobilized cells)	Raffinose in beet juice, soya bean milk	STR	Raffinose-free solutions

Production of Fructose Syrup

The most important application of immobilized enzymes in the industry is in the conversion of glucose syrups to high-fructose syrups by the enzyme glucose isomerase. It is evident that most of the commercial preparations use either the adsorption or the cross-linking technique. Application of glucose isomerase technology has gained considerable importance especially in tropical countries like India, where sugar cane cultivation is abundant. The high-fructose syrups can be obtained by a simpler process of hydrolysis of sucrose using invertase.

Use of Immobilized Invertase

Compared to sucrose, invert sugar has a higher humectancy, higher solubility and osmotic pressure. Historically, invertase is perhaps the first reported enzyme in an immobilized form. A large number of immobilized invertase systems have been patented. The possible use of whole cells of yeast as a source of invertase was demonstrated by D'Souza and Nadkarni as early as 1978.

Use of Lactose in Dairy Industry

One of the major applications of immobilized biocatalyst in dairy industry is in the preparation of lactose-hydrolysed milk and whey using β-galactosidase. A large population of lactose intolerants can consume lactose-hydrolysed milk. This is of great significance in a country like India where lactose intolerance is quite prevalent. Lactose hydrolysis also enhances the sweetness and solubility of sugars and can find future potentials in the preparation of a variety of dairy products.

Lactose-hydrolysed whey-based beverages, leavening agents, feedstuff may be fermented to produce ethanol and yeast, thus converting an inexpensive by-product into a highly nutritious, good quality food ingredient. The first company to commercially hydrolyse lactose in milk by immobilized lactase was Centrale Del Latte of Milan, Italy, utilizing the Snamprogetti technology. The process makes use of a natural lactase from yeast entrapped in synthetic fibres. An immobilized preparation obtained by cross-linking β-galactosidase in hen's egg white has been used for the hydrolysis of lactose. A major problem in the large-scale continuous processing of milk using immobilized enzyme is the microbial contamination which has necessitated the introduction of intermittent sanitation steps. A co-immobilizate obtained by binding of glucose oxidase on the microbial cell wall using con A has been used to minimize the bacterial contamination during the continuous hydrolysis of lactose by the initiation of the natural lacto-peroxidase system in milk.

Use of Immobilized Enzymes in Pharmaceutical Industry

One of the major applications of immobilized enzymes in pharmaceutical industry is the production of 6-aminopenicillanic acid (6-APA) by the deacetylation of the side chain either in penicillin G or V, using penicillin acylase (penicillin amidase). More than 50% of 6-APA produced today is produced enzymatically using the immobilized route. One of the major reasons for its success is that a purer product is obtained, thereby minimizing the purification costs. Currently, most of the pharmaceutical giants make use of this technology. A number of immobilized systems have been patented or commercially produced for penicillinic acylase which makes use of a variety of techniques either using the isolated enzyme or the whole cells. This is also one of the major applications of the immobilized enzyme technology in India. Similar approach has also been used for the production of 7-amino deacetoxy cephalosporanic acid, an intermediate in the production of semisynthetic cephalosporins.

Use of Glucose Oxidase in Enzyme Electrodes

Immobilized oxidoreductases are gaining considerable importance in biotechnology to carry out synthetic transformations. Of particular significance in this regard are oxidoreductase-mediated asymmetric synthesis of amino acids, steroids and other pharmaceuticals and a host of specialty chemicals. They play a major role in clinical diagnosis and other analytical applications like the biosensors.

Future applications of oxidoreductases can be in areas as diverse as polymer synthesis, pollution control and oxygenation of hydrocarbons. Immobilized glucose oxidase can find application in the production of gluconic acid, removal of oxygen from beverages and in the removal of glucose from eggs prior to dehydration in order to prevent Maillard reaction.

Production of Amino Acids

Immobilized D-amino acid oxidase has been investigated for the production of keto acid analogues of the amino acids, which find application in the management of chronic uremia. Keto acids can be obtained using either L- or D-amino acid oxidases. The use of D-amino acid oxidase has the advantage of simultaneous separation of natural L-isomer from DL-racemases along with the conversion of D-isomer to the corresponding keto acid which can then be transaminated in the body to give the L-amino acid. Of the several microorganisms screened, the triangular yeast *Trigonopsis variabilis* was found to be the most potent source of D-amino acid oxidase with the ability to deaminate most of the amino acids. The permeabilized cells entrapped in radiation-polymerized acrylamide-calcium alginate or gelatins have shown promise in the preparation of a keto acid.

Detergent Industry

Lipases catalyse a series of different reactions. Although they were designed by nature to cleave the ester bonds of triacylglycerols (hydrolysis), lipases are also able to catalyse the reverse reaction under microaqueous conditions, viz. formation of ester bonds between alcohol and catalytic acid moieties. These two basic processes can also be combined in a sequential fashion to give rise to a set of reactions generally termed as inesterification. Immobilized lipases have been investigated for both these processes. Lipases possess a variety of industrial potentials starting from use in detergents; leather treatment; controlled hydrolysis of milk fat for the acceleration of cheese ripening; hydrolysis, glycerolysis and alcoholysis of bulk fats and oils; production of optically pure compounds, flavours, etc. Lipases are spontaneously soluble in aqueous phase but their natural substrates (lipids) are not. Although use of proper organic solvents as emulsifiers help in overcoming the problem of intimate contact between the substrate and the enzyme, the practical use of lipases in such pseudohomogeneous reactions pose technological difficulties.

Immobilization of Glucoamylase

This is an example of an immobilized enzyme that probably is not competitive with the free enzyme and hence has not found large-scale industrial application. This is mainly because soluble enzyme is cheap and has been zused for over two decades in a much optimized process without technical problems. Immobilization has also not found to significantly enhance the thermostability of amylase. Immobilized rennin or other proteases might allow for the continuous coagulation of milk for cheese manufacture. Efforts have been made to minimize these problems by attaching the enzymes through spacer arms.

Immobilized Enzymes in Organic Solvents

Hydrolytic enzymes can catalyse esterification and transesterification reactions in monophasic organic solvents and in water organic biphasic systems. Conventional immobilized enzymes are principally used to facilitate catalyst recovery. Even though enzymes by themselves are insoluble in organic solvents in addition to others, the prime importance of the use of enzymes in organic media is the necessity to avoid enzyme deactivation or denaturation. A number of enzymes which are used in organic solvents have been immobilized using a variety of techniques with a view to stabilizing them. In systems containing enzymes immobilized on solid supports and working in organic media, the support has a significant influence on the total enzyme activity and can also displace the reaction equilibrium due to the interaction of the support with the water molecules. Thus the choice of suitable support material, proper water content, and the selection of the organic solvents are crucial for the use of immobilized enzymes in organic solvents.

Immobilization of Enzymes for the Fabrication of Biosensors

Stable membranes have been prepared by binding glucose oxidase to cheese cloth for the fabrication of a glucose biosensor. Enzymes entrapped inside the reversed micelle have also shown promise in the fabrication of biosensor. Cross-linked enzyme crystals (CLCs) provide their own support and so achieve enzyme concentration close to the theoretical packing limit in excess of even highly concentrated enzyme solutions. In view of this, CLCs are particularly attractive in biosensor applications where the largest possible signal per unit volume is often critical. Sensors based on small transducers or thinner enzyme-immobilized membranes are also emerging.

The development of molecular devices incorporating a sophisticated and highly organized biological information processing function is a long-term goal of bioelectronics. For this purpose, it is necessary in the future to develop suitable methods for micro-immobilizing the potential enzymes into an organized array or pattern, as well as designing molecular structures that are capable of performing the required function. A typical example is the micro-immobilization of proteins into an organized pattern on silicone water, on a specific binding reaction between streptavidin and biotin combined with photolithography techniques.

Immobilized enzymes have also been used for various other analytical purposes. A recent development has been in obtaining a stable, dry immobilized enzyme like acetylcholinesterase, on polystyrene microtitration plates for mass screening of its inhibitors in water and biological fluid.

Immobilized Multienzymes and Enzyme Cell Co-Immobilization

A common feature of metabolic pathways in intermediary metabolism is that the product of one enzyme in sequence is the substrate for the next and so forth. One of the advantages of such an arrangement is that a favourable high local concentration of intermediates in the microenvironment of an enzyme system can be created. Such an approach perhaps will be useful in the development of enzymatic processes requiring multiple enzymes. The enzymes in question can be simultaneously bound to the same support or could be simultaneously immobilized through entrapment. The kinetic advantages of this system over mixed soluble enzymes and immobilized and then mixed systems include reduction in the lag and significant enhancement in the final product, and these have been demonstrated.

A practical approach is to immobilize the deficient enzyme directly on the cell wall. This approach which is termed as enzyme-cell co-immobilizate permits the tailoring of a whole cell for a specific complex chemical conversion by combining the biochemical potential of both the cell and the additional enzyme from another source. Enzymes can be immobilized on the microbial cell surfaces coated with polyethyleneamine through adsorption followed by cross-linking.

Immobilized Enzymes for the Regeneration of Cofactors

Enzyme technology has dealt mainly with simple reactions which do not require coenzymes. One of the major challenges for enzyme technologists is in the application of cofactor-dependent enzymes. Over 25% of the known enzymes require cofactors like NAD/NADH, ATP/ADP, etc. which, unlike the enzymes, undergo stoichiometric changes during a biochemical reaction. The living cells have circumvented this problem either by introducing certain cofactor-recycling enzyme systems in their metabolic pathways or through their recycling via the electron transport system in the case of aerobic metabolism.

In vitro applications of such enzymes will largely depend on developing efficient *in vitro* cofactor-recycling systems. These can be obtained by co-immobilizing another enzyme so that it can recycle the cofactors, e.g. alcohol dehydrogenase in the presence of ethanol can be used for the recycling of NAD to NADH in a bioprocess catalysed by an NADH-dependent enzyme. Choice of the coupling enzyme for such application has been made based on its ability to use an economical substrate like ethanol or formate (formate dehydrogenase) as well as one which results preferably in a volatile by-product like CO_2 or acetaldehyde so as to minimize the downstream processing problems. In addition to the recycling of the cofactor, the retention of the cofactor is also an important criteria for the biochemical process.

Currently, methods are available for the covalent binding of the cofactors along with the enzymes on polymeric supports or directly on enzymes or on soluble high molecular weight polymers like dextran or polyethylene glycol. Membrane reactors containing continuous cofactor-regenerating systems have been used in the large-scale production of sugar derivatives, and optically active α-hydroxy acids and α-amino acids.

Medical/Clinical Applications

Immobilized enzymes find applications in clinical analysis. For example, glucose content in urine can easily and accurately be determined with immobilized glucose oxidase (GOD) and peroxidase (POD). In this method, a strip of cellulose acetate paper coated with immobilized glucose oxidase (GOD) and peroxidase (POD) is immersed in the urine sample mixed with a chromogenic substrate, *o*-toluidine. The following reactions take place;

$$\text{D-Glucose} + H_2O + O_2 \xrightarrow{\text{GOD}} \text{D-Gluconic acid} + H_2O_2$$

$$H_2O_2 \xrightarrow{\text{POD}} H_2O + [O]$$

$$\underset{\text{(Colourless)}}{\text{Orthotoluidine}} \longrightarrow \underset{\text{(Blue colour)}}{\text{Oxidized orthotoluidine}}$$

Aspartic acid is widely used in medicines and as a food additive. The enzyme aspartase catalyses a one-step stereospecific addition of ammonia to the double bond of fumaric acid. This enzyme has been immobilized using the whole cells of *Escherichia coli.* This is the first reported industrial application of an immobilized microbial cell. L-aspartic acid is also used in the production of low-calorific sweetener, aspartame. Immobilized fumarase have been used widely for the production of malic acid, an acidulant in many pharmaceutical preparations. Recently the market of malic acid has been increased as food acidulant in competition with citric acid.

Applications in Food Processing

One of the most important industrial applications of immobilized enzyme is in the conversion of glucose syrups to high-fructose syrups by immobilized glucose isomerase. This reaction is of special interest especially in non-tropical countries where sugar cultivation is not abundant. After the isomerization process, pH is lowered to 4–5, purified by ion-exchange chromatography and decolorized with activated charcoal. Then it is concentrated by evaporation to about 70% dry solid.

This method has been extended to produce sucrose from starch.

Sucrose is becoming increasingly expensive and high-fructose syrups, which contain almost equal concentration of glucose and fructose is getting more popular in the market. Fructose is twice as sweet as glucose and this interconversion of glucose to fructose is economically feasible. Industrial process for the production of the high-fructose syrups uses glucose isomerase adsorbed on to DEAE–cellulose and porous glass. Enzymatic inversion of sucrose by immobilized invertase also produces high-fructose syrups, which avoids the high colour, high salt ash, relatively low inversion and batch variability problems associated with the conventional acid hydrolysis.

Immobilization of Raffinase

Immobilized raffinase finds application in the sugar industry. Raffinase hydrolyses raffinose, a trisaccharide, into sucrose and galactose. When sugar beet juice is treated with immobilized raffinase, much of the raffinose in the former is hydrolysed to sucrose and galactose. The galactose released is destroyed in the alkaline conditions during the juice purification while sucrose is recovered. This process gives ~3% increase in productivity. Immobilized raffinase is also used to remove raffinose and stachyose from soya bean milk. These sugars are responsible for the flatulence that may be caused when soya bean milk is used as a substitute for milk. Immobilized glucose oxidase can find application in the production of gluconic acid, an acidulant, in the removal of oxygen from beverages, in the removal of glucose from eggs prior to dehydration in order to prevent the Maillard reaction and in the isolation of fructose from honey.

REVIEW QUESTIONS

1. What do you mean by immobilization of enzymes?
2. Discuss the necessity of immobilizing an enzyme.
3. Describe how immobilization alters the characteristics of the enzyme.
4. Compare and contrast the different methods available for immobilization of enzymes.
5. Give an account of the applications of immobilized enzymes.

13

IMMOBILIZATION BIOREACTORS

Long before anyone understood the concept of bioreaction, human beings were taking advantage of its results. Bread, cheese, wine, and beer were all made possible through what was traditionally known as fermentation, a little understood process, successful more by chance than design. By definition, a bioreactor is a system in which a biological conversion is effected. Bioreactors differ from conventional chemical reactors in that they support and control biological entities. As such, bioreactor systems must be designed to provide a high degree of control over process upsets and contaminations, since the organisms are more sensitive and less stable than chemicals.

KEY ISSUES IN BIOREACTOR DESIGN AND OPERATION

The goal of an effective bioreactor is to control, contain and positively influence the biological reaction. To accomplish this, the chemical engineer must take into consideration two areas. One is the suitable reactor parameters for the desired biological, chemical and physical (macrokinetic) system. The macrokinetic system includes microbial growth and metabolite production.

The other area of major importance in bioreactor design involves the bioreaction parameters, including:

- Controlled temperature
- Optimum pH
- Sufficient substrate (carbon source, e.g. sugars, proteins, fats, etc.)
- Water availability
- Salts for nutrition
- Vitamins
- Oxygen (for aerobic process)
- Gas evolution and
- Product and by-product removal

In addition to controlling these, the bioreactor must be designed to both promote formation of the optimal morphology of the organism and eliminate or reduce contamination by unwanted organisms or mutation of the organisms.

The various types of bioreaction systems covered here include batch, continuous, semi-continuous and fed-batch bioreactions.

BATCH BIOREACTIONS

The majority of bioreactions are batchwise. In batch cultures, a bioreactor is filled with fresh medium and inoculated. At the end of fermentation,

- the contents are removed for downstream processing,
- the reactor is cleaned and sterilized,
- it is refilled for next fermentation

Overall, batch bioreaction systems provide a number of advantages, including:

- Reduced risk of contamination or cell mutation, due to a relatively brief growth period
- Lower capital investment when compared to continuous processes for the same bioreactor volume
- More flexibility with varying product/biological systems
- Higher raw material conversion levels resulting from a controlled growth period

The disadvantages include:

- Lower productivity levels due to time taken for filling, heating, sterilizing, cooling, emptying and cleaning the reactor
- Increased focus on instrumentation due to frequent sterilization
- Greater expense incurred in preparing several subcultures for inoculation
- Higher costs for labour and/or process control for this non-stationary process

Common applications for batch bioreactors include:

- Products can be produced with minimal risk of contamination or organism mutation
- Operations in which only small amounts of product are produced
- Processes using one reactor to make various products
- Processes in which batch or semicontinuous product separation is adequate

CONTINUOUS BIOREACTIONS

The defining characteristic of continuous bioreaction is a perpetual feeding process of the media and, simultaneously, bioreactor fluid is continuously removed, i.e., the cells continuously propagate on the fresh medium entering the reactor and, at the same time, products, metabolic waste products and cells are removed in the effluent. The reaction variables and control parameters remain consistent, establishing a time constant state within the reactor. The result is continuous productivity and output.

These systems provide a number of advantages, including:

- Increased potential for automating the process
- Reduced labour expense due to automation
- Less non-productive time expended in emptying, filling and sterilizing the reactor
- Consistent product quality due to invariable operating parameters
- Decreased toxicity risks to staff, due to automation
- Reduced stress on instruments due to sterilization

The disadvantages of continuous bioreactors include:

- Minimal flexibility, since only slight variations in the process are possible (throughput, medium composition, oxygen concentration and temperature)
- Higher investment costs in control and automation equipment, and increased expenses for continuous sterilization of the medium
- Greater processing costs with continuous replenishment of non-soluble, solid substrates such as straw
- Higher risk of contamination and cell mutation, due to the relatively brief cultivation period

Continuous bioreaction is frequently used for processes with high volume production; for processes using gas, liquid or soluble solid substrates; and for processes, involving microorganisms with high mutation stability. Typical end products include vinegar, baker's yeast and treated waste water.

Types of Continuous Cultivation

The types of continuous culturing are as follows:

1. *Chemostat* In this process the medium is pumped continuously, and the volume is constant because the excess medium overflows.

2. *Turbidostat* It employs feedback control of pumping rate to maintain a fixed turbidity of the culture. The controller of a turbidostat slows feeding when the cell concentration is below the set point so that growth can restore the turbidity. If the set point is exceeded, the pump speeds up to dilute the cell concentration.

3. *Auxostat* It is a method of feeding fresh nutrient based on the concentration of metabolic product and related parameter changes like pH and dissolved oxygen. When the product level is increased above the set point by either increasing or decreasing pH and dissolved oxygen, the fresh medium is added to retore them.

SEMI-CONTINUOUS BIOREACTIONS

This is a hybrid of batch and continuous operation, found in many types of processes. One of the more frequently used is initiating the bioreaction in the batch mode, until the growth-limiting substrate has been consumed. Then, the substrate is fed to the reactor as specified (batch) or is maintained by an extended culture period (continuous). Like batch reactors, semicontinuous reactors are non-stationary.

These systems provide a number of advantages, including:

- Higher yield, resulting from a well-defined cultivation period during which no cells are added or removed.
- Increased opportunity for optimizing environmental conditions of the microorganism with regard to the phase of growth or production and age of the culture.
- Nearly stationary operation, important for slightly mutating microorganisms and for those at risk of contamination.

The disadvantages include:

- Lower productivity levels due to time-consuming procedures for filling, heating, sterilizing, cooling, emptying and cleaning the reactor
- Greater expenses in labour and/or dynamic process control for the process

Semicontinuous bioreactors are typically used when continuous methods are not feasible, for example, those in which slight mutation or contamination of the microorganism occurs. Such bioreactors are also used when batch methods do not offer the desired productivity levels.

FED-BATCH BIOREACTIONS

In the fed-batch system, a fresh aliquot of the medium is continuously or periodically added, without the removal of the culture fluid. The fermenter is designed to accommodate the increasing volumes. The system is always at a quasi-steady state. The two types of fed-batch are:

1. *Fixed volume fed-batch* In this type, the limiting substrate is fed without diluting the culture.

2. *Variable volume fed-batch* Here, the volume changes with the fermentation time due to the substrate feed. The way

this volume changes is dependent on the requirements, limitations and objectives of the operator.

APPLICATION OF ENZYME IMMOBILIZATION IN INDUSTRIAL BIOREACTORS

The development and use of bioreactors of a new generation with immobilized enzymes remains a topical problem of the modern fine biotechnology. On the one hand, it is necessary to optimize the immobilization procedure, because the known methods of enzyme incorporation into a gel decreases the catalytic performance of the immobilized substance. This is due to the hindered diffusion of the substrate and product in a gel phase; furthermore this immobilization procedure is inapplicable to all the substrates. On the other hand, the bioreactor design should allow continuous specific hydrolysis of macromolecular viscous substrates and colloids whose solutions cannot be filtered.

In an enzyme reactor, the highest specific enzyme activity is desirable. It is considered advantageous if the support aids in separation. One approach is to use a molecular sieve as the support and pulse the reactor bed with the alternating passage of substrate solution and water. The result is that bands of unused substrate and product progress down the column. If it happens so, then this technique will be useful otherwise it is better to immobilize the enzymes on a porous support. For an industrial reactor, it is preferable to use supports that are non-biodegradable, e.g. glass, silica, celite, bentonite, alumina or titanium oxide.

LARGE-SCALE IMMOBILIZED CELL BIOREACTORS

Large-scale immobilized cell bioreactors include: stirred tank reactors, continuous stirred tank reactor (CSTR), airlift reactor systems, packed bed reactor, fluidized bed reactor and recycle reactor.

Stirred Tank Reactors

The most common type of aerobic bioreactor in use today is the stirred tank reactor, which may feature a specific internal configuration designed to provide a specific circulation pattern. It is composed of a reactor and a mixer such as a stirrer, a turbine wing or a propeller. This reactor is useful for substrate solutions of high viscosity and for immobilized enzymes with relatively low activity.

The operating principles of the stirred tank bioreactor are relatively simple. As shown in Figure 13.1, the sterile medium and inoculum are introduced into a sterilized tank, and the air

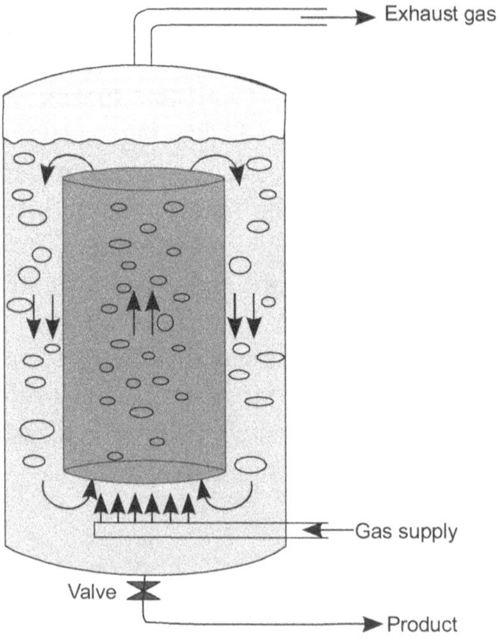

Figure 13.1 Stirred tank reactor

supply typically enters at the bottom. For optimal mixing, the tank features not only with an agitator system but also baffles, which help to prevent a whirlpool effect that could impede proper mixing. In the early stages of the process, warm water may be circulated through the baffles to heat up the system; later, cool water may be circulated inside of them to prevent overheating. The number of baffles typically ranges from four to eight. As the bioreaction progresses, the bubbles produced by the air supply are broken up by the agitator as they travel upward. Many types of agitators are currently used, with the most common one being the four-bladed disc turbine. At the top of the tank, exhaust gas is discharged and the product flows back down, where it is drained from the tank.

Continuous Stirred Tank Reactor (CSTR)

Continuous bioreactors have been designed to carry out submerged continuous fermentation. In continuous reactors, fresh medium is continuously added and bioreactor fluid is continuously removed. As a result, cells continuously receive fresh medium and are simultaneously used for processing. In a CSTR, one or more fluid reagents are introduced into a tank reactor equipped with an impellor, while the reactor effluent is removed (Figure 13.2). The impellor stirs the reagents to ensure proper mixing. In CSTR, at steady state, the inflow rate must equal the mass outflow rate, otherwise, the tank will overflow or go empty (transient state). The reaction proceeds at the reaction rate associated with the final (output) concentration. In situations where the optimal rate of product formation may occur at temperature or pH different from that of growth, two or more stirred tank reactors operated in series permit optimized growth and product formation conditions in each reactor. When mixed substrates are used, the preferential substrate is consumed first. The use of two CSTRs in series enables the preferred substrate to be completely consumed in the first reactor and the second substrate in the second reactor, minimizing the required reactor volume. The reactor can be operated for a long time. Also, continuous reactors can be more productive than batch reactors due to the fact that continuous reactors do not have to be shutdown regularly and the growth rate of cells can be more easily controlled and optimized. In addition, cells can be immobilized in continuous reactors to prevent wastage, which further increases the productivity of these reactors.

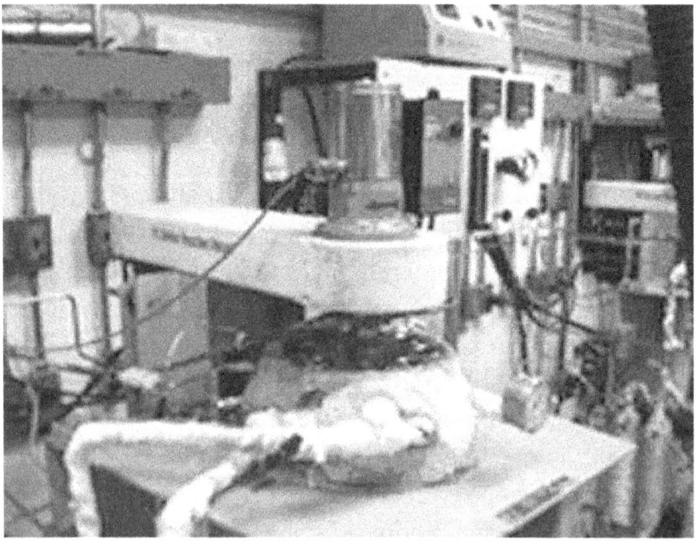

Figure 13.2 Continuous stirred tank reactor

Airlift Bioreactor

Airlift bioreactor is also known as tower reactor. It can be described as a bubble column containing a draught tube. Greater air throughput and higher pressures are needed, particularly for large-scale operation. When compared with a stir tank fermenter, an airlift fermenter lacks an agitating device. When the filtered air with high pressure is passed through the sparger present at the bottom of the reactor, the medium is raised along with the airstream which adequately agitates the medium enough. Besides, it provides adequate aeration also. An airlift reactor is divided into three parts: the air riser (ascending column), downcomer (descending column) and disengagement zone. The air riser is the region into which bubbles are sparged. It may be inside or outside of the draft tube. The latter design is preferred in large-scale fermenters as it provides better heat-transfer efficiencies. The bubbles rising in the air riser cause the liquid to flow in vertical direction. To counteract these upward forces through the ascending column, when the liquid is driven up due to pressure or air, the raising pressure is reduced gradually. When the liquid reaches the disengagement zone, it is free from pressure and flows in downward direction in the downcomer. This leads to liquid circulation and improved mixing efficiencies as compared to bubble columns. The enhanced liquid circulation causes bubbles to move in uniform direction at relatively uniform velocity.

Many types of airlift bioreactors are currently in use today. Air is typically fed through a sparger ring into the bottom of a central draught tube that controls the circulation of air and the medium. Air flows up the tube, forming bubbles, and exhaust gas disengages at the top of the column. The degassed liquid then flows downward and the product is drained from the tank. The tube can be designed to serve as an internal heat exchanger, or a heat exchanger can be added to an internal circulation loop (Figure 13.3).

Airlift systems provide some advantages over conventional bioreactors.

- Simple design with no moving parts2 or agitator shaft seals, for less maintenance, less risk of defects and easier sterilization

- Large, specific, interfacial contact area with low energy input
- Well-controlled flow and efficient mixing
- Well-defined residence time for all phases
- Increased mass transfer due to enhanced oxygen solubility achieved in large tanks with greater pressures

Figure 13.3 AirLift Bioreactor

The main disadvantages are:

- Higher initial capital investments due to large-scale processes
- Greater air throughput and high pressures are needed, particularly for large-scale operation
- Low friction with an optimal hydraulic diameter for the riser and downcomer
- Lower efficiency of gas compression
- Inherently impossible to maintain consistent levels of substrate, nutrients and oxygen with the organisms circulating through the bioreactor and conditions changing.
- Inefficient gas/liquid separation when foaming occurs.

However, these disadvantages can and must be minimized in designing airlift systems.

Packed Bed Reactor

Packed bed reactors are widely used for immobilized enzymes and microbial cells. In these systems, it is necessary to consider the pressure drop across the packed bed or column and the effect of the column dimensions of the reaction rate. There are three substrate flow possibilities in a packed bed:

- Downward flow method
- Upward flow method
- Recycling method

The recycling method is advantageous when the linear velocity of the substrate solution affects the reaction flow rate. This is because the recycling method allows the substrate solution to pass through the column at a desired velocity. For industrial applications, upward flow is generally preferred over downward flow because upward flow does not compress the beds in enzyme columns as downward flow does. When gas is produced during an enzyme reaction, upward flow is preferred (Figure 13.4).

Packed bed reactor has the following advantages:

- Automatic, easy control and operation
- Reduction of labour costs
- Stabilization of operating conditions

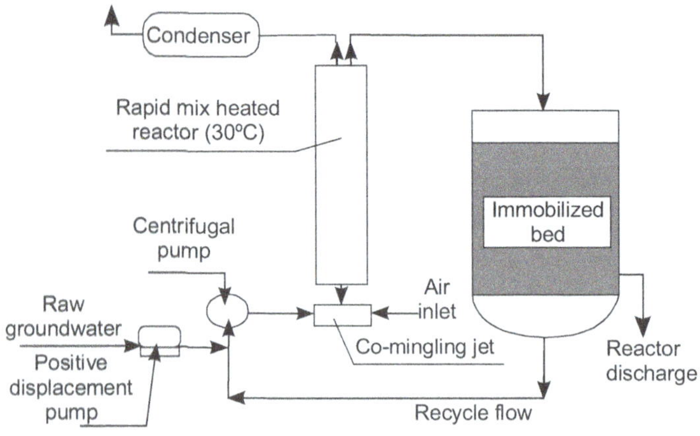

Figure 13.4 Packed bed reactor

Fluidized Bed Reactor

Fluidized bed reactor is a combination of the two most common, packed bed and stirred tank, continuous flow reactors (Figure 13.5). In a fluidized bed reactor, the substrate is passed upward through the immobilized enzyme bed at a high enough velocity to lift the particles. However, the velocity must not be so high that the enzymes are swept away from the reactor entirely. This type of reactor is ideal for highly exothermic reactions because it eliminates the local hot spots, due to its mass and heat transfer characteristics mentioned before. It is most often applied in immobilized enzyme catalysis where viscous, particulate substrates are to be handled.

Packed bed reactor has the following advantages:

- Uniform particle mixing
- Uniform temperature gradients
- Ability to operate reactor in continuous state

The disadvantages are:

- Increased reactor vessel size
- Pumping requirements and pressure drop
- Particle entertainment
- Erosion of internal components

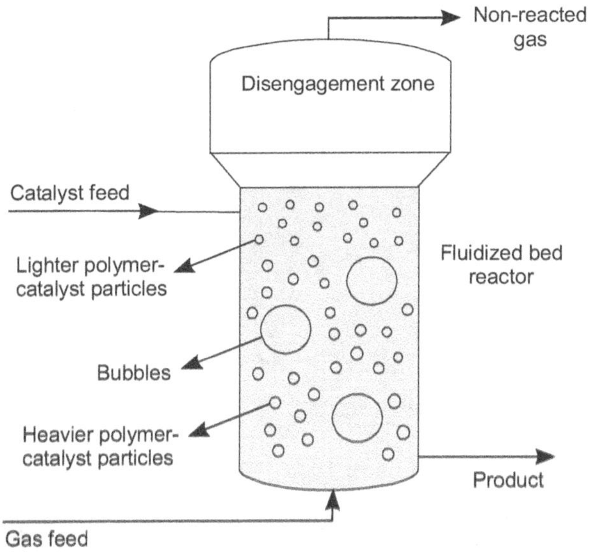

Figure 13.5 Fluidized bed reactor

Recycle Reactor

A recycle reactor is simply a reactor, such as CSTR or fluidized bed reactor, with a recycle system (Figure 13.6). This type of reactor is not seen very often, but is very important to consider when studying immobilized enzymes. In a recycle reactor, a portion of the product stream is recycled and mixed with the inlet flow to the reactor. If the entire product stream is recycled

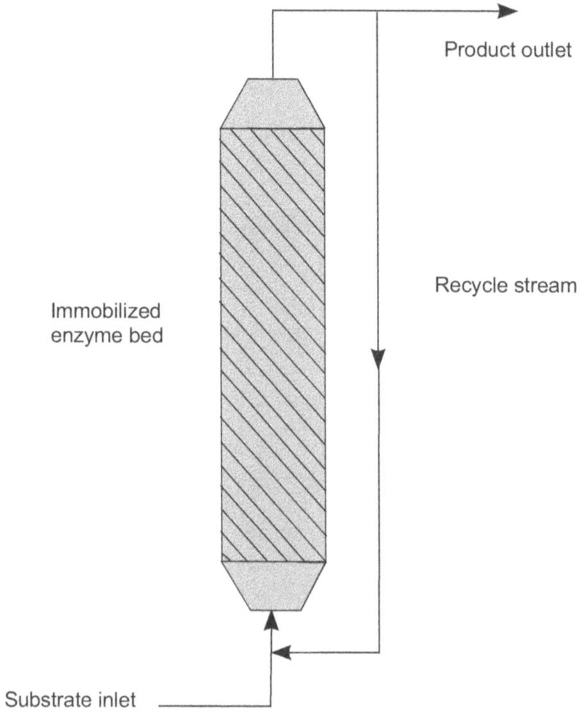

Figure 13.6 Recycle reactor

back to the inlet stream, then it is called a total recycle reactor. This type of reactor is used when you have a substrate that cannot be completely processed on a single pass, such as with an insoluble substrate. These reactors continue to move the same substrate through the reactor so that the effective contact time is high enough to allow the substrate to be processed. Recycle reactors also allow the reactor to operate at high fluid velocities. This is important because it minimizes the bulk mass transfer resistance to the transport of the substrate.

Thus, the knowledge of bioprocessing is an integration of biochemistry, microbiology and engineering science applied in industrial technology. Thus fermentation products and the ability to cultivate large amounts of organisms are the focus of bioprocessing and such achievements may be obtained by using bioreactors.

REVIEW QUESTIONS

1. Discuss the application of enzyme immobilization in industrial bioreactors.
2. Describe the different types of bioreactors used in this technology.
3. Discuss the key issues to be kept in mind in designing a reactor.

14

RIBOZYMES

INTRODUCTION

Until about 25 years ago, all known enzymes were proteins. Although nucleic acids (DNA and RNA) could serve as blueprints for the transcription and translation that would produce the proper amino acid sequences to make up an enzymatic protein, it was the protein that did the actual work. But then it was discovered that some RNA molecules also have enzymatic property, i.e., catalysing covalent changes in the structure of substrates. These ribonucleic acids with enzymelike activity were named ribozymes. The first ribozymes were discovered in the 1980s by Thomas R. Cech and Sidney Altman. These ribozymes were found in the intron of an RNA transcript, which removed itself from the transcript, as well as in the RNA component of the RNase P complex, which is involved in the maturation of pre-tRNAs. In 1989, Thomas R. Cech and Sidney Altman won the Nobel Prize in chemistry for their "discovery of catalytic properties of RNA." The term "ribozyme" was first introduced by Kelly Kruger *et al.* in 1982 in a paper published in *Cell*.

The few known naturally occurring ribozymes are listed below.

- Peptidyl transferase 23S rRNA
- RNase P
- Group I and Group II introns
- Leadzyme
- Hairpin ribozyme
- Hammerhead ribozyme
- HDV ribozyme
- Mammalian CPEB3 ribozyme
- VS ribozyme
- *glmS* ribozyme

CLASSES OF RIBOZYMES

Catalytic RNAs are broadly grouped into two classes based on their size and reaction mechanisms: large and small ribozymes. Currently recognized groups are the group I intron, the group II intron, the RNA subunit of ribonuclease P, the hammerhead ribozyme, the hairpin ribozyme, and the hepatitis delta virus (HDV) ribozyme. The group I and II intron ribozymes and the ribonuclease P ribozyme are relatively large ribozymes (hundreds of nucleotides).

The hammerhead, hairpin, and HDV ribozymes are substantially smaller (less than one hundred nucleotides). The shortened form of the *Tetrahymena* group I intron is the best characterized of the larger ribozymes. Among the small ribozymes, the hammerhead is the best characterized. Large ribozymes consist of several hundreds up to 3000 nucleotides and they generate reaction products with a free 3´-hydroxyl and 5´-phosphate group. In contrast, small catalytically active nucleic acids which range from 30 to ~150 nucleotides in length generate products with a 2´, 3´-cyclic phosphate and a 5´-hydroxyl group.

HAMMERHEAD RIBOZYMES

Hammerhead ribozyme is the smallest of the naturally occurring ribozymes. It has been found in several satellite RNAs of plant viruses, viroids and transcripts of a nuclear satellite DNA of newt. Like protein-based enzymes, ribozyme catalysis obeys Michaelis–Menten principles, with some exceptions. The reaction is composed of 3 steps: binding of the ribozyme to the target, cleavage of the target, and release of the cleavage products (Figure 14.1).

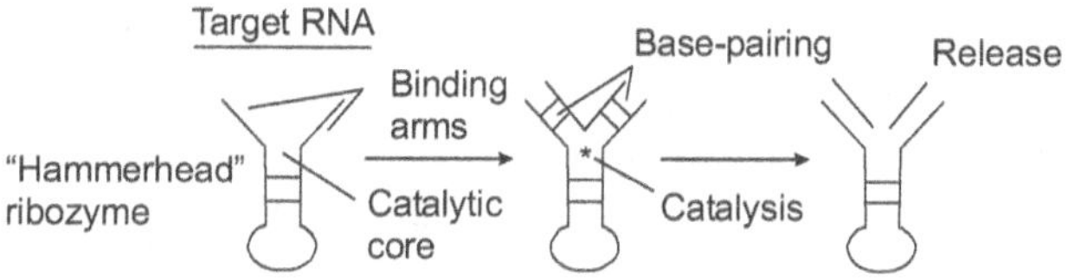

Figure 14.1 Mechanism of ribozyme catalysis

Hammerhead ribozyme (Figure 14.2) has an absolute requirement for divalent metal ions. Mg^{2+} is the referred metal, but Mn^{2+}, Co^{2+}, Ca^{2+}, Sr^{2+}, and Ba^{2+} also support cleavage activity.

Standard reaction conditions include 10 mM metal, pH 7.5, at temperatures between 27°C and 0°C. Under these conditions chemical cleavage rates of 1 min^{-1} are typical. In some cases additives that stabilize higher order nucleic acid structures, for example, the polycations spermine or spermidine, or sodium ions, can minimize the divalent metal ion requirement. These observations imply that the metal may serve a structural as well as a catalytic role. The hammerhead ribozyme's response to changes in pH provides additional clues about the cleavage mechanism. This indicates that a single deprotonation is required for cleavage.

Moreover, pH rate profiles with a series of divalent metals correlate with the metal's pK_a values which shows that the hammerhead acts as the base in the cleavage reaction. The crystal structure of the hammerhead ribozyme was the first example of an atomic structure of a catalytic RNA molecule to be determined. The hammerhead complex consists of 3 helical stems and includes 11 consensus nucleotides that form the catalytically active core. Nine of the 11 conserved nucleotides are nominally single-stranded. The hammerhead ribozyme is capable of cleaving *in cis* or *in trans*. The *in trans* construct allows for multiple turnover and enables virtually any sequence to be targeted for cleavage. With the exception of U7, 11 substitutions led to a dramatic decrease in cleavage activity. Substitutions at G5 and A14 were especially detrimental, and totally destroyed the cleavage activity.

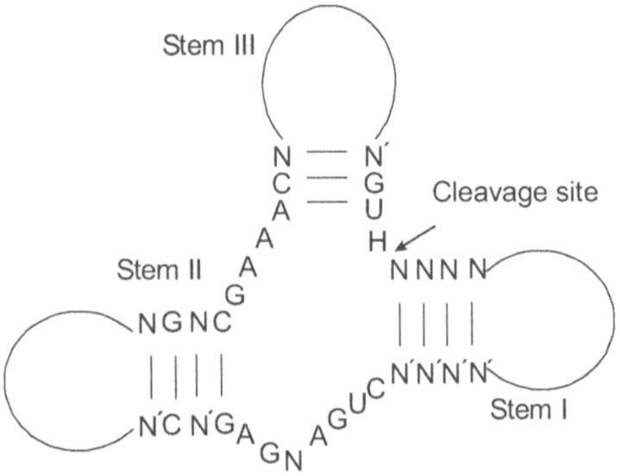

Figure 14.2 The minimal consensus structure of the hammerhead ribozyme

The sequences shown as outlines are required for efficient cleavage. Residues indicated by **N** can be any base and **N′** indicates the complement of the base with which it is paired. The **H** at the cleavage site means any base but G.

The lines at the ends of **stems I, II and III** indicate that any number of bases may connect the two halves of each stem, or they need not be connected at all.

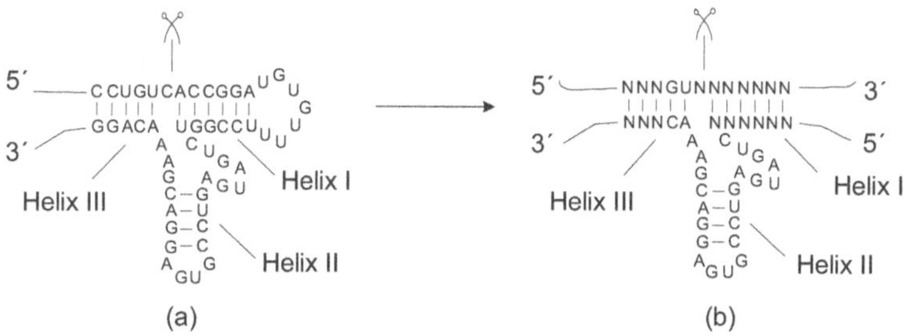

Figure 14.3 Hammerhead ribozyme (a) *cis*-acting hammerhead ribozyme (b) *trans*-acting hammerhead ribozyme

In Figure 14.3, the structure of the wild-type, *cis*-acting hammerhead ribozyme is shown. The three helices and the cleavage site are indicated. The self-splicing happens after the sequence GUC between helix I and III, and the G and U are conserved and crucial for the catalysis. Except for itself, a hammerhead ribozyme is not catalytic in cells because of the nature of its intramolecular reaction. However, most ribozymes can be chemically engineered to cleave RNA in *trans* by separating the catalytic domain from ribozyme and attaching recognition (i.e., antisense) arms to the catalytic centre, in order to target the substrate. These *trans*-acting ribozymes can be very useful in the study of molecular biology and pharmaceuticals. In comparison to the conventional antisense RNAs, ribozymes provide the potential of turnover with a single molecule being able to inactivate multiple-target RNAs.

Binding of the enzyme and substrate results in the catalytically active structure. After cleavage, the two product strands dissociate and the ribozyme strand can go into the next cycle, which earns the fidelity to cleave a unique target. Both the turnover and specificity are affected by the length of the binding arm (helix I and III). If the length of the binding arms is very short, the rate of dissociation of the target from the ribozyme may exceed the rate of cleavage, resulting in poor efficiency (Figure 14.4).

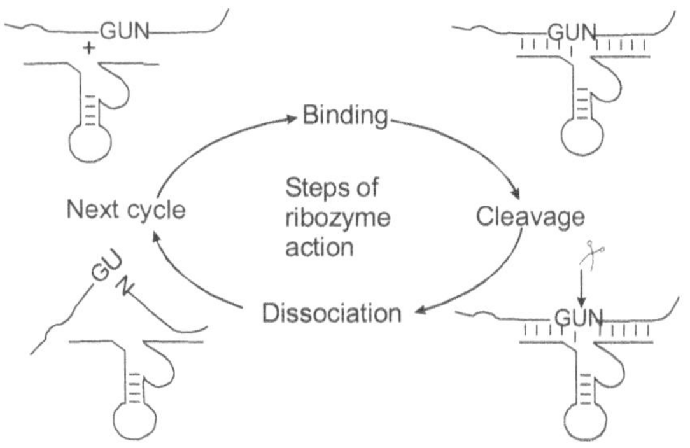

Figure 14.4 Turnover cycle of RNA cleavage by a hammerhead ribozyme

HAIRPIN RIBOZYMES

Another catalytic RNA domain found in satellite RNAs of pathogenic plant viruses is the hairpin motif. Three different hairpin ribozymes have so far been found in nature, of which, the one from satellite RNA associated with tobacco ring spot virus (TRSV) is the best characterized. The other two hairpin motifs, isolated from different satellite viruses, show sequence variations. The hairpin ribozyme, like the hammerhead, effectively uses Mg^{2+}, Sr^{2+}, and Ca^{2+} metal ions. However, unlike the hammerhead, the hairpin ribozyme is inhibited by Mn^{2+} and O^{2+} unless spermine is also present.

Catalytic activity of the hairpin ribozyme results from distortion and precise orientation of the substrate RNA and general acid–base catalysis by neighbouring nucleotides without involvement of metal ions in catalysis. The hairpin shares many of the preferred reaction conditions with the hammerhead ribozyme, i.e., 10 mM metal, pH 7.5, 27–50°C, and the chemical cleavage rate is comparable. The second critical difference is the hairpin ribozyme's response to phosphorothioate substitution. The hairpin ribozyme effectively cleaves a phosphorothioate at the target site, which is inconsistent with interaction between a metal ion and either non-bridging phosphate oxygen at the scissile phosphate. Moreover the hairpin ribozyme effectively catalyses cleavage in the presence of EDTA and inert cobalt (III) complexes.

As with the hammerhead ribozyme, the hairpin ribozyme can function both *in cis* and *in trans*. Although the hammerhead and hairpin ribozymes perform similar functions, they do not share a common structural motif. The shortened form of the hairpin ribozyme is larger than the minimum hammerhead sequence (50 nucleotides vs ~30 nucleotides). To the other small ribozymes, hairpin catalysts cleave concatemeric precursor molecules into mature satellite RNA during rolling-circle replication, giving rise to a 2´, 3´-cyclo phosphate and a free 5´-OH terminus. Depending on reaction conditions, the hairpin ribozyme may also favour RNA ligation over cleavage.

The hairpin contains four helical regions and two single-stranded loops. It consists of two domains, each harbouring two helical regions separated by an internal loop, connected by a hinge region. One of these domains results from the association of 14 nucleotides of a substrate RNA with the ribozyme via base-pairing. In this respect, the hairpin ribozyme presents a mechanistically distinct class of catalytic RNA. Hairpin ribozymes have been developed mainly for antiviral applications. Intracellularly expressed ribozymes conferred resistance to incoming HIV-1. When $CD4^+$ lymphocytes from HIV-1 infected donors were transduced with an anti-HIV ribozyme vector and subsequently expanded, viral replication was delayed by two to three weeks.

INTRONS

Introns are non-coding sequences that interrupt the coding sequences of most eukaryotic genes. These must be removed or "spliced", after transcription, to allow expression of functional mRNA, rRNA, or tRNA molecules. Group I, II, and III introns are self-splicing introns and are relatively rare compared to spliceosomal introns.

Group I Introns

Group I introns are the only class of introns whose splicing acquires a free guanine nucleoside. They possess a secondary structure different from that of group II and III introns. In the case of group I introns, excision is mediated by the autocatalytic activity of the intronic sequences themselves. There are several hundred examples of group I introns, including those found in plant and fungal mitochondria, bacteriophage, eubacteria, and chloroplast tRNA. Despite considerable variability in size and sequence, group I introns have phylogenetically conserved secondary structures and a common reaction mechanism.

Group II and III introns are similar and have a conserved secondary structure. Ribozymes derived from group I introns generally consist of hundreds of nucleotides and adopt catalytically competent structures that are not yet well-defined at the molecular level. An example of one of the smallest group I ribozymes is the *Anabaena* ribozyme. The first confirmation of self-splicing activity came from work on ribosomal RNA genes from *Tetrahymena thermophila. In vivo*, these reactions occur with the assistance of protein factors that, in some cases, are encoded within the intron itself.

The enzymatic activity of group I introns involves a two-step transesterification with a requirement for a divalent cation, such as magnesium, and a guanosine co-factor. The first step involves a nucleophilic attack of the guanosine co-factor 39-hydroxyl group on the 59-splice site forming a free 39-hydroxyl on the 59-exon. In a second step, this free hydroxyl group makes a nucleophilic attack on the 39-splice site, releasing the intron as a circular molecule and leaving a ligated exon. Substrate specificity is determined by a sequence within the intron, the internal guide sequence (IGS), and there is a requirement for a U at position 21, relative to the cleavage site, paired with a conserved G in the IGS.

The ability of group I ribozymes to perform *trans*-splicing reactions *in vitro* suggests the possibility of therapeutic modification of disease-relevant RNA targets *in vivo*. The *Tetrahymena* ribozyme employs an external guanosine nucleoside or nucleotide co-factor as a nucleophile. The guanosine binds at a saturable binding site in the core of the ribozyme through at least three different interactions. The 3´-hydroxyl of the guanosine acts as the nucleophile, which results in the extension of the intron sequence by one nucleotide after cleavage. The *Tetrahymena* ribozyme requires either Mg^{2+} or Mn^{2+} ions for activity.

Shortened forms of the ribozyme bind to and catalyse sequence-specific cleavage of both DNA and RNA oligonucleotides. The substrate binds by base-pairing to an internal guide sequence on the ribozyme and, in the case of RNA, also by interactions with specific 2´-hydroxyl groups. Nucleophilic attack of the target phosphodiester by the 3´-hydroxyl group of the external guanosine co-factor proceeds with inversion of configuration at phosphorus, which infers an in-line, S_N^2-type reaction. The 3´-cleaving group of a substrate for the *Tetrahymena* ribozyme has also been replaced with a sulphur atom, thus producing a 3´-bridging phosphorothioate. Compared to the native phosphodiester sequence, the bridging phosphorothioate was cleaved 1000 times more slowly in the presence of Mg^{2+} ions. Taken together, the data from the cleavage and ligation reactions support the model of a two metal ion active site for the group I ribozyme.

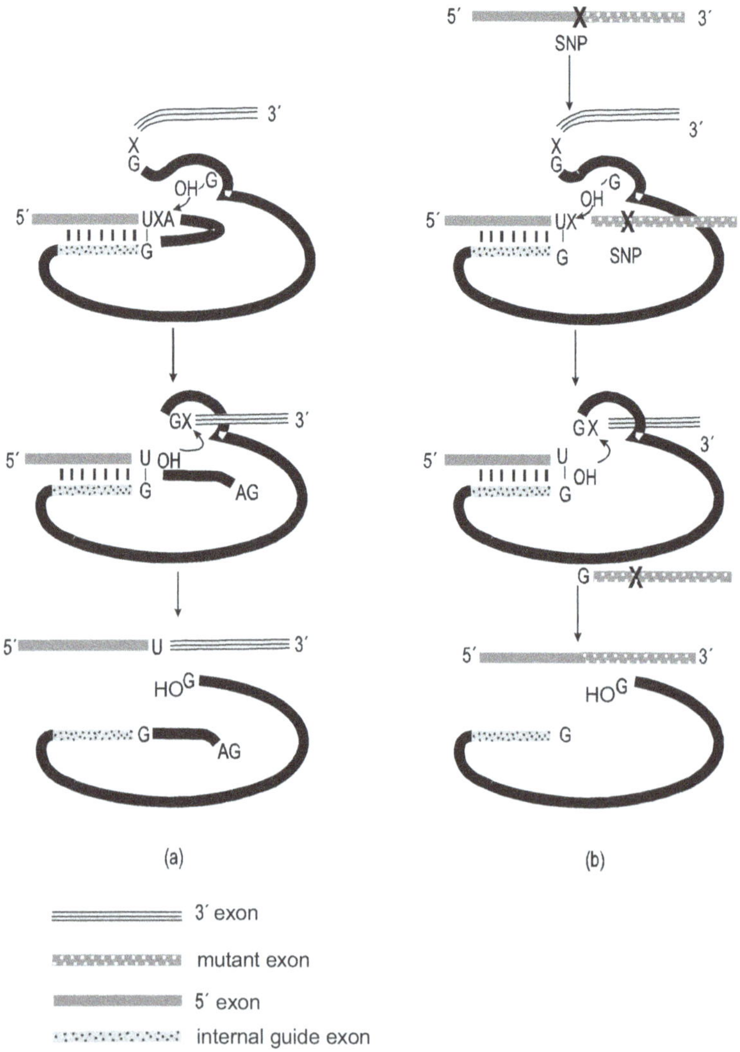

Figure 14.5 Schematic representation of the catalytic action of group I introns

Figure 14.5(a) illustrates the *cis*-splicing reaction. The reaction is initiated by the attack of an external guanosine cofactor on the 5′-splice site (up), followed by joining of the exons by nucleophilic attack of the upstream exon on the 3′-splice site (middle), and dissociation of the joined exons from the intron (down). Figure 14.5(b) illustrates the therapeutic application of a *trans*-splicing group I intron. A mutant exon containing a single nucleotide polymorphism (SNP) is exchanged for a corrected form of the same exon.

Group II Introns

Group II introns are commonly found in mitochondrial genes in plants, fungi, yeast, and other lower eukaryotes. Like the group I introns, they are generally large fragments of RNA consisting of a series of helical domains and nominally single-stranded regions of RNA. Catalytic competency requires the correct folding and the binding of requisite metal cofactors.

As with the group I intron, the group II introns catalyse two sequential reactions that result in release of the intron and ligation of the flanking exon sequences. However, the group II introns do not require a free guanosine co-factor. Instead, the group II introns exploit the 2′-hydroxyl of a specific internal adenosine nucleotide that is part of the intron sequence. Transesterification results in a branch point, adenosine and upon cleavage a lariat structure is formed which contains a 2′, 5′-phosphodiester bond at the branch site.

Group II introns bear no apparent structural similarity to group I introns. Efficient *in vitro* cleavage requires high Mg^{2+} concentrations (up to 100 mM) and monovalent cations (up to 1.5 M K^+ or NH_4^+), and under physiological salt conditions a hydrolytic mechanism, in which a water molecule is positioned for attack (rather than the 2′-OH of the branch point, adenosine), tends to dominate *in vitro* analyses. In addition to transesterifications (with a 2´-OH nucleophile) or hydrolysis (H_2O or metal-coordinated H_2O as the nucleophile), both resulting in a 3´-OH leaving group, group II introns can also catalyse the reverse reaction (debranching) in which the nucleophile is the 3´-OH and the leaving group is the 2´-OH. The intron can recognize both the 5´–3´ linkage as well as the 2´–5´–3´ linkage, and in some cases even a triphosphate. These activities suggest a certain degree of flexibility in the active site, and the nature of the active-site residues.

In a two-dimensional representation, group II introns can be organized into a set of six helical domains about a central wheel. The only interdomain contact that has been suggested based upon phylogenetic and mutational studies is the interaction of the GAAA tetraloop of domain five (V) with the stem loop region of domain one (I). The catalytic activity of the group II intron is centred about domain V. Despite its small size (34 nucleotides) domain V mediates the activity of the much larger intron sequence, and many of the phylogenetically conserved nucleotides are located within the D5 stem/loop sequence. Mutations in this portion of the sequence can block the splicing reaction.

A subgroup of this class of ribozymes is able to insert itself into an intron-less allele on the DNA level by reverse splicing and reverse transcription, a process called **retrohoming**. It has been demonstrated that group II introns can be redirected to insert themselves into therapeutically relevant DNA target sites in human cells. This emerging technology may ultimately be used to disrupt deleterious genes on the DNA level or to insert new genomic information specifically into the target genes.

RNase P

RNase P is a ubiquitous enzyme that acts as an endonuclease to generate the mature 5´-end of the tRNA precursors.

In bacteria, RNase P exists as a ribonucleoprotein complex, consisting of a long RNA, typically 300–400 nucleotides in length, and a small protein of approximately 14 kDa. The discovery that the RNA component of the enzyme alone possesses catalytic activity *in vitro* provided the first example of an RNA-based catalyst that acts *in trans* on multiple substrates. RNase P can be considered to be the only known true, naturally occurring *trans*-cleaving RNA enzyme. Further RNase P is one of two known multiple turnover ribozymes in nature, the discovery of which earned Professor Sidney Altman the Nobel Prize in Chemistry in 1989.

Sidney Altman characterized RNase P and its activity in processing the 5´ leader sequence of the precursor tRNA back in the 70's. It has been shown that human nuclear RNase P is required for the normal and efficient transcription of various small non-coding RNA genes, such as tRNA, 5S rRNA, SRP RNA and U6 snRNA genes, which are transcribed by RNA polymerase III, one of three major nuclear RNA polymerases in human cells. RNA cleavage is via nucleophilic attack on the phosphodiester bond leaving a 59-phosphate and 39-hydroxyl at the cleavage site, and there is an absolute requirement for divalent metal ions. The RNA subunit of bacterial ribonuclease P (RNase P) specifically cleaves precursor sequences from the 5´-ends of pre-tRNA molecules to produce mature tRNAs.

RNase P is unique in that it consists of both RNA and protein subunits, but all of the catalytic activity resides in the RNA subunit. The protein subunit is necessary for *in vivo* activity, but under high Mg^{2+} and monovalent salt concentrations the RNA portion itself is sufficient for catalysis. The high salt requirement in the absence of protein is thought to be necessary to allow the proper folding of the RNA by screening unfavourable electrostatic interactions of the folded, polyanionic RNA. Unlike other ribozymes, the primary mechanism of phosphodiester cleavage is a hydrolysis rather than transesterification. Multiple Mg^{2+} ions appear to be required for cleavage activity, both to labilize a water molecule for attack at the scissile phosphodiester, as well as to stabilize the resulting transition-state structure. The ubiquitous nature of RNase P, the lack of a requirement for specific nucleotide sequences for cleavage, and the inherent efficiency of utilizing a cellular enzyme, have created interest in directing RNase P-mediated cleavage to the therapeutic target mRNAs.

HEPATITIS DELTA VIRUS RIBOZYME

The hepatitis delta virus (HDV) is a single-stranded circular RNA virus of approximately 1700 nucleotides, which causes significant pathology in man. HDV infection requires the presence of the hepatitis B virion either coincident with HDV infection or as a pre-existing infection. Hence, HDV is designated as a satellite virus of hepatitis B. The virus is thought to replicate via a double rolling-circle mechanism where both the positive and negative polarity strands promote self-cleavage. Shortened forms of the HDV ribozyme are 85 nucleotides in length and, as with the other ribozymes described above, it is possible to design *trans*-acting constructs.

Several secondary-structure models have been proposed of which pseudoknot model is well-supported by mutagenesis experiments and chemical probing studies. It shows four helical regions, in many of which the conserved nucleotides are located in helix II and its hairpin. The structural motif resembles neither the hammerhead nor the hairpin ribozymes. There is a strong evidence that the catalytic mechanism of the hepatitis delta virus ribozyme involves the action of a cytosine base within the catalytic centre as a general acid–base catalyst. The hepatitis delta ribozyme displays high resistance to denaturing agents like urea or formamide.

HDV contains two ribozymes, both required for RNA replication—one on the genomic, infectious RNA strand, the other in the complementary region of the antigenomic strand. The self-cleaving activity of HDV RNA is enhanced in the presence of denaturants, which suggests that the active structures are present in nascent transcripts, not the mature, folded, genomic RNA. Substrate recognition requires formation of the P1 stem, comprising a GU wobble pair plus six nucleotides hybridized to the target, with cleavage occurring just 59 to the wobble base pair. The minimal ribozyme sequence is 85 nucleotides, almost entirely 39 with respect to the cleavage site, useful for the generation of discrete RNA transcript termini or processing ribozyme multimers to monomers in *cis*-reactions. *Trans*-cleaving HDV ribozymes have been designed that are active against oligonucleotide substrates; these include a P4–2 ribozyme, a P1 ribozyme, a circular ribozyme and a hybrid of both genomic and antigenomic HDV ribozymes.

However, the modest activity of *trans*-cleaving HDV ribozymes against long substrates limits their utility for therapeutic applications. As with the hairpin, the cleavage rate of the HDV ribozyme is unaffected by pH between pH 5 and 9.1. The HDV ribozyme requires a divalent metal for activity and effectively uses both Ca^{2+} and Mg^{2+}. The metals Mn^{2+} and Sr^{2+} also support cleavage, and to a lesser extent Co^{2+}, Pb^{2+}, and Zn^{2+}. An apparently unique feature of the HDV ribozyme is its remarkable tolerance for denaturants and high temperatures. Full cleavage activity is maintained in the presence of 20 M formamide or 10 M urea, and at temperatures up to 80°C, which speaks to the extremely stable secondary structure of this ribozyme. In the case of a *trans*-acting HDV ribozyme when targeted against a 13-mer substrate, the log of the cleavage rate increased linearly between pH 4.0 and 6.0 and showed a maximum cleavage rate at pH 7.

THE VARKUD SATELLITE (VS) RIBOZYME

The Varkud satellite (VS) ribozyme is a 154-nucleotide-long catalytic entity that is transcribed from a plasmid discovered in the mitochondria of certain strains of *Neurospora*. The VS ribozyme is the largest of the known nucleolytic ribozymes. The global structure has been determined by solution methods, particularly by fluorescence resonance energy transfer (FRET) method, which revealed a formal H-shape of the five helical segments. Binding of the substrate is determined primarily by tertiary interactions.

In Figure 14.6, cleavage sites are indicated by arrows. Solid line in Varkud satellite ribozyme denotes tertiary interactions between loops. Of hammerhead and hairpin ribozymes, *trans*-cleaving forms are depicted. The DNAzyme (E) consists of deoxyribonucleotides.

VIROIDS

Viroids are RNA molecules that infect plant cells as conventional viruses do, but are far smaller (one has only 246 nucleotides). They are naked, that is, they are not encased in a capsid. Some viroidlike molecules get into the cell as passengers inside a conventional plant virus. These are called virusoids or viroidlike satellite RNAs. In both cases, the molecules consist of single-stranded RNA whose ends are covalently bonded to form a circle. There are several regions where base-pairing occurs across adjacent portions of the molecule. New viroids and virusoids are synthesized by the host cell as long precursors in which the viroid structure is tandemly repeated. These repeats must be cut out and ligated to form the final product.

Most virusoids and at least one viroid are self-splicing, that is, they can cut themselves out of the precursor and ligate

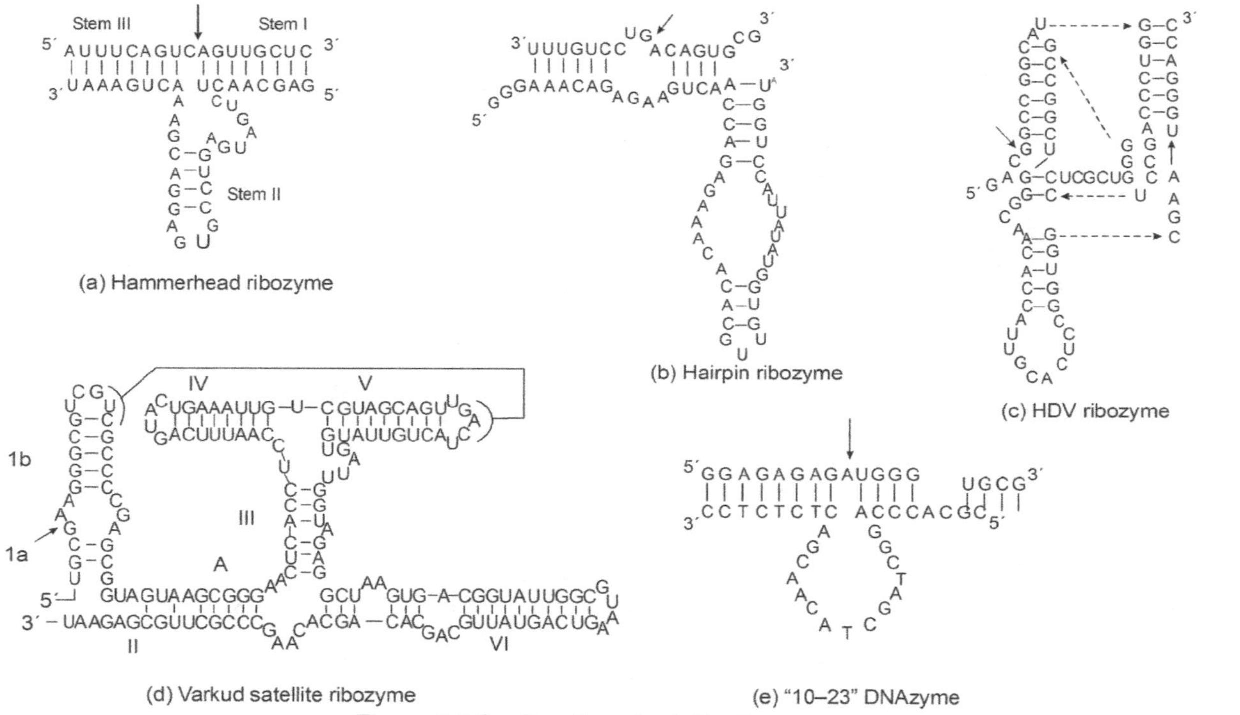

Figure 14.6 Small nucleolytic ribozymes

their ends without the aid of any host enzymes. Thus they represent another class of ribozyme. These ribozymes replicate themselves in copies attached to their own genome. The viroids then undergo self-cleavage, sending fragments to colonize other areas of the plant. The viroids harm the plants by rapidly proliferating and using nucleotide materials that the plant itself needs. Further damage is caused as the viroid bundles interfere with the plant's internal structures much like a tumour. They have no protein-producing capabilities.

Their cleavage from the mother strand is completely self-controlled and self-initiated. They are catalytic in their own replication and processing. A specific site in these viroids is responsible for the process of self-cleavage. The site is less than 30 nucleotides long and consists of three stems, coming of a central loop. This secondary structure, called a "hammerhead", is capable of cleaving very specific sequences of RNA in order to release viable daughter strands of RNA. Synthetic hammerheads, consisting of only 19 nucleotides, have already been produced, which can act as highly specific catalysts. Similar synthetic ribozymes are being designed to break up RNA viruses and RNA involved in the transcription and translation of mutant DNA.

ARTIFICIAL RIBOZYMES

Since the discovery of ribozymes that exist in living organisms, there has been interest in the study of new synthetic ribozymes made in the laboratory. For example, artificially produced self-cleaving RNAs that have good enzymatic activity have been produced. Zang and Breaker isolated self-cleaving RNAs by *in vitro* selection of RNAs originating from random-sequence RNAs. Some of the synthetic ribozymes that were produced had novel structures, while some were similar to the naturally occurring hammerhead ribozyme.

The techniques used to discover artificial ribozymes involve Darwinian evolution. This approach takes advantage of RNA's dual nature as a catalyst and as an informational polymer, making it easy to produce vast populations of RNA catalysts using polymerase enzymes. The ribozymes are mutated by reverse-transcribing them with reverse transcriptase into various cDNA and amplified with mutagenic PCR. The selection parameters in these experiments often differ.

One approach for selecting a ligase ribozyme involves using biotin tags, which are covalently linked to the substrate. If a molecule possesses the desired ligase activity, a streptavidin matrix can be used to recover the active molecules.

Leadzyme

Leadzyme is a small ribozyme that was artificially made using *in vitro* selection techniques. Leadzyme is able to cleave RNA in the presence of lead. The structure of leadzyme has been determined by X-ray crystallography. It has been proposed that a naturally occurring leadzyme occurs in the 5S rRNA and that this may be an important mechanism in lead toxicity.

Ligase Ribozymes

RNA ligase ribozymes were the first of several types of synthetic ribozymes produced by *in vitro* evolution and selection techniques. They are an important class of ribozymes because they catalyse the assembly of RNA fragments into RNA polymers by forming phosphodiester bond, a reaction required of all extant nucleic acid polymerases and thought to be required for any self-replicating molecule in a pre-biotic RNA World. The ligase ribozyme family includes type 1 ligases; type 2 ligases; from ligase to polymerase; the L1 ligase: a simple type 1 ligase. Michael Robertson and Andrew Elington evolved a ligase ribozyme that performs the desired 5´–3´ RNA assembly reaction, and called this the L1 ligase.

Allosteric Ribozymes

A number of enzymatic processes are subject to allosteric regulation, in which the catalytic activity of a key enzyme in a metabolic pathway is induced or inhibited upon binding of an allosteric effector molecule. Natural ribozymes and ribozymes that have been isolated by *in vitro* selection are not known to operate as true allosteric enzymes. The activity of natural catalytic RNAs, however, can be greatly affected by specific proteins that support proper structural folding. Small antibiotic molecules have been shown to inhibit as well as enhance the catalytic rates of certain ribozymes.

As with allosteric effectors of proteins, there is no true similarity between the effector molecule and the substrate of the ribozyme. Substrate and effector occupy different binding sites, yet conformational changes upon effector binding result in functional changes in the neighbouring catalytic domain. The characteristics of the conjoined aptamer–ribozyme complexes parallel the characteristics that are typically ascribed to natural allosteric enzymes that are made of protein. Allosteric ribozymes could be exploited as rate-regulated catalysts or as "molecular switches" that could be used to activate or deactivate various biological and chemical processes. Novel biosensors and controllable therapeutic ribozymes are significant examples.

THE "10-23" RNA-CLEAVING DNA ENZYME

The 10-23 DNA enzyme or deoxyribozyme was named from its origin as the 23rd clone of the 10th cycle of *in vitro* selection.

10-23 DNAzymes consist of a catalytic core of 15 nucleotides and two substrate-binding arms of variable length and sequence. One of the most active DNAzymes is the RNA-cleaving "10-23" deoxyribozyme. The 10-23 DNAzyme cleaves its RNA substrate in a reaction dependent on divalent ions to yield a 2′, 3′-cyclo phosphate and a free 5′-hydroxyl group. This enzyme has a number of applications both *in vitro* and *in vivo*. These include its ability to cleave almost any RNA sequence with high specificity provided it contains a purine–pyrimidine dinucleotide.

Kinetic efficiency The ability of the 10-23 deoxyribozyme to cleave purine–pyrimidine junctions meant that the AUG start codon of any gene could be used as a target. The kinetic efficiency of deoxyribozyme-catalysed cleavage varied substantially from one substrate sequence to the next. This sequence-dependent variability seemed to be closely associated with the thermodynamic stability of the enzyme–substrate heteroduplex.

Sequence specificity With the potential to bind any RNA sequence and cleave purine–pyrimidine junctions, the 10-23 DNA enzyme has unprecedented target site flexibility. In this system, deoxyribozymes have the advantage for two reasons: one, they were able to get closer to the junction because of their superior target flexibility and another, their activity was more easily perturbed by mismatch and hence were less reactive with the RNA sequence from the wild type *abl* gene.

Biological activity The ability of the 10-23 deoxyribozyme to specifically cleave RNA with high efficiency under simulated physiological conditions has useful biological application in a gene inactivation strategy. Another advantage of ribozymes consisting of DNA is the relative ease of synthesis and handling compared to RNA ribozymes. Moreover, DNA molecules are less susceptible to degradation by nucleases in tissue and cell culture, making them considerably more stable for exogenous delivery in biological applications.

Applications The 10-23 DNAzyme has been employed as an antiviral agent either directly targeting HIV-1 RNA or preventing virus entry by down regulation of the CCR5 co-receptor. Furthermore, DNAzymes have been developed to target cancer-related genes like the chimeric *BCR-ABL* oncogene or protein kinase C.

KINETICS OF RIBOZYMES

Under multiple turnover conditions, the substrate is in excess of the ribozyme so that the ribozyme can catalyse the cleavage of several substrate molecules. The catalytic rate constant or turnover number, k_{cat}, is a measure of the rate-limiting step, which can be the cleavage, conformational transitions of the ribozyme–substrate complex, or product release. When a ribozyme with six bases in each arm was used in kinetic analysis, k_{cat} and K_M were within the typical values of 1–2 min and 20–200 nM, respectively. Extension of the helical arms generally results in an increased stability and markedly decreased k_{cat}. Under single-turnover conditions, the ribozyme is in excess of the substrate, and such conditions are normally used for cleavage of long mRNA substrate. Cleavage rates are generally several orders lower in magnitude when compared with the rates obtained for short substrate. In the case of hammerhead and hairpin ribozymes, the presence of MgCl is essential for the cleavage, and the convincing evidence suggested that MgCl not only assists in RNA folding but also participates directly in the cleavage mechanism.

MECHANISM OF RIBOZYME-MEDIATED RNA CLEAVAGE

Ribozymes are RNA molecules which adopt three-dimensional structures that allow them to catalyse chemical reactions, usually with the participation of one or more metal ions. As such, they are generally regarded as a distinct class of metalloenzymes.

The formal mechanisms by which RNA cleavage occurs remain elusive although they can be categorized into two general classes based upon the nature of the products obtained. In most cases the reactions proceed as transesterifications with the attacking nucleophile being either a 2´-hydroxyl or a 3´-hydroxyl. With the hammerhead, hairpin, and hepatitis delta ribozymes, the 2´-hydroxyl adjacent to the scissile phosphodiester is employed as an internal nucleophile, while the group I introns employ an external guanosine co-factor. The group II introns employ an external-like mechanism, and although the 2´-hydroxyl recruited as a nucleophile is located outside of the catalytic pocket, it is technically part of the intronic sequence. In these latter cases, both the group I and group II introns as with the RNase P ribozyme, a water molecule can be employed as the attacking nucleophile such that cleavage occurs as a hydrolytic event rather than a transesterification.

Involvement of Metal Ions

All known catalytic RNAs have an absolute requirement for divalent metal ions, usually Mg^{2+}, but similar metals can often be substituted. It is well-established that metal cations bind to specific sites of an RNA sequence to stabilize specific structures. X-ray crystallographic analysis of tRNA reveals that Mg^{2+} can stabilize sharp turns of the polyanionic backbone, stabilize hairpin loops, and link two single-stranded regions.

The metal ions could serve to deprotonate a nucleophile, activate an electrophile, stabilize a transition state, or protonate a leaving group.

Catalytic RNAs utilize metals both for proper folding and for active-site chemistry. Some metal ions can perform both functions, while others can either promote proper folding or participate only in active-site chemistry. Protein enzymes also employ metal ions to cleave phosphodiester bonds. Protein residues bind and correctly orient the metal ions for catalysis; no protein side chain is implicated in the actual cleavage chemistry. In this case the first metal, metal A, facilitates the formation of the attacking OH^- with its lone pair of electrons oriented toward the phosphorus atom and situated for in-line attack. Metal B is hypothesized to assist departure of the 3´-leaving group by acting as a Lewis acid and also by interacting with the non-bridging oxygen to stabilize the transition state. Other protein enzymes that are thought to use divalent ions to facilitate phosphoryl-transfer reactions include alkaline phosphatase, the RNase H domain of HIV's reverse transcriptase, P1 nuclease, and phospholipase. It has been proposed that catalytic RNAs could use a similar "two-metal" mechanism.

Internal Transesterification

The reaction pathway of internal transesterification involves an adjacent 2´-OH group. The nucleophilic 2´-OH group attacks the phosphorus centre to produce a pentacoordinate transition state or intermediate. The pentacoordinate transition state breaks down to yield two fragments: a 2´, 3´-cyclic phosphate product and a 5´-hydroxyl product. This is the same general pathway that exists for the non-enzymatic, random cleavage of RNA linkages in alkali solutions and/or in the presence of metal ions. A metal coordinated to the non-bridging phosphate oxygen could also help to stabilize the pentacoordinate transition state. An acid catalyst could help to neutralize the developing charge on the 5´-leaving group, thus lowering the energy of the transition state. Any RNA linkage is susceptible to metal-catalysed cleavage, especially by metal ions whose attached water molecules have a low pK_a, such as Pb^{2+}. The folded structures of these RNAs contain metal-binding pockets that position the metal ions to facilitate cleavage of a specific phosphodiester bond

External Transesterification

External transesterification involves the recruitment of nucleophiles from outside the cleavage site. This is the strategy employed by all of the known large ribozymes, such as the group I and II introns and the RNA subunit of RNAse P. These RNAs catalyse splicing function as well, so they are inherently more complex. *In vivo*, proteins or other RNAs generally help to facilitate the cleavage, but these are not necessarily required. All of these ribozymes activate an exogenous nucleophile to facilitate cleavage, yielding a 3´-OH and a 5´-phosphate. Nucleophiles can include water or nucleotide hydroxy functionalities.

The reaction pathway involves nucleophilic attack on phosphorus to generate a pentacoordinate transition state or intermediate. Displacement and protonation of the 3´-oxy species gives rise to a 3´-hydroxy fragment and a 5´-phosphate mono- or diester. For instance, metal ions coordinating to non-bridging oxygen of the scissile phosphate moiety can activate the phosphorus atom toward attack by increasing the electrophilicity of the phosphorus centre.

In addition, bases can help to deprotonate the nucleophile. In the transition state, a coordinating metal ion can help to delocalize developing negative charge on the 3′-leaving group and Bronsted acids can protonate the 3′-oxyanion. Again, metal ions can help to stabilize the pentacoordinate transition state, and with all of these ribozymes, metal ions are essential.

APPLICATIONS

RIBOZYMES AS THERAPEUTIC AGENTS

Ribozymes have been proposed for treatment of a variety of diseases, including infectious diseases and cancer. To use ribozymes as therapeutic agents, several factors must be considered. The first is design in which the critical properties of ribozymes such as specificity and turnover should be considered. Specificity is the ability to cleave at unique sites, whereas turnover is the ability to cleave multiple-substrate strands. Both are affected by the lengths of binding arms I and III. If the lengths of the binding arms are too short, the rate of dissociation of the target from the ribozyme will exceed the rate of cleavage, resulting in poor turnover. If the lengths of the substrate-binding arms are too long, the target can be cleaved efficiently, but the ribozyme turnover will be slow because of the rate-limiting release of the cleavage products. In long substrates, the facilitators have the potential to preform the substrate for the ribozyme attack to increase both the rate of ribozyme–substrate association and the rate of the cleavage step.

Once designed, the second consideration is delivery. In order to deliver the ribozymes, they have to undergo chemical modification to enhance stability. This can assume a variety of forms: chemical modification by methylation of certain bases, substitution of sulphur for oxygen in the phosphate backbone (phosphorothioate linkages), or construction of whole sequences out of DNA instead of RNA. The last approach offers several advantages, including the following: 1) DNA is cheap and easy to synthesize; 2) DNA is more resistant to nucleases than is RNA and can be additionally protected by modification; 3) DNA arms hybridize less easily to the target substrate than do those of RNA, and they may have greater specificity because the hybridizing arms can be longer while still maintaining a competitive substrate dissociation rate; and 4) DNA ribozymes are less prone to self-hybridization than are RNA ribozymes and form inactive structures.

Methods to Deliver Ribozymes

One of the most challenging parts in ribozyme applications is the delivery of catalytic oligonucleotides into the desired subcellular compartment of the targeted cell population within an organism. Different cellular RNAs may be localized in different subcellular compartments as nucleus, nucleoli, or cytoplasm. The targeted delivery of ribozymes is thus a complex issue, and several methods have been described for various applications.

In general, two ways of transfer can be discerned. In exogenous delivery, pre-synthesized ribozymes are taken up by the target cells, whereas endogenous delivery denotes the enforced synthesis of catalytic RNAs directly within cells for gene-therapeutic approaches. The most frequently applied method to enhance uptake of antisense oligonucleotides and ribozymes by cells is the complexation with formulations of phospholipids containing positively charged headgroups. The cationic lipid reagents then assemble into oligonucleotide-containing liposomes that adsorb to cell membranes and are taken up by endocytosis. To facilitate the subsequent release of the nucleic acid from endosomes or lysosomes, helper lipids are often added that interfere with endosomal membranes.

Specific targeting of defined cells can be achieved by coupling the ribozyme to antibodies or ligands that are recognized by cellular receptors. Taken together, there is no simple rule for the uptake and subcellular localization of ribozymes in cells. Many parameters, including the delivery system and cell cycle, are involved in the outcome of a ribozyme-targeting approach. Surprisingly, simple injection of chemically stabilized ribozymes in a saline solution has led to efficient uptake and biological effects in *vivo.* A different strategy to deliver ribozymes to their areas of action is the expression of the ribozyme gene of vectors inside cells. Two challenges have to be met—efficient transfection of target cells and expression from suitable promoters.

A variety of viral transduction methods has been employed to introduce ribozyme-coding genes into cells. They randomly integrate the ribozyme-coding sequence into the host genome leading to sustained expression, but holding the risk of insertional mutagenesis. Adenoviral vectors have also been engineered to deliver ribozyme genes. Subsequent expression, however, is transient since their genetic material is not integrated into the host genome. In addition, adenoviruses are known to be immunogenic. An alternative may be the usage of adeno-associated viral vectors. Adeno-associated viruses are non-pathogenic, integration of the ribozyme gene into the host genome occurs at defined sites, and the transduction of non-dividing cells is also possible.

Ribozymes can either be transcribed from polymerase II to polymerase III promoters. Transcription by polymerase II leads to the addition of a 5´-cap and a poly(A) tail, resulting in enhanced stability and cytoplasmic localization. In addition, pol II promoters can be used for tissue-specific transcription. Ribozymes may also be inserted into RNAs with defined subcellular localizations including tRNAs, small nuclear RNAs like U1 and small nucleolar RNAs.

Advantages of Ribozyme in Clinical Application

Ribozymes, including hammerhead ribozymes, can be used in the study of gene function and gene therapy for diseases. Those ribozymes which are designed to target specific gene transcripts (hnRNA, mRNA) would bind and cleave the substrate RNA *in vitro*, or *in vivo*. The overall effect is the down-regulation of the expression on that target gene, similar to the gene knock-out. The potential advantage of inactivation by ribozyme over gene knock-out is that, the binding and catalysis efficiency of ribozymes on the target RNA can be flexible through the design and selection of different kinds of ribozymes, while gene knock-out decreases the gene expression to zero.

The property of ribozymes makes them suitable in the research of genetics and developmental biology. The other important application of ribozymes is to develop new drugs and protocols for gene therapy. Ribozymes have been proposed as potentially efficacious therapeutics for treatment of variety of infectious agents. Even targeting of prokaryotic pathogens is possible through the use of phage or other delivery systems. Vectors expressing ribozymes have been used for cleaving essential gene products and work as tentative therapeutics on diseases, such as HIV infection.

To date, the achievement of ribozymes for human clinical trials has centred on viral pathogens, and ribozymes appear to offer some unique advantages when compared to the traditional antisense protein expression strategies. By protecting cells against viral cytopatheic effects, ribozymes may promote survival of populations of functional transduced cells *in vivo*, leading to increased therapeutic effects over time. By producing only therapeutic RNAs, ribozymes are likely to avoid entirely or at least largely the problems of immune eradication of transduced cells or production of autoimmune antibodies. Ribozyme therapy may also be applied on DNA virus infections.

By inhibiting expression of the early genes of many DNA viruses (e.g. herpes simplex virus), it should be possible to block virus replication and the production of progeny virions. This suggests a possible role of ribozymes as antineoplastic agents by targeting the replication of the oncogenic DNA viruses.

The gradual maturation of ribozyme technology from the bench to clinical application involves several major challenges, many of which still need to be resolved. These include extra- and intracellular stability of the ribozyme, delivery of ribozymes to target cells, target accessibility, co-localization of ribozyme and target within cells, and optimal catalytic activity and specificity of the ribozyme. Despite their enzymatic activity, ribozymes must be delivered to target cells in amounts sufficient to affect both a significant proportion of the cell population and the target mRNA.

Although therapeutic application of ribozymes may be achieved by exogenous delivery of chemically produced ribozyme, for example, ribozyme complexed to a cationic lipid which is used in chronic disease or in antiviral therapeutic settings, they may require permanent availability to be of therapeutic benefit. This can be achieved by transfer of genes coding for ribozymes using viral vector systems. The preclinical and clinical application of ribozyme-based gene therapy has been primarily focused on AIDS, cancer, and other viral infections.

New Therapeutic Approach

A first step in the development of new therapeutic approaches is the identification and validation of suitable target molecules. Ribozyme libraries with randomized substrate recognition arms have been employed for this purpose.

In one study, an inverse genomic approach based on a randomized hairpin ribozyme library was applied to identify genes regulating the expression of *BRCA 1* which is down-regulated in many cases of breast and ovarian cancer. The ribozyme gene library was introduced into human ovarian cancer-derived cells that stably expressed a reporter gene under control of the *BRCA* 1 promoter. Treated cells with an increased reporter gene expression were selected and the dominant negative transcriptional regulator Id4 was identified as a modulator of *BRCA* 1 expression. The identification of genes involved in regulation of tumour-relevant genes might facilitate the development of new therapeutic options.

Cells expressing ribozymes that enhance the invasive properties of NIH 3T3 fibroblasts were selected in a filter-based invasion assay. Genes that were likely to enhance the metastatic properties of tumour cells were then identified by sequence analysis of the ribozyme genes. This approach has subsequently been transferred to an *in vivo* model. Weakly metastatic melanoma cells that had been treated with a ribozyme library were injected intravenously into mice. Ribozymes were isolated from cells converted into strongly metastatic cells and sequenced to identify the targeted genes relevant to metastasis. Again, several candidates were found which are involved in the complex mechanism of tumour metastasis and might be new targets for cancer therapy.

These examples and several further studies demonstrate that ribozymes are valuable tools to identify new targets for therapeutic approaches. Ribozyme approaches have also widely been used in cancer research. Telomerase is an interesting target for the development of new anti-cancer drugs.

Clinical Trials with Ribozymes

Several clinical studies with *trans*-cleaving ribozymes have been carried out to treat infectious diseases or cancer.

Both strategies, i.e., endogenous expression of ribozymes or exogenous delivery of chemically pre-synthesized ribozymes, were employed (Figure 14.7). Tumours are rapidly growing tissues with a high demand for oxygen and nutrition. Therefore, the formation of new blood vessels, a process called angiogenesis, is required for sustained tumour growth. The furthest developed, chemically synthesized ribozyme "ANGIOZYME" was designed to inhibit tumour angiogenesis. It is directed against the receptor, tyrosine kinase Flt-1 (VEGF-R1), the high-affinity receptor for the vascular endothelial growth factor (VEGF).

A phase I clinical trial to test ANGIOZYME's safety, tolerability and pharmacokinetics in healthy volunteers began in late 1998. The ribozyme was found to be maintained in the plasma for several hours following single-intravenous infusion. The ribozyme was then tested in a multidose phase I/II trial. Again, it was well tolerated by all subjects and has subsequently been investigated in phase II trials in patients with breast cancer as well as in combination with chemotherapy in humans with metastatic colorectal carcinoma. The effect of ANGIOZYME could not be separated from the chemotherapeutic drug, but it may result in decreased level of the VEGF receptor and improved clinical outcome. Although any kind of virus can be treated with ribozymes by cleaving essential RNAs, RNA viruses are particularly suited to be addressed by this class of therapeutics.

HERZYME is the ribozyme that was studied in a clinical trial by RPI. It is directed against the mRNA of the human epidermal growth factor receptor-2 (HER-2), which is over-expressed in aggressive breast cancer. This ribozyme was based on the chemistry developed by Zinnen *et al.*

In a phase I trial it was found to be well tolerated with no significant systemic adverse events observed according to the company's information. All these clinical studies demonstrated that the ribozymes were well tolerated and that their application in humans is safe. The efficacy of ribozymes to prevent tumour growth or virus replication has not yet been proven convincingly. Consequently, the above-mentioned company RPI as a major player in clinical applications of ribozymes now focuses on RNA-interference approaches under the name **Sirna therapeutics**.

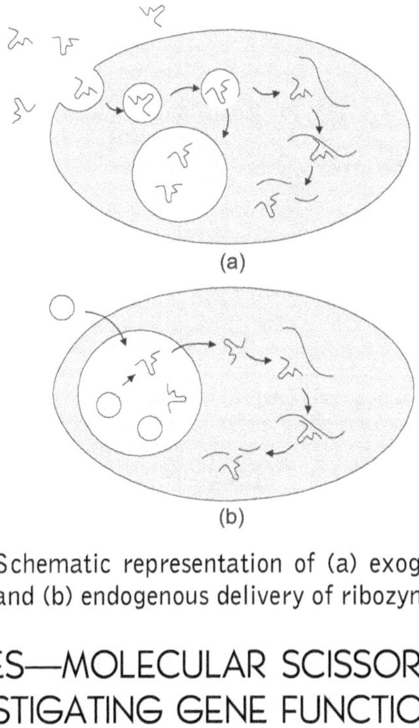

Figure 14.7 Schematic representation of (a) exogenous delivery and (b) endogenous delivery of ribozymes

RIBOZYMES—MOLECULAR SCISSORS FOR INVESTIGATING GENE FUNCTION

Ribozymes are molecular scissors that cut RNA, the molecular messages given by genes in order to produce proteins. These molecular scissors provide a very useful means of studying gene function since by cutting the RNA with a ribozyme a gene can be effectively turned off (Figure 14.8). However, using ribozymes as a molecular scissors is not as easy as it sounds.

The biggest problem is that the cell is producing a large number of RNAs from a huge number of different genes. When the ribozyme is introduced into the cell, the problem of specificity is something that has to be considered very carefully prior to introducing the ribozyme into the cell otherwise anything that then happens in the cell will be too difficult to interpret.

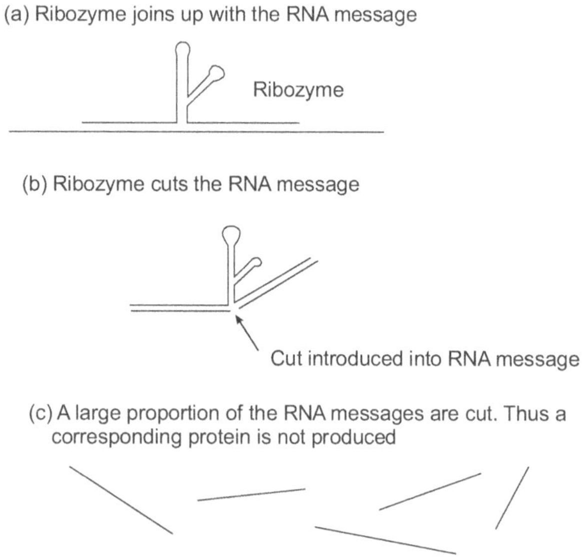

Figure 14.8 Ribozymes as molecular scissors

RIBOAPTDB—A COMPREHENSIVE DATABASE OF RIBOZYMES AND APTAMERS

Catalytic RNA molecules are called ribozymes. The aptamers are DNA or RNA molecules, possessing desirable affinity, selected by Systemic Evolution of Ligands by Exponential (SELEX) enrichment method. This SELEX method is an *in vitro* iterative process that isolates binding aptamers from the random pool and amplifies each sequence by the polymerase chain reaction after each round of isolation. The selected oligonucleotide sequences (~200 bp in length) have the ability to recognize specific ligands by forming binding pockets and can bind to nucleic acids, proteins or small organic, inorganic chemical compounds and even small organisms like viruses. Aptamers are a promising class of compounds, both for target validation and therapy. As designer drugs, they exhibit high specificity, high affinity, and modifiable bioavailability. The ability to generate inhibitors with such properties against a variety of target proteins will be invaluable as the human genome and proteome are deciphered.

Availability and Requirements

RiboaptDB was created and is maintained in the Department of Biological Sciences at the University of Southern Mississippi. It is publicly available at the http://mfgn.usm.edu/ebl/riboapt/.

The RiboaptDB is not only extremely useful for identifying available aptamers and artificial ribozymes but also for acquiring information about *in vitro* selection experiments such as the type of the nucleic acid, type of the target and conditions of the experiment as a whole and for better understanding the distribution of functional nucleic acids in the given sequence space. Like other types of sequences, the amount of sequences generated by *in vitro* selection experiments has been accumulating exponentially. The sheer number and diversity of selection experiments has risen to the point where it is now essential to gather all the sequence data into a comprehensive, continuously updated database. The general sequence databases like GenBank, EMBL and DDBJ do not maintain the complete collection of artificial nucleic acid sequences like aptamer and ribozyme. Another database, "Aptamer database" also contains lot of information on this type of data but not regularly updating with new data.

Structure and Implementation

The design of the RiboaptDB database schema follows the three-level schema architecture. The database has been implemented using an open source relational database MySQL server version 4.1. The user-friendly web interface has been developed by using PHP language. The web server is an open source Apache web server version 2.0. The "sequence" table is the key table in the database to which all other tables are related directly or indirectly. This table contains the sequence ID and relates directly with its child tables, "aptamer" and "ribozyme", which contains the corresponding sequence information.

The other important tables in the database are "publication" and "experiment" which store the citation information like title, journal name, authors, pubmed ID and experiment details like template type and experiment conditions respectively.

The target-specific information, the target name and its category (organic, inorganic, nucleic, peptide, protein and other) are obtained from the "target" table. If any information about non-canonical base pair is available, it can be retrieved through the "non-canonical" table.

Content

RiboaptDB is a relatively small database but is, nonetheless, essentially complete. The data was sourced from a previous compilation and exhaustive searching of the primary literature. The current size of the database is 4212 sequences from 423 citations. In this, there are 370 artificial ribozyme sequences and 3842 aptamer sequences in the total 4212 sequences. The database is updated every month as new literature comes on aptamers and artificial ribozyme sequences (Figure 14.9). The initial collection of data is done through searching the NCBI-Pubmed for the literature with keywords like "artificial ribozymes", "ribozyme", "aptamers", "SELEX," etc. The usefulness of a database is governed by the accuracy of the data it contains. The data in this database is compiled manually from previously published, peer-reviewed articles, and verified.

Utility and Discussion

RiboaptDB provides users with an easy-to-use web interface with flexibility to select either ribozyme or aptamer sequences to browse the corresponding information. Beginning at the welcome page the user can navigate via the top menu or the browse database tables on the side menu. A brief description about the navigation is given below.

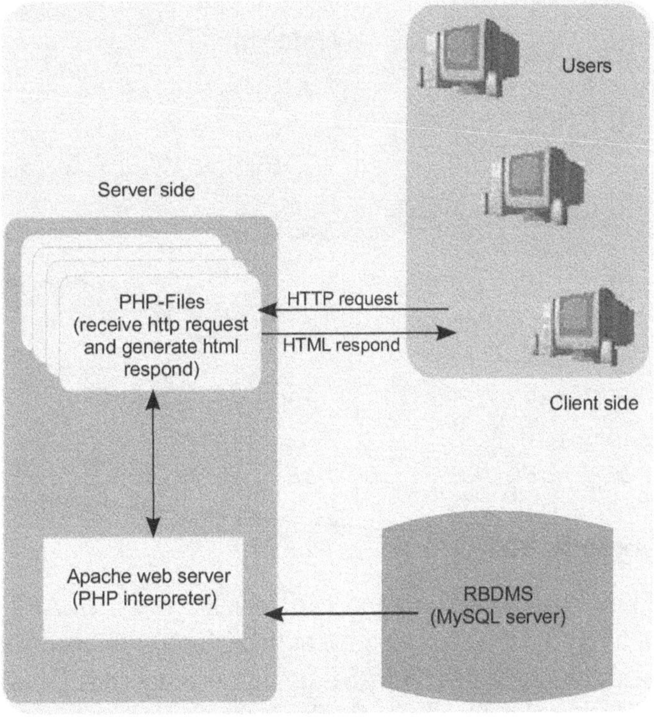

Figure 14.9 RiboaptDB Three-tier architecture

Search

The general complete search option provides an interface for a variety of queries to the database. It can be used to search the database for sequence, experiment, target, author, publication and non-canonical along with either ribozyme or aptamer or both and also with either natural or artificial type of sequences.

Local BLAST

The Local BLAST option can be used to do BLAST search against the local archived data to perform sequence-similarity searches using the BLAST family of programs. This will be useful to the user to know the most similar sequences to the submitted sequences and also to get further information about its target and the experiment details.

Home | About | Search | Blast | Submit Data | Contact

Experiment List

exp id	2001
template	GCCTGTTGTGAGCCTCCTCTCGAA-N40-TTGAGCGTTTA TTCTTGTCTCCC
tem desc	DNA N40
nucleic type	dna
exp cond	selection buffer (30 mM Tris-HCl pH 8.5, 40 mM $(NH_4)_2SO_4$, 20 mM $MgCl_2$, 0.01% 2-mercaptoethanol)
target id	4001

Sequence list matching to this experiment: 1300, 1296, 1299, 1298, 1297, 1294, 1295, 1293, 1292, 1291, 1290, 1289, 1288, 1287, 1286, 1285, 1284, 1283, 1282, 1281, 1280, 1279, 1278, 1277, 1276, 1275, 1274, 1273, 1272, 1271, 1270, 1269, 1268, 1301, 1302, 1303, 1304,

Submit Data

It facilitates online sequence submission to the database. It allows users to fill in a form containing new sequence information along with user details. Related information can also be submitted through uploading a text file. The data which is then saved into a directory on the server side and an e-mail is sent automatically to the curator who then checks the data to make sure there are no errors and then the information is loaded into the database automatically.

Alternate to the general search option on the top menu, there is a search option on the side menu of the home page to search the whole database on a specific keyword. Also, specific table search is available on the side menu of each related page.

The user can also retrieve the selected sequences into a text file for further studies. The idea behind the combining of ribozymes and aptamers data into one database is increasing the chance of generating ribozymes with modified and novel properties.

One example is combining both the "target identification" of aptamer and "catalytic activity" of ribozymes into a commercial "riboswitch" application.

Results

The comprehensive sequence information on aptamers and ribozymes that have been generated by *in vitro* selection methods are included in this RiboaptDB database. Such types of unnatural data generated by *in vitro* methods are not available in the public "natural" sequence databases such as GenBank and EMBL. The amount of sequence data generated by *in vitro* selection experiments has been accumulating exponentially. There are 370 artificial ribozyme sequences and 3842 aptamer sequences in the total 4212 sequences from 423 citations in this RiboaptDB. We included general search feature, individual feature wise search, user submission form for new data through online, and also local BLAST search.

This database, besides serving as a storehouse of sequences that may have diagnostic or therapeutic utility in medicine, provides valuable information for computational and theoretical biologists. The RiboaptDB is extremely useful for garnering information about *in vitro* selection experiments as a whole and for better understanding the distribution of functional nucleic acids in sequence space. The database is updated regularly and is publicly available at http://mfgn.usm.edu/ebl/riboapt/.

REVIEW QUESTIONS

1. What are ribozymes?
2. Enlist the classes of ribozymes and bring out their functions.
3. Write notes on:
   i. RNase P
   ii. Hepatitis delta virus ribozyme
   iii. The Varkud Satellite (VS) ribozyme
   iv. Artificial ribozymes
   v. Allosteric ribozymes
4. Discuss how ribozymes can be used as therapeutic agents.
5. Describe how ribozymes are used for target identification.
6. Discuss RiboaptDB (Database of Ribozymes and Aptamers).

15

ABZYMES

INTRODUCTION

Abzymes are antibodies, which are capable of catalysing a chemical reaction. In the early 1940s Linus Pauling first proposed the concept that antibodies may catalyse chemical reactions in the early 40's. In his words, "One way to do this is to prepare an antibody to a haptenic group which resembles the transition state of a given reaction."

He suggested that an enzyme lowers the energy barriers of a reaction by stabilizing preferentially the transition state of the substrate during the reaction rather than the substrate in its ground state. Twenty years later, William Jencks proposed a strategy to prepare new biocatalysts. This strategy is based on the production of antibodies directed against stable molecules resembling the transition-state structure of a specific chemical transformation.

The technique of developing monoclonal antibody by Kohler and Milstein provided an excellent method for producing monoclonal antibodies with a single antigenic-defined specificity.

Bernard Green and his group reported for the first time that antibodies may act as catalysts. Californian laboratories of Richard Lerner and Peter Schultz demonstrated that tailor-made antibodies could catalyse specifically chemical reactions. Strategies like hapten design, immunization or methods for screening and selection were developed to improve the production and efficiency of abzymes. An **abzyme** (from antibody and enzyme), also called catmab (catalytic monoclonal antibody), is a monoclonal antibody with catalytic activity. Molecules which are modified to gain new catalytic activity are called **synzymes**. Abzymes are usually artificial constructs, but are also found in normal humans (anti-vasoactive intestinal peptide autoantibodies) and in patients with autoimmune diseases such as systemic lupus erythematosus, where they can bind to and hydrolyse DNA. Abzymes are potential tools in biotechnology, e.g. to perform specific actions on DNA.

KINDS OF ABZYMES

1. Antibodies directed against transition-state analogues
2. The anti-idiotypic approach—from enzymes to abzymes
3. Naturally occurring abzymes
4. Metallo- and haemo-abzymes

Antibodies Directed Against Transition State Analogues—the Substrate-based Approach

Based on the concept of L. Pauling, first abzymes developed by W. Jencks was produced by mimicking enzymatic reactions. This method consisted of immunizing mice with stable molecules resembling the transition-state structure of the chemical reaction that was expected to be catalysed.

The first successful results were obtained for hydrolysis of ester and carbonate bonds by the groups of R. Lerner and P. Schultz. They used organophosphate compounds as immunogens that were thought to mimic the tetrahydral structure of the carbon atom during the hydrolytic process. For example, the 28B4 abzyme catalyses periodate oxidation of *p*-nitrotoluene methyl sulphide to sulphoxide, as shown below, where electrons from the sulphur atom are transferred to the more electronegative oxygen atom.

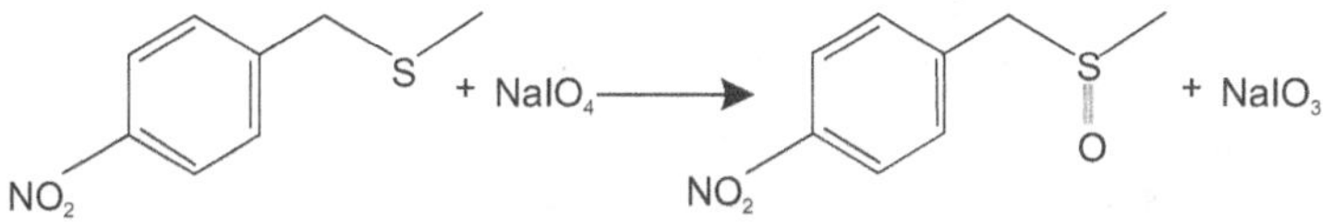

p-nitrotoluene methyl sulphide *p*-nitrotoluene methyl sulphoxide

The rate of this reaction is promoted by enzyme catalysts that stabilize the transition state of this reaction, thereby decreasing the activation energy and allowing for more rapid conversion of substrate to product. In this case, the transition state is thought to involve a transient positive charge on the sulphur atom and a double-negative charge on the periodate ion as shown in Figure 15.1.

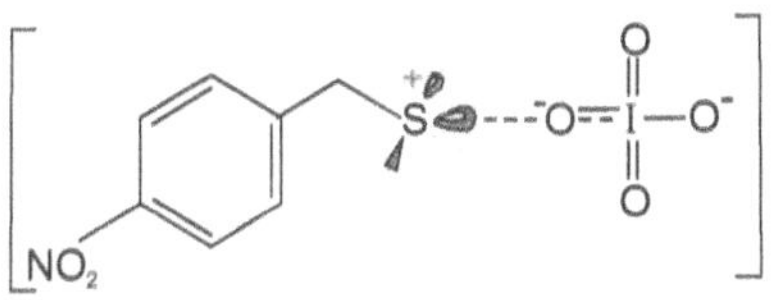

Figure 15.1 Transition state

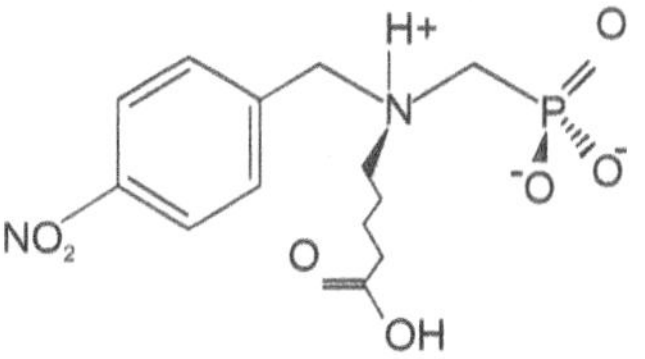

Figure 15.2 Hapten

In order to generate abzymes complementary in structure to this transition state, mice were immunized with an aminophosphonic acid hapten, as shown in Figure 15.2. Obviously, its structure mirrors the structure and electrostatic properties of the sulphoxide transition state. Of the hapten-binding monoclonal antibodies produced with this hapten, many were found to catalyse sulphide oxidation but with a wide range of binding affinities and catalytic efficiencies. In particular, abzyme 28B4 binds hapten with high affinity (K_d=52 nM) and exhibits a correspondingly high degree of catalytic efficiency (k_3/K_M = 190,000 $M^{-1}s^{-1}$).

Elucidation of the molecular structure of abzyme 28B4 bound to the hapten reveals much about the nature of its catalytic action. Highly specific structural and electrostatic interactions create a remarkable degree of structural complementarity between the antigen-binding site and the sulphoxide transition-state analogue as illustrated in the following series of three-dimensional views of the antibody–hapten complex.

Since that time, more than 70 different chemical reactions were described to be catalysed by antibodies. These reactions include hydrolysis of chemical bonds (esters, carbonates, amides, phosphates), stereospecific synthesis of compounds (esters, amides, Diels–Alder addition) as well as reactions of isomerization, decarboxylation, oxidation and reduction.

The Anti-idiotypic Approach—from Enzymes to Abzymes

In 1974, Niels Jerne advanced the theory that regarded the immune system as a network of interacting idiotypes. A major postulate of the idiotypic network was that for each immunoglobulin (Ab1) generated against an antigenic determinant, there existed a complementary antibody (Ab2) directed against the idiotypic determinants of Ab1.

When the idiotypic determinants superimpose with the binding site of Ab1, some of the Ab2 may mimic the antigen's determinants and are designed as "internal images" of the original antigen.

A second set of antibodies (Ab2) is produced against the Ab1 combining site. Among these second-generation, or anti-idiotypic antibodies, some of them may represent a structural internal image of the original enzymatic site, and in some cases may exhibit a catalytic activity.

Using this approach, antibodies with esterase and amidase activities were characterized. These abzymes usually bear efficient catalytic activities, with a relaxed specificity when compared with the model enzymes.

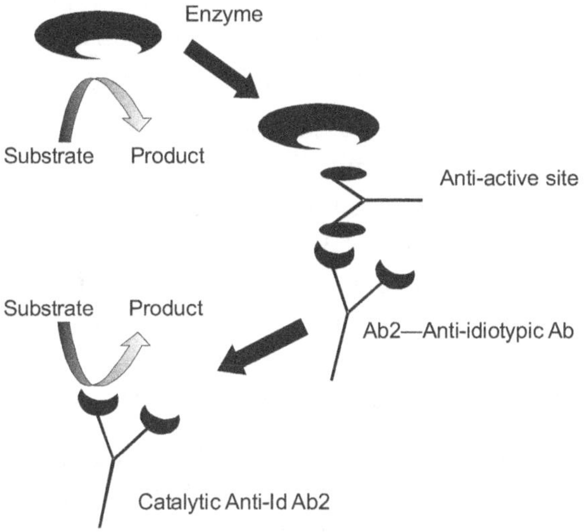

Figure 15.3 Anti-idiotypic approach

Figure 15.3 shows the experimental process, which is used to produce abzymes. For producing catalytic antibodies, a first antibody Ab1 is raised that recognizes the active site of an enzyme so that the combining site of Ab1 has structural features complementary to those of the enzyme.

Naturally Occurring Abzymes

Antibodies with catalytic activities were isolated from sera of patients with different diseases. The first natural abzymes were obtained by means of antibody purification from human serum.

Antibodies with protease activity against vasoactive intestinal peptide (VIP) were first isolated in the serum of patients with asthma. Surprisingly, antibodies exhibiting the same catalytic activity were obtained by immunizing mice with VIP in its ground state. These monoclonal antibodies have allowed to demonstrate that the catalytic activity is borne by the isolated light chain of the antibody molecule.

This VIPase activity was also shown to be present on Bence Jones proteins, that are monoclonal human light chains found in urine of patients with multiple myeloma.

Other protease activities were characterized for cleaving of thyroglobulin in the serum of patients with Hashimoto thyroiditis, or for hydrolysing factor VIII in haemophilia patients infused with homologous factor VII. DNA-hydrolysing auto-antibodies were also isolated from the sera of patients with systemic lupus erythematosus or rheumatoid arthritis. The DNA-hydrolysing activity could be correlated to the presence of high levels of antitopoisomerase I antibodies in the sera of patients.

Through the use of protein engineering, abzyme catalysis can be improved even further, perhaps even to surpass the activity of natural enzymes. Molecular biologists have developed methods to clone the array of genes that encode IgG molecules. The millions of gene products from an immunized animal are screened for the production of antibodies with desirable catalytic activities. Once candidates are isolated, these so-called "recombinant antibodies" can be produced in bacteria in large amounts. In this way, an antibody gene can be "immortalized" for unlimited study.

Metallo- and Haemo-Abzymes

The idea to produce antibodies that mimic metalloenzymes or haemoenzymes was early advanced to generate antibodies able to cleave stable chemical bonds or to catalyse peroxidase-like reactions.

In 1989, an antibody catalysing the hydrolysis of a glycine–phenylalanine bond was obtained by using the antigen as hapten that not only induces the selective recognition of the gly–phe sequences, but also induces the generation of residues that are able to complex a metal ion. When complexed with zinc, this antibody exhibits a good catalytic activity.

The generation of antibodies directed against different metalloporphyrins has allowed demonstrating that antibodies may mimic metalloproteins. The haemo-abzymes obtained are able to catalyse peroxidase-like activities and could be very interesting tools for the enantio-selective oxidation of molecules.

STRATEGY FOR ABZYME PRODUCTION

Haldane Pauling Hypothesis of Transition State Stabilization as a Primary Effector of Catalysis

A major focus was on the development of a strategy to produce antibodies that efficiently form and break **carbon–carbon bonds.** Much of this work is based on the chemistry of enamines and the development of antibodies that use covalent catalysis. The specific reactions with prospects for abzyme production are aldol and a variety of decarboxylation reactions.

Designing Abzyme to Catalyse Aldol Reaction

Reactive immunization approach was applied for the development of the aldol-reaction-catalysing abzyme. In this, 1,3-dicarbonyl hapten was designed to act as a chemical and entropic trap to select antibodies with a reactive lysine residue in their active site. Antibodies displaying the appropriate chemical reactivity, transform the hapten to form a covalently bound enaminone. The reaction mechanism that results in the formation of the hapten–antibody complex can then be recruited to catalyse aldol reactions. In this way, antibodies were programmed to mimic natural class I aldolase enzymes into antibodies.

Product of these Studies

The product of these studies is antibody 38C2, the world's first commercially available catalytic antibodies.

38C2 has been shown to catalyse the aldol addition of a wide variety of aliphatic open chain and aliphatic cyclic ketones to various aromatic and aliphatic aldehydes. More than 100 different substrate combinations, cross aldol and also intramolecular aldol reactions have been identified. Solution of the X-ray crystal structure of the antibody with the Wilson laboratory has confirmed the approach to mechanistic programming of biocatalysts.

The use of 38C2 has been demonstrated as an efficient catalyst for the retro-aldol reaction, for allowing the kinetic resolution of racemic secondary aldols. By using both the forward (or synthetic) aldol and retro-aldol reactions, both aldol enantiomers become accessible.

INDUSTRIAL AND MEDICAL APPLICATIONS

Two medical targets readily offer themselves to antibody applications. The first is in the clearance of toxic substances from the bloodstream by specific abzyme hydrolysis, e.g. cocaine hydrolysis. And second is the conversion of a prodrug into active agent in a process that is designed to improve either its pharmacokinetic, pharmacological, or toxicological profile.

13 years later Richard Lerner and Peter Schultz introduced catalytic antibodies simultaneously. The first antibody to be commercialized is the abzyme with an aldolase activity developed in the Lerner's group and sold by Aldrich chemical group. A handful of firms, notably Med-Immune, Advanced Biotech Ltd and Prolifaron are developing catalytic antibodies for commercialization, along with researchers at The Scripps Research Institute, Hebrew University of Jerusalem (Israel) and Columbia University (New York City).

The use of abzymes at industrial scale for specific synthesis of molecules is still at the laboratory step, even if some firms like Novartis (Switzerland) or Bristol-Myer Squibb (USA) have shown their interest for using the aldolase abzyme, produced in the Lerner's group, for synthesis of Epothilone A, a new anti-cancer compound.

Antigen-Directed Abzyme Prodrug Therapy (ADAPT)

In an attempt to correct the immunostimulatory effects of ADAPT's bacterial enzyme component, ANTIBODY-directed abzyme prodrug therapy has been proposed.

This therapy strategy will employ catalytic antibodies instead of the antibody–enzyme conjugate. ADAPT offers the similar advantage that there is no equivalent molecule in the organism thus avoiding incidental activation by circulating endogenous enzymes (Tellier, 2002). Additionally, humanized prodrug-activating abzymes can be prepared to lessen the immunogenic response (Dubowchik and Walter, 2002 and Tellier, 2002), thereby allowing for the repeated administration of the drug without the need for concomitant administration of immunosuppressant drugs (Tellier, 2002). Furthermore, several abzymes have been found to catalyse reactions not known to be carried out in natural systems. It would be medicinally beneficial if a highly specific drug-releasing abzyme could be prepared to exploit this unique ability (Dubowchik and Walter, 2002).

Different laboratories have also proposed to use catalytic antibodies for medical applications. One application could concern the use of hydrolytic properties of abzymes to activate prodrugs. By targeting this activity in the vicinity to tumour cells, prodrugs could be transformed into cytotoxic compounds directly on tumour cells.

This anti-cancer therapy is designed as Antibody-directed Abzyme Prodrug Therapy (ADAPT) (Figure 15.4). Antibody-directed Enzyme Prodrug therapy (ADEPT) is a cancer therapeutic programme pioneered in the 1980s. It builds on the discovery that tumours have specific antigens associated with their surface and that these are capable of eliciting an immune response. An effector function, such as an enzyme, can be concentrated in the vicinity of the tumour by simply conjugating it to antibodies which recognize the antibody-specific surface epitopes.

Administration of a cytotoxic agent in the form of a prodrug of greatly diminished activity reduces the adverse effects that arise from destruction of fast dividing, healthy cells in the body. If the effector function, typically an enzyme, can convert the prodrug into the parent cytotoxic agent at the locus of the tumour, then a high concentration of that drug will be delivered selectively to primary and secondary tumour cells.

Problems have been encountered with this system because the enzyme component, usually of bacterial origin, is foreign to the mammalian host which brings into operation its immune system to try to neutralize the enzyme. The objective of our

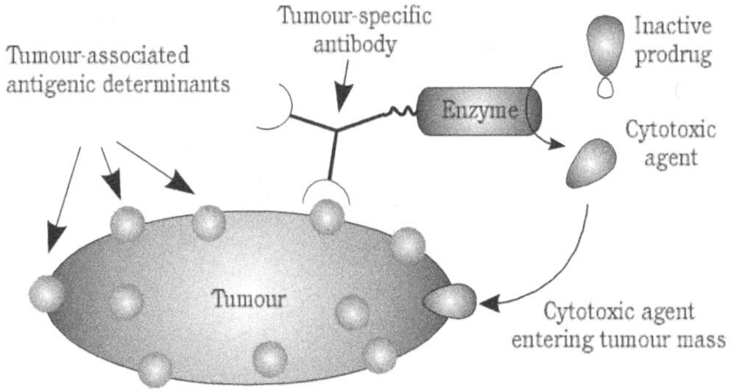

Figure 15.4 ADEPT—Mab-enzyme conjugates targeted at the tumour surfaces convert less toxic prodrugs into more active cytotoxic agents

research is the production of a non-immunogenic biocatalyst, specifically a humanizable catalytic antibody, capable of activating a prodrug of choice to release a toxic, aryl nitrogen mustard agent.

An antibody-directed abzyme prodrug therapy (ADAPT) overcomes the problems of ADEPT and is much superior in its action.

Catalytic Antibodies—Targeting Cancer with Prodrugs

Not only can catalytic antibodies be used to synthesize anti-cancer drugs, they can be used to deliver them in a highly specific fashion to the cancerous cell itself. Chemotherapeutic regimes are typically limited by non-specific toxicity. To solve this problem, a novel and broadly applicable drug masking chemistry that operates in conjunction with the unique, broad-scope catalytic antibody 38C2 has been developed. This masking chemistry is applicable to a wide range of drugs since it is compatible with virtually any heteroatom. It has been demonstrated that generic drug masking groups can be selectively removed by sequential retro-aldol-retro-Michael reactions catalysed by antibody 38C2 (Figure 15.5).

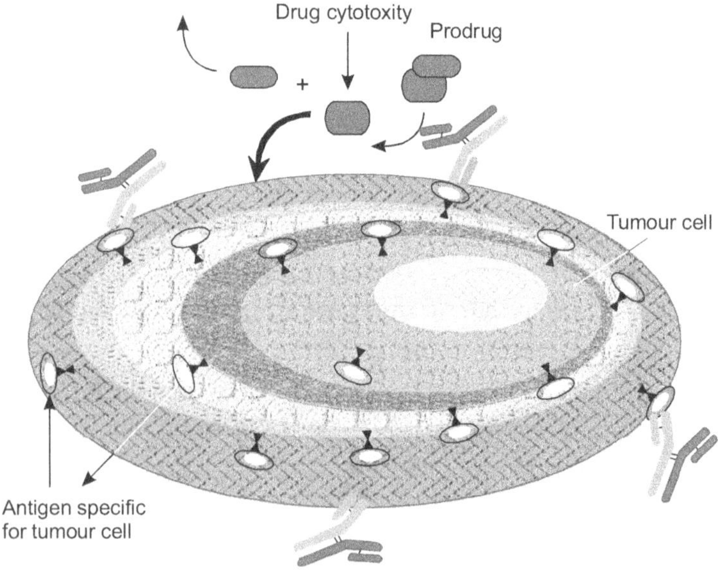

Figure 15.5 Decarboxylation prodrug activation via a tandem retro-aldol-retro-Michael reaction catalysed by antibody 38C2

This reaction cascade is not catalysed by any known natural enzyme. Application of this masking chemistry to the anti-cancer drugs doxorubicin and camptothecin produced prodrugs with substantially reduced toxicity. These prodrugs are selectively unmasked by the catalytic antibody when it is applied at therapeutically relevant concentrations.

Though the first obtained abzymes were characterized by low efficiency, improvements were made for efficient abzyme production. Improvements include:

- Better hapten design
- Improvement in the strategies used for immunization

- Site-directed modification of antibodies (mutagenesis)
- Improvement in the methods developed for screening (cat ELISA) and selection (phage display)

These improvements allow to obtain catalysts with efficiency sufficient for an industrial or a medical application.

On the other hand the resolution of an increasing number of tri-dimensional structures of abzymes brings a better understanding of the appearance and evolution of the catalytic function, not only in antibody-binding sites but also in enzyme active sites.

Another example of medical application concerns antibodies that specifically hydrolyse cocaine. A commercialization agreement between Columbia University and Ixsys Inc. could allow to use abzymes for treating cocaine overdose and addiction.

A third example is developed in Israel where Advanced Biotech is producing catalytic antibodies for the treatment of gram-negative sepsis, targeting the endotoxin LPS.

HIV Treatment

In a June 2008 issue of Journal of Autoimmunity Reviews, researchers S. Planque Sudhir Paul and Yasuhiro Nishiyama of the University of Texas Medical School of Houston have engineered an abzyme that degrades the superantigenic region of the gp120 CD4-binding site. This is one part of the HIV virus outer coating that does not change, because it is the attachment point to T lymphocytes, the key cell in cell-mediated immunity. Once compromised, patients produce antibodies to the more changeable parts of the viral coat. The antibodies are ineffective given the virus' ability to change that coat rapidly. Because this protein, gp120, is necessary for the HIV virus to attach, it does not change, and is vulnerable across the entire range of the HIV variant population.

The abzyme does more than bind to the site, rendering the HIV virus inert, it actually destroys the site, then can attach to other viruses. A single abzyme can destroy thousands of HIV viruses. Human clinical trials will be the next step in treatment and perhaps even in the production of preventive vaccines and microbicides.

Finally all researches aimed to produce antibodies with a sequence-specific protease activity could open new ways for anti-virus therapy and for the conception of new vacancies.

REVIEW QUESTIONS

1. What are abzymes?
2. Discuss the types of abzymes.
3. Write an account of enzymes acting as abzymes.
4. Write a note on naturally occurring abzymes.
5. Discuss how naturally occurring abzymes help in treatment of diseases.
6. Discuss the strategies of abzyme production.
7. Differentiate and discuss ADAPT and ADEPT.
8. How is abzyme used in the treatment of HIV?
9. Discuss how catalytic antibodies are used in targeting cancer.

16

ENZYME ENGINEERING

Enzymes have fascinated scientists since their discovery and, over some decades, one aim in organic chemistry has been the creation of molecules that mimic the active sites of enzymes and promote catalysis. Nevertheless, even today, there are relatively few examples of enzyme models that successfully perform Michaelis–Menten catalysis under enzymatic conditions (i.e., aqueous medium, neutral pH, ambient temperature) and for those that do, very high-rate accelerations are seldom seen.

Enzymes serve as catalysts for nearly all the chemical reactions that define cellular metabolism. The rate of acceleration and specificity of the enzyme make them valuable outside the cell. As a consequence, enzymes are being used increasingly in research, industry and medicine. The properties of enzymes are determined by their precise three-dimensional structures, but the detailed rules that govern protein folding and the structure–function relationships in proteins are not known. Although it is not yet practical to design enzymes from scratch, existing protein scaffolds can be readily modified with recombinant techniques, by site-selective chemical modification or even through chemical (semi) synthesis.

In fact, site-directed mutagenesis has become an indispensable tool for studying enzyme mechanism and for altering enzyme selectivity. Fundamentally, new activities can be conferred on proteins in this way, as well. Conversion of a protease into a peroxidase is illustrated in Figure 16.1. As shown in Figure 16.1, treating the serine protease subtilisin

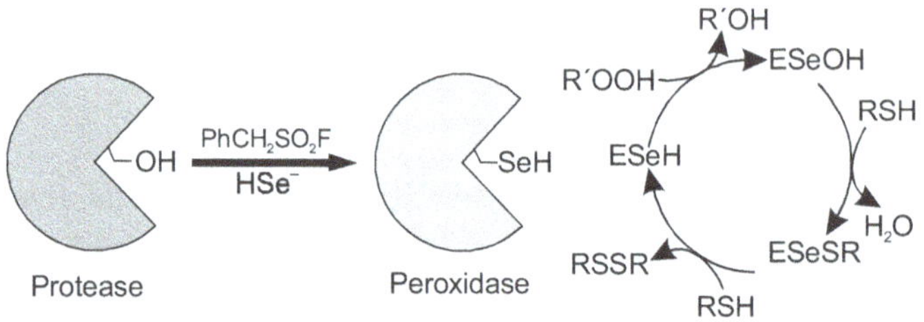

Figure 16.1 Site-selective chemical conversion of a serine protease into a peroxidase

sequentially with phenyl methane sulphonyl fluoride and hydrogen selenide selectively converts Ser 221 at the enzyme active site into a selenocysteine. The resulting artificial selenoenzyme, selenosubtilisin, efficiently catalyses the oxidation of thiols by alkyl hydroperoxides, mimicking the action of glutathione peroxidase, an important natural enzyme that protects cells from oxidative damage. In addition to its redox activity, selenosubtilisin also promotes the hydrolysis and aminolysis of activated esters. Indeed, acyl transfer to amines is four orders of magnitude more efficient than with native subtilisin. This result, and the fact that the modified enzyme does not hydrolyse peptides, suggests that it could be a practical peptide ligase (Figure 16.2).

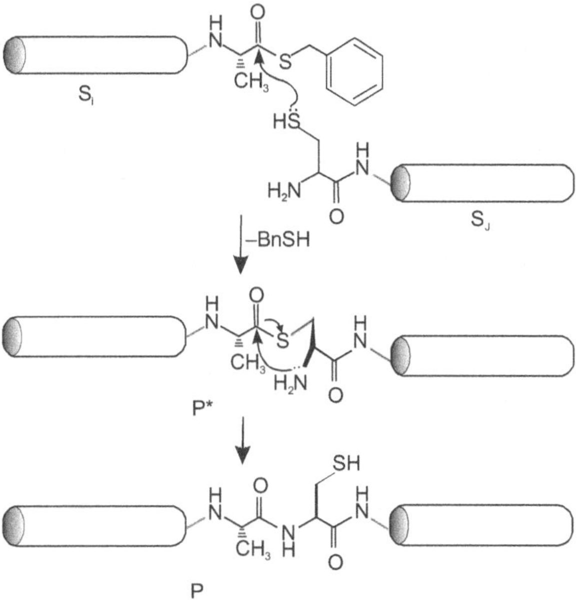

Figure 16.2 A de novo-designed peptide ligase and mechanism. (Peptide backbones are shown as *cylinders*)

PRINCIPLES OF ENZYME ENGINEERING

The structure and function of an enzyme molecule, for that matter of any protein molecule, are chiefly determined by its amino acid sequence, i.e., its primary structure. Therefore, any change in the properties of an enzyme is reflected in its primary structure.

Conversely, a change in the amino acid sequence should alter the properties of the enzyme. But this is not always the case because the enzymatic properties are changed only when amino acid changes are introduced in certain critical regions of the protein. However, the present knowledge of the relationships between amino acid sequence, and the three-dimensional structure and properties of the enzymes, obtained from a large database, is only partially operative. It allows an explanation of the changes in structure and function on the basis of the changes in amino acid sequence, but it does not allow a dependable prediction of the influences of specific amino acid changes on the structure and function of the enzymes. It may, however, be reasonable to anticipate that as more elaborate databases and improved software become available, it should become possible to predict with a far greater confidence, the structural and functional changes in enzymes produced by the specified changes in the amino acid sequences. The effectiveness of enzyme engineering will be greatly enhanced then, and, this activity may have a tremendous influence on enzyme technology.

STRATEGIES FOR ENZYME ENGINEERING

Enzyme engineering accelerates evolutionary events and can bring relief to existing bottlenecks in enzyme activity, stability and substrate specificity, thereby creating new useful catalysts which were not existing in nature before. Cloning the genes and recombinant enzyme expression is an indispensable step for intended biocatalytic applications on large-scale and opens the way for cost-effective production. The availability of a gene allows for recombinant protein expression in suitable host organisms. Moreover, the gene at hand is a prerequisite for enzyme engineering on a molecular level, resulting in protein production in a high-yield quality.

Two main strategies for enzyme engineering are: 1) Since de novo introduction of new enzymatic functions in existing scaffolds was rarely resulting in efficient enzymes, it is generally accepted that biodiversity offers a good starting point for enzyme development and discovery. 2) If at least some enzymatic activity is present, structure-guided design and directed evolution can be regarded as powerful strategies to improve enzyme fitness, and enzymes that can be readily expressed in *E. coli*, can be trimmed to surprising technological fitness. However, both approaches are finally depending on the powerful expression of the improved mutein. However, limitations in protein production affects the expression of muteins from archaea, actinomycetes and eukaryotic organisms.

ENZYME ENGINEERING TOOLS

Enzyme engineering is the use of genetic and chemical techniques to change the structure and function of a protein, thus producing a novel product with specific, desired properties.

Enzymes are generally engineered to achieve the following goals:

- Increase in V_{max}
- Decrease in K_m
- Change in pH optimum
- Elimination of an inhibition site
- Alteration of specificity of reaction
- Elimination of a residue that confers instability
- Improvement in thermostability and stability in organic solvents

Enzyme engineering group also focuses on the evaluation of kinetic mechanisms for activation/inhibition of the enzymes by metal ions. The quality of engineered enzyme molecules is evaluated in the light of kinetic and thermodynamic parameters for catalysis and stability.

The enzyme engineering tools are:

1. Chemical modification of enzymes
2. Metal binding

3. Enzyme immobilization
4. Mutagenesis of microbes
5. Genetic engineering

Enzymes have excellent features (activity, selectivity, specificity) for designing processes of synthesis of very complex products under very mild and environmental friendly conditions. Enzymes are biomolecules, which have significant application in a variety of industrial processes. Many industries such as paper and pulp industry, food industry, leather industry, pharmaceutical industry, alcohol and beverage industry, and animal feed and detergents industry, are routinely using enzymes. Furthermore, enzymes have been widely used in textile industry for desizing, stain removing, fabric softening, de-pilling, pilling prevention as anti-redepositors, colour care agents, stone washing, bio-polishing, bio-finishing and smooth surfacing of cotton fabric. Moreover, enzymes have great ecological and commercial importance like, amelioration of municipal, forestry, agricultural and industrial wastes to control environmental pollution, biocomposting to produce natural organic fertilizers.

Enzymes have been optimized, via evolution, to fulfil their biological function and to catalyse reactions in very complex metabolic routes submitted to many levels of regulation. Therefore, enzymes seldom have the features adequate to be used as industrial catalysts. Thus, enzymes are soluble catalysts, quite unstable and inhibited by high concentrations of substrate or products. Therefore, thermostable and efficient enzymes are always required by industry to withstand robust industrial conditions.

Site-directed mutagenesis has been used for thermo stabilization of enzymes, but dramatic results have never been found, because the enzyme molecules are fairly tolerant to single amino acid replacements. Thermostabilization of enzymes has been generally related to the changes basically occurring on their surface because the interior is already optimally packed. The structural elements responsible for higher enzyme thermostability have been found to be: increase in hydrophobic interactions; reduction of: conformational strain, the entropy of unfolding and solvent accessible on hydrophobic surface; compact packing; surface loop stabilization; hydrogen bonds; disulphide bridges; metal binding; glycosylation; resistance to covalent degradation and aromatic–aromatic interactions. Engineering of enzymes by carboxyl and amino group modifications of the surface amino acids could result into a dramatic increase in thermostability.

APPLICATION OF GENETIC ENGINEERING TECHNIQUES TO ENZYME TECHNOLOGY

The most exciting development over the last few years is the application of genetic engineering techniques to enzyme technology. There are a number of properties, which may be improved or altered by genetic engineering including the yield and kinetics of the enzyme, the ease of downstream processing, and various safety aspects. Enzymes from dangerous or unapproved microorganisms and from slow-growing or limited plant or animal tissue may be cloned into safe, high-production microorganisms. In the future, enzymes may be redesigned to fit more appropriately into industrial processes, for example, making glucose isomerase less susceptible to inhibition by the Ca^{2+} present in the starch saccharification processing stream.

The amount of enzyme produced by a microorganism may be increased by increasing the number of gene copies that code for it. This principle has been used to increase the activity of penicillin G-amidase in *Escherichia coli*. The cellular DNA from a producing strain is selectively cleaved by the restriction endonuclease *Hind* III. This hydrolyses the DNA at relatively rare sites containing the 5´-AAGCTT-3´ base sequence to give identical "staggered" ends.

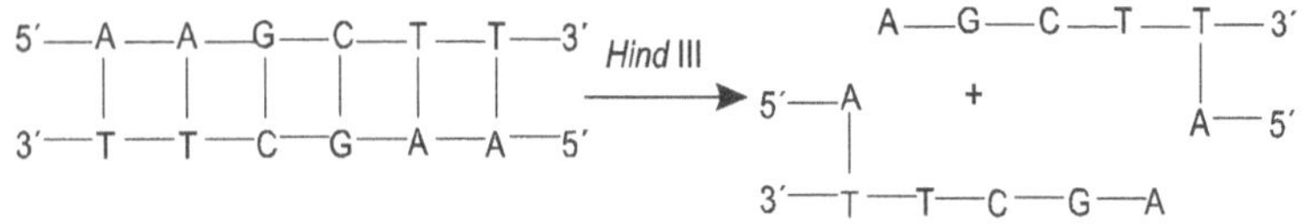

The total DNA is cleaved into about 10,000 fragments, only one of which contains the required genetic information. These fragments are individually cloned into a cosmid vector and thereby returned to *E. coli*. These colonies containing the active gene are identified by their inhibition of a 6-amino penicillanic acid-sensitive organism. Such colonies are isolated and the penicillin G-amidase gene transferred on to pBR322 plasmids and recloned back into *E. coli*. The engineered cells, aided by the plasmid amplification at around 50 copies per cell, produce penicillin G-amidase constitutively and in considerably higher quantities than does the fully induced parental strain.

Protein Engineering

Another extremely promising area is protein engineering. New enzyme structures may be designed and produced in order to improve on existing enzymes or to create new activities. Such factitious enzymes are produced by site-directed mutagenesis. In site-directed mutagenesis, the preferred pathway for creating new enzymes is by the stepwise substitution of only one or two amino acid residues out of the total protein structure. Although a large database of sequence-structure correlations is available, and growing rapidly together with the necessary software, it is presently insufficient to predict accurately three-dimensional changes as a result of such substitutions. The main problem is assessing the long-range effects, including solvent interactions, on the new structure.

Protein engineering, therefore, is presently rather a hit or miss process which may be used with only little realistic likelihood of immediate success. Apparently quite small sequence changes may give rise to large conformational alterations and even affect the rate-determining step in the enzymic catalysis. However it is reasonable to suppose that, given a sufficiently detailed database plus suitable software, the relative probability of success will increase over the coming years and the products of protein engineering will make a major impact on enzyme technology.

Artificial Enzymes

Many attempts have been made to take nature's blueprint and adapt it for industrial applications. For example, scientists have tried to introduce transition metals into existing enzymes, or to alter natural enzymes step by step (a process called directed evolution) to optimize or change their functionality. In contrast to these approaches, the ARTIZYMES aim to develop a new class of highly selective transition metal catalysts resembling natural enzymes. The so-called artizymes are based on unnatural phospholine-containing amino acids, which will be built into natural proteins. The ability of these enzymes to orient substrate molecules in a chemical reaction will be combined with the catalytic properties of metal ions such as rhodium, palladium and nickel. Tweaking the amino acids in the metalloenzymes so that they have "unnatural" metal-binding side chains, and using the aforementioned unusual metal ions will produce novel enzymes capable of catalysing the industrially important reactions such as CO insertions, and asymmetric C–C coupling on specific substrate molecules. The artizymes could be eco-friendly by transferring the elegance of catalytic systems in nature to the catalytic systems employed in industry and academia should lead to important breakthroughs. Along with waste-free (i.e., selective) synthesis, these artificial enzymes could make many chemical transformations energy-efficient (they do not require high temperature), non-toxic (using water as a solvent) and non-hazardous.

APPLICATIONS OF ENZYME ENGINEERING

To Degrade Explosives

The use of microbial enzymes for the bioremediation of explosives-contained land, either in their native or heterologous hosts, offers a powerful and low-cost route to eliminate these undesirable hazards from the environment. Explosives offer an abundance of nitrogen in their chemical structure. It is therefore not surprising that microorganisms have adapted to utilize these xenobiotic compounds as a source of nitrogen for their growth. As many of these explosives have only been present in the environment for many decades and with no known similar molecules existing in nature, the origin of enzymes apparently specialized for the breakdown of explosives are of particular interest.

Screening of environmental isolates resulted in the discovery of flavoproteins capable of denitrating the explosives pentaerythritol tetranitrate (PETN) and nitroglycerin (GTN). These nitrate ester reductases are related in sequence and structure to Old Yellow Enzyme *Saccharomyces carlsbergensis*. All the members of this family have Tim-barrel structures, FMN as a prosthetic group and display various chemistries with electrophilic substrates. Activity towards substrates with α, β-unsaturated carbonyl structure is a common property of the family. The nitrate ester reductases are, however, unusual in that they display activity towards the highly recalcitrant aromatic explosive 2,4,6–trinitrotoluene (TNT) via a reductive pathway resulting in nitrogen liberation.

To Treat Lysosomal Storage Disease

Liposomes, first developed by Bangham are biodegradable lipid vesicles with concentric lipid bilayers alternating with aqueous compartments. A number of inherited metabolic disorders lead to the accumulation of metabolic intermediates in the lysosomes, due to the absence of some key enzymes responsible for their degradation. The treatment of such disorders arising from enzyme deficiencies can be approached by enzyme engineering in two ways: by correcting the malfunctioning gene, or by an exogenous replacement of the missing enzyme. Both the treatments involve the delivery of appropriate molecules (DNA fragments or intact enzymes) to appropriate intracellular sites. In enzyme therapy various approaches have been used to compensate for the defective enzyme.

REVIEW QUESTIONS

1. What do you mean by enzyme engineering?
2. Discuss the principles of enzyme engineering.
3. Describe the strategies for enzyme engineering.
4. Discuss the enzyme engineering tools in detail.
5. Briefly explain the applications of enzyme engineering.
6. Discuss how application of genetic engineering techniques can be used in enzyme technology.
7. Write notes on:
   i. Protein Engineering
   ii. Artificial enzymes

17

BIOSENSOR TECHNOLOGY

A biosensor is an analytical tool consisting of biologically active material used in close conjunction with a device that will convert a biochemical signal into a quantifiable electrical signal. The device incorporates a biological sensing element with a traditional transducer. The biological sensing element can be an enzyme, antibody, DNA sequence or even microorganism (Table 17.1). The biochemical component selectively recognizes a particular biological molecule through a reaction, specific adsorption, or other physical or chemical process, and the transducer converts the biochemical event into a measurable signal, thus providing the means for detecting and quantifying it. Common transduction systems are optical, electro-optical, or electrochemical; this variety offers many opportunities to tailor biosensors for specific applications. For example, the glucose concentration in a blood sample can be measured directly by a biosensor by simply dipping the sensor into the sample.

The most widespread example of a commercial biosensor is the blood glucose biosensor, which uses the enzyme, glucose oxidase, to break down blood glucose. In doing so it first oxidizes

Table 17.1 Some common biosensor materials

Analytes	Examples
Respiratory gases	O_2, CO_2
Anaesthetic gases	N_2O, halothane
Toxic gases	H_2S, Cl_2, CO, NH_3
Flammable gases	CH_4
Ions	H^+, Li^+, K^+, Na^+, Ca^+, phosphates, heavy metal ions
Metabolites	Glucose, urea
Trace metabolites	Hormones, steroids, drugs
Toxic vapours	Benzene, toluene
Proteins and nucleic acids	DNA, RNA
Antigens and antibodies	Human Ig, anti-human Ig
Microorganisms	Viruses, bacteria, parasites

glucose and uses two electrons to reduce the FAD (a component of the enzyme) to $FADH_2$. This in turn is oxidized by the electrode (accepting two electrons from the electrode) in a number of steps. The resulting current is a measure of the concentration of glucose. In this case, the electrode is the transducer and the enzyme is the biologically active component.

GENERATIONS OF BIOSENSORS

There are three generations of biosensors. First-generation biosensor is where the normal product of the reaction diffuses to the transducer and causes the electrical response. Second-generation biosensor involves specific mediators between the reaction and the transducer in order to generate improved response. Third-generation biosensor is where the reaction itself causes the response and no product or mediator diffusion is directly involved. The different types of biosensors are listed in Table 17.2.

Table 17.2 Types of biosensors

Transducers	Measurement	Applications
Ion-selective electrodes	Potentiometric	Ions (in biological media)
Gas sensing electrodes	Potentiometric	Gases. Enzyme, organelle/ cell/ tissue electrodes for substrates and inhibitors, enzyme immunoelectrodes
Field effect transistors (FET)	Potentiometric	Ions, gases, enzyme substrates, immunological analytes
Enzyme electrodes	Amperometric	Enzyme substrates, immunological systems
Conductimeter	Conductance	Enzyme substrates
Piezoelectric crystals, acoustic waves	Mass charge	Volatile gases and vapours, immunosensors
Thermistors	Calorimetric	Enzyme/organelle/whole cell/ tissue sensors for products/substrates/inhibitors, gases, pollutants, antibiotics; vitamins; immunosensors (TELISA)
Luminescence, fluorescence, SPR, waveguides	Optical	pH, enzyme substrates, immunological analysis

COMPONENTS OF BIOSENSORS

A biosensor has three components—the sensitive biological element, the transducer and the associated electronics (Figures 17.1 and 17.2).

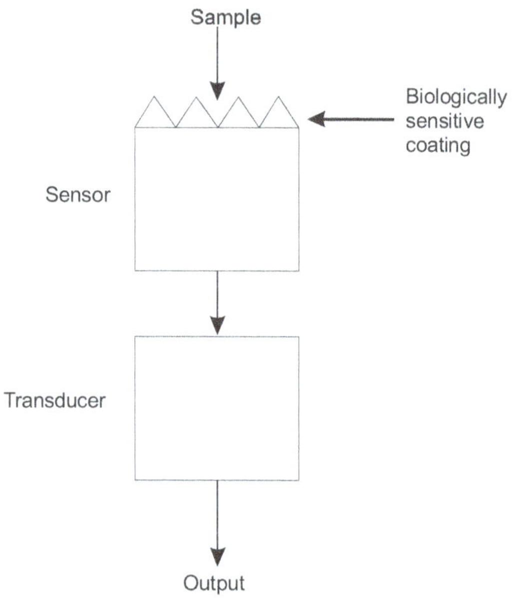

Figure 17.1 Structure of a biosensor

1. Sensitive Biological Element or Receptor Element

The sensitive biological element is biological material (e.g. tissue, microorganisms, organelles, cell receptors, enzymes, antibodies, nucleic acids, etc.) and a biologically derived material or biomimic. The sensitive elements can be created by biological engineering.

2. Transducer or Detector Element

The transducer or the detector element works in a physico-chemical way (optical, piezoelectric, electrochemical, etc.), and transforms the signal resulting from the interaction of the analyte with the biological element into another signal

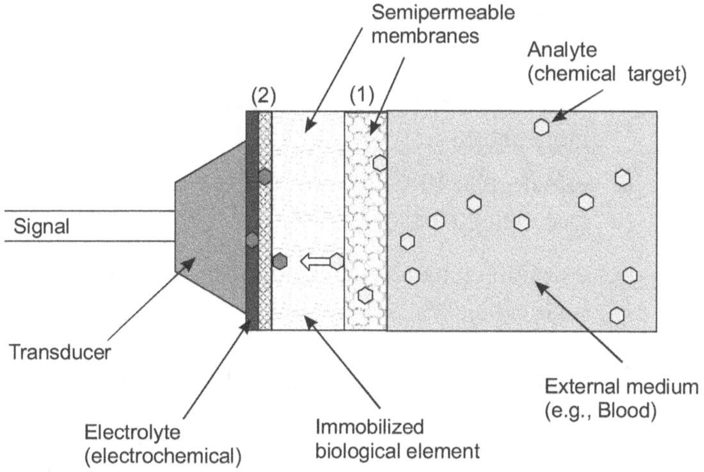

Analyte—chemical/biological target

Semipermeable membrane—(1) allows preferential passage of analyte (limits fouling)

Detection element (Biological)—provides specific recongnition/detection of analyte

Semipermeable membrane (2)—(some designs) preferential passage of by-product of recognition event

Electrolyte—(electrochemical-based) ion conduction medium between electrodes

Transducer—converts detection event into a measurable signal

Figure 17.2 Components of biosensors

(i.e., transducers) that can be more easily measured and quantified. The detector is not selective. It makes use of a physical change accompanying the reaction. This may be:

- the heat output (or absorbed) by the reaction (calorimetric biosensors)
- changes in the distribution of charges causing an electric potential to be produced (potentiometric biosensors)

- movement of electrons produced in a redox reaction (amperometric biosensors)
- light output during the reaction or a light absorbance difference between the reactants and products (optical biosensors)
- effects due to the mass of the reactants or products (piezoelectric biosensors)

For example, it can be a pH electrode, an oxygen electrode or a piezoelectric crystal.

3. Associated Electronics or Signal Processors

Associated electronics or signal processors are primarily responsible for the display of the results in a user-friendly way (Figure 17.3).

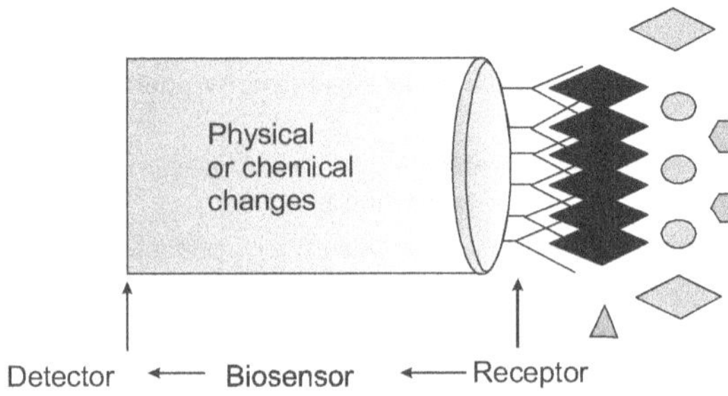

Figure 17.3 Schematic diagram of a biosensor device

USE OF ENZYMES IN ANALYSIS

Enzymes are natural catalysts. They usually operate approximately at neutral pH and mild temperatures, generate no by-products and are highly selective. Enzyme-catalysed reactions can be selective for one substrate or a group of substrates. These characteristics have resulted in frequent use of enzymes in analytical applications.

Enzymes make excellent analytical reagents due to their specificity, selectivity and efficiency. They are often used to determine the concentration of their substrates (as analytes) by means of the resultant initial reaction rates. If the reaction conditions and enzyme concentrations are kept constant, these rates of reaction (V) are proportional to the substrate concentrations ([S]) at low substrate concentrations. When [S] < 0.1, K_m, simplifies to give

$$V = (V_{max}/K_m)[S]$$

The rates of reaction are commonly determined from the difference in optical absorbance between the reactants and products.

Example

D-galactose dehydrogenase (EC 1.1.1.48) assay for galactose which involves the oxidation of galactose by the redox coenzyme, nictotinamide adenine dinucleotide (NAD^+).

$$\text{D-galactose} + NAD^+ \longrightarrow \text{D-galactono 1,4-lactone} + NADH + H^+$$

The drawbacks to this type of analysis become apparent when a large number of repetitive assays need to be performed. Then, they are seen to be costly in terms of expensive enzyme and coenzyme usage, time-consuming, labour-intensive and in need of skilled and reproducible operation within properly equipped analytical laboratories. For routine or on-site operation, these disadvantages must be overcome. This is being achieved by the production of biosensors which exploit biological systems in association with advances in micro-electronic technology.

TYPES OF BIOSENSORS

Electrochemical Affinity Biosensors

Affinity sensors detect the coupling reaction between the selective binding unit (SBU) (e.g. avidin, antibody, single-stranded DNA, lectin) and its complementary component (e.g. biotin, antigen, sugar sequence). Affinity sensor's design ensures that binding of SBU and complement takes place on the transducer surface. The sensors are thus implicitly heterogeneous. The transducer converts the binding event into a measurable response.

The sensors can be divided into two categories: Non-labelled and labelled. Non-labelled affinity sensors directly detect the affinity complex by measuring physical changes at the transducer induced by complex formation. In contrast, labelled affinity sensors incorporate a sensitively detectable label, and the presence of the affinity complex is then determined through measurement of the label. Heinman pioneered the use of alkaline phosphatase antibody conjugates to perform sandwich immunoassays. In these assays, aminophenyl phosphate is used as a substrate (in place of nitrophenyl phosphate) and the aminophenyl product is detected anodically with a flow injection analysis system.

Biocatalysis-based Biosensors

Biocatalysis-based biosensors for environmental applications depend universally on the use of enzymes. These enzyme-based biosensors primarily rely on two operational mechanisms. The first mechanism involves the catalytic transformation of a pollutant (typically from a non-detectable form to detectable form). The second mechanism involves the detection of pollutants that inhibit or mediate the enzyme's activity. Inherent limitations for this type of biosensor are primarily those imposed by the nature of the enzyme itself and include the limited number of environmental pollutants which happen to be substrates for the enzyme and the relatively high detection limits for environmental pollutants.

The detection limits for these sensors are determined by the enzyme's catalytic properties and are defined by K_m and V_{max} values. The other primary mechanism used in biocatalytic biosensors for environmental applications is inhibition of the enzyme by the pollutant. Inherent advantages for these formats involve the larger number of environmental pollutants, usually of a particular chemical class, that inhibit the enzyme and the low concentrations needed to affect the enzyme activity. Detection limits for biosensors based on irreversible inhibitors are usually within the range required for a variety of environmental applications.

The irreversible nature of many analyte–enzyme interactions that result in increased sensitivity also renders the biosensor inactive after a single measurement. This may not be a problem, however, for those systems that can be reactivated or that employ disposable sensing elements. Another potential limitation to this type of biosensor mechanism involves the sometimes diverse classes of pollutants that inhibit a specific enzyme. In most circumstances, this would not be expected to create problems. In some cases, e.g. the co-contamination of environmental samples with organics and heavy metals, such interferences may lead to unexpected results.

The interface of these devices to the contaminated matrix is critical to the successful application of biocatalysis-based biosensors, particularly for *in situ* applications. These matrices may range from pristine drinking water to highly contaminated industrial sludge. Although a considerable amount of work has been done with respect to use of biosensors in biological matrices, little work has focused on the development of membrane barriers for the direct sampling of groundwater or pore water in sludge and saturated sediments.

Bioaffinity-based Biosensors

Bioaffinity-based biosensors for environmental applications primarily depend on the use of antibodies because of the availability of monoclonal and polyclonal antibodies directed toward a wide range of environmental pollutants as well as the relative affinity and selectivity of these recognition proteins for a specific compound or closely related groups of compounds (Figure 17.4).

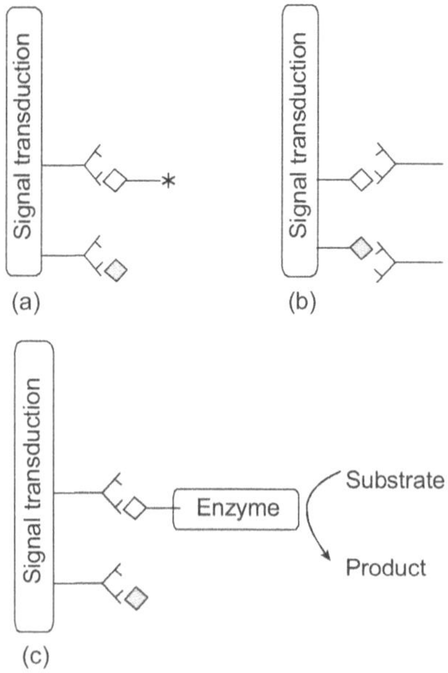

Figure 17.4 Bioaffinity-based biosensor formats (a) direct competitive format using labelled analyte tracer; (b) direct competitive format using immobilized analyte tracer; (c) indirect competitive format using enzyme-labelled analyte tracer.

For affinity-based biosensors, this is typically accomplished in one of several ways. In one type of format, antigen-tracer competes with analyte for immobilized antibody-binding sites. This format is often used in fluorescence-based systems. In another commonly used format, the antigen is immobilized to the signal transducer (operationally becoming the analyte-tracer) while free binding sites on the antibody, which has been previously exposed to the analyte, bind to the surface-immobilized antigen. Because the antibody is relatively a large molecule, its binding to the surface can be detected by signal transduction methods such as surface plasmon resonance, acoustic systems, and optical systems that measure changes in the refractive index and thus, do not require an optical tag.

The third commonly used format requires an indirect competitive assay and relies on the use of an enzyme-labelled antigen-tracer. In this format, the assay is completed in two steps. First, the enzyme-tracer competes with analyte for immobilized antibody-binding sites. Then, after removal of the unbound tracer (by means of a washing step), a non-detectable substrate is catalytically converted to an electrochemically or optically detectable product. This assay format is used almost universally with electrochemical signal transduction.

Microorganism-based Biosensors

Microorganism-based biosensors tend to use one of three primary mechanisms. For the first mechanism, the pollutant is a respiratory substrate. Biosensors that detect biodegradable organic compounds measured as biological oxygen demand (BOD) are the most widely reported of the microorganism-based biosensors. Biological oxygen demand is widely used as an indicator of the amount of biodegradable organic compounds found in industrial waste water.

The standard procedure (termed BOD_5) involves a 5-day incubation of the environmental or industrial water sample with an inoculum of microorganisms typically present in the waste treatment system to yield an endpoint measurement for oxygen consumption. The use of indicator microorganisms interfaced to signal transducers allows the measurement of the rate of organic compound metabolism rather than an endpoint. Thus, data can be acquired in a significantly shorter time frame (e.g. 15 min to 1 hr), rendering this technology highly advantageous for process control applications. Although these biosensors appear to work well for *in situ* monitoring of industrial waste waters that result in high BOD values, they currently require improvements in several areas. The primary limitations for these methods involve the variability encountered in calibration of the biosensor response to BOD_5 values. This arises from the fact that specific microbial species (used in biosensors) have characteristic substrate spectra which may or may not correspond well with the spectrum of compounds present in the sample.

Additional variability results from the presence of polymers (such as protein, starch, and lipid) which must be broken down to monomers before they can be metabolized; this changes the linearity of the response over time and can make interpretation of the result problematic. Another mechanism used for microorganism-based biosensors involves the inhibition of respiration by the analyte of interest. Microbial respiration and its inhibition by various environmental pollutants have been measured both optically and electrochemically.

Inherent advantages of these techniques apply primarily to the use of microorganisms as compared to the isolated enzymes. Microorganism-based biosensors are relatively inexpensive to construct and can operate over a wide range of pH and temperature. General limitations involve the long assay times including the initial response and return to baseline. These characteristics are primarily determined by the cellular diffusion characteristics that can be modified by using genetically engineered microorganisms.

Biosensors have also been developed using genetically engineered microorganisms (GEMs) that recognize and report the presence of specific environmental pollutants. The microorganisms used in these biosensors are typically produced with a constructed plasmid in which genes that code for luciferase or galactosidase are placed under the control of a promoter that recognizes the analyte of interest. Because the organism's biological recognition system is linked to the reporting system, the presence of the analyte results in the synthesis of inducible enzymes which then catalyse reactions resulting in the production of detectable products. With respect to environmental applications, the primary disadvantage for this type of biosensor is the limited number of GEMs which have been constructed to respond to specific environmental pollutants.

Potentiometric Biosensors

Potentiometric biosensors make use of ion-selective electrodes in order to transduce the biological reaction into an electrical signal. In the simplest terms this consists of an immobilized enzyme membrane surrounding the probe from a pH-meter where the catalysed reaction generates or absorbs hydrogen ions. The reaction occurring next to the thin sensing glass membrane causes a change in pH which may be read directly from the pH-meter's display. Typical of the use of such electrodes is that the electrical potential is determined at very high impedance allowing effectively zero current flow and causing no interference with the reaction.

Figure 17.5 shows a potentiometric biosensor in which a semi permeable membrane surrounds the biocatalyst entrapped next to the active glass membrane of a pH probe. The electrical potential is generated between the internal Ag/AgCl electrode bathed in dilute HCl and an external reference electrode.

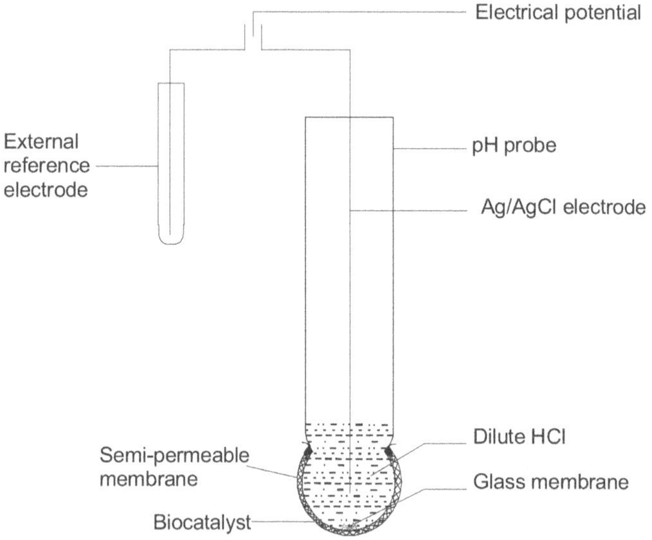

Figure 17.5 A simple potentiometric biosensor

There are three types of ion-selective electrodes which are of use in biosensors:

1. Glass electrodes for cations (e.g. normal pH electrodes) in which the sensing element is a very thin hydrated glass membrane which generates a transverse electrical potential due to the concentration-dependent competition between the cations for specific binding sites. The selectivity of this membrane is determined by the composition of the glass. The sensitivity to H^+ is greater than that achievable for NH_4^+.

2. Glass pH electrodes coated with a gas-permeable membrane selective for CO_2, NH_3 or H_2S. The diffusion of the gas through this membrane causes a change in pH of a sensing solution between the membrane and the electrode which is then determined.
3. Solid-state electrodes where the glass membrane is replaced by a thin membrane of a specific ion conductor made from a mixture of silver sulphide and a silver halide. The iodide electrode is useful for the determination of I^- in the peroxidase reaction and also responds to cyanide ions.

Reactions involving the release or absorption of ions that may be utilized by potentiometric biosensors are as follows:

i. H^+ *cation*

$$\text{D-Glucose} + O_2 \xrightarrow{\text{Glucose oxidase}} \text{D-Glucono 1,5-lactone}$$
$$+ H_2O_2 \xrightarrow{H_2O} \text{D-Gluconate} + H^+$$

$$\text{Penicillin} \xrightarrow{\text{Penicillinase}} \text{Penicilloic acid} + H^+$$

$$H_2NCONH_2 + H_2O + 2H^+ \xrightarrow[\text{(pH 6.0)(a)}]{\text{Urease}} 2NH_4^+ + CO_2$$

$$H_2NCONH_2 + 2H_2O \xrightarrow[\text{(pH 9.5)(b)}]{\text{Urease}} 2NH_3^+ + HCO_3^- + H^+$$

$$\text{Neutral lipids} + H_2O \xrightarrow{\text{Lipase}} \text{Glycerol} + \text{Fatty acids} + H^+$$

ii. NH_4^+ *cation*

$$\text{L-amino acid} + O_2 + H_2O \xrightarrow{\text{L-amino acid oxidase}} \text{Keto acid}$$
$$+ NH_4^+ + H_2O_2$$

$$\text{L-asparagine} + H_2O \xrightarrow{\text{Asparaginase}} \text{L-aspartate} + NH_4^+$$

$$H_2NCONH_2 + 2H_2O + H \xrightarrow[\text{(pH 7.5)}]{\text{Urease}} 2NH_4^+ + HCO_3^-$$

iii. I^- *anion*

$$H_2O_2 + 2H^+ + 21 \xrightarrow{\text{Peroxidase}} I_2 + +2H_2O$$

iv. CN^- *anion*

$$\text{Amygdalin} + 2H_2O \xrightarrow{\text{D-glucosidase}} \text{2Glucose} + \text{Benzaldehyde} + H^+ + CN^-$$

(a) Can also be used in NH_4^+ and CO_2 (gas) potentiometric biosensors.

(b) Can also be used in an NH_3 (gas) potentiometric biosensor.es80l166bp

The response of an ion-selective electrode is given by

$$E = E_0 + \frac{RT}{zF} \ln([I])$$

where,

E = measured potential (in volts),

E_0 = characteristic constant for the ion-selective/external electrode system,

R = gas constant,

T = absolute temperature (K),

z = signed ionic charge,

F = Faraday, and

[I] = concentration of the free uncomplexed ionic species (strictly, [I] should be the activity of the ion but at the concentrations normally encountered in biosensors, this is effectively equal to the concentration).

This means that, for example, if there is an increase in the electrical potential of 59 mV for every second then there will be an increase in the concentration of H^+ at 25°C. The logarithmic dependence of the potential on the ionic concentration is responsible for both the wide analytical range and the low accuracy and precision of these sensors.

Their normal range of detection is 10^{-4}–10^{-2} M, although a minority are tenfold more sensitive. Typical response time is between one and five minutes allowing up to 30 analyses every hour. A recent development from ion-selective electrodes is the production of ion-selective field effect transistors (ISFETs) and enzyme-linked field effect transistors (ENFETs). Enzyme membranes are coated on the ion-selective gates of these electronic devices, the biosensor responding to the electrical potential change via the current output.

Thus, these are potentiometric devices although they directly produce changes in the electric current. The main advantage of such devices is their extremely small size (<< 0.1 mm^2) which allows cheap mass-produced fabrication using integrated circuit technology. As an example, a urea-sensitive FET (ENFET containing bound urease with a reference electrode containing bound glycine) has been shown to show only a 15% variation in response to urea (0.05–10.0 mg ml^{-1}) during its active lifetime of a month. Several analytes may be determined by miniaturized biosensors containing arrays of ISFETs and ENFETs. The sensitivity of FETs, however, may be affected by the composition, ionic strength and concentrations of the solutions analysed.

Amperometric Biosensors

Amperometric biosensors function by the production of a current when a potential is applied between two electrodes.

They generally have response times, dynamic ranges and sensitivities similar to the potentiometric biosensors. The simplest amperometric biosensors in common usage involve the Clark oxygen electrode. This consists of a platinum cathode at which oxygen is reduced and a silver/silver chloride reference electrode. When a potential of –0.6 V, relative to the Ag/AgCl electrode is applied to the platinum cathode, a current proportional to the oxygen concentration is produced. Normally both electrodes are bathed in a solution of saturated potassium chloride and separated from the bulk solution by an oxygen-permeable plastic membrane (e.g. Teflon, polytetrafluoroethylene). The following reactions occur:

Ag anode $$4Ag^{0} + 4Cl^{-} \rightarrow 4AgCl + 4e^{-}$$

Pt cathode $$O_2 + 4H^{+} + 4e^{-} \rightarrow 2H_2O$$

The efficient reduction of oxygen at the surface of the cathode causes the oxygen concentration there to be effectively zero. The rate of this electrochemical reduction therefore depends on the rate of diffusion of the oxygen from the bulk solution, which is dependent on the concentration gradient and hence the bulk oxygen concentration. It is clear that a small, but significant, proportion of the oxygen present in the bulk is consumed by this process; the oxygen electrode measures the rate of a process which is far from equilibrium, whereas ion-selective electrodes are used close to the equilibrium conditions. This causes the oxygen electrode to be much more sensitive to changes in the temperature than potentiometric sensors.

Figure 17.6 shows a simple amperometric biosensor in which a potential is applied between the central platinum cathode and the annular silver anode. This generates a current (I) which is carried between the electrodes by means of a saturated solution of KCl. This electrode compartment is separated from the biocatalyst (here shown glucose oxidase, GOD) by a thin plastic membrane, permeable only to oxygen. The analyte solution is separated from the biocatalyst by another membrane, permeable to the substrate(s) and product(s). This biosensor is normally about 1 cm in diameter but has been scaled down to 0.25 mm diameter using a Pt wire cathode within a silver-plated steel needle anode and utilizing dip-coated membranes.

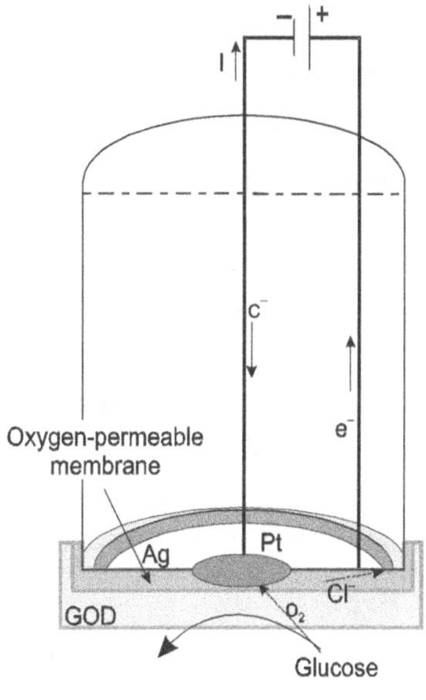

Figure 17.6 Schematic diagram of an immunosensor device

A typical application for this simple type of biosensor is the determination of glucose concentrations by the use of an immobilized glucose oxidase membrane. The reaction results in a reduction of the oxygen concentration as it diffuses through the biocatalytic membrane to the cathode, this being detected by a reduction in the current between the electrodes. Other oxidases may be used in a similar manner for the analysis of their substrates (e.g. alcohol oxidase, D- and L-amino acid oxidases, cholesterol oxidase, galactose oxidase, and urate oxidase).

Calorimetric Biosensors

The calorimetric biosensor, first reported in 1974, was designed by attaching the biological component to a heat-sensing transducer, the thermistor; alternatively, the biological component was immobilized on a column in which the thermistor was embedded. These biosensors have a large number of applications, given that most enzyme-catalysed reactions are accompanied by heat production of 25–100 kJ/mol. A wide range of clinical analytes can be measured this way, e.g. cholesterol, glucose, ATP, urea, triglycerides, and ascorbic acid.

The construction of microcalorimetric biosensors composed of miniaturized thin-film thermistors, and the incorporation of immobilized antibodies for determination of antigens, indicate the possibility for producing devices of reasonable size and simplicity. The latter technique, known as thermometric ELISA, has been successfully used in immunological analysis for determining concentrations of human pro-insulin (measured in media from genetically engineered *Escherichia coli* with a detection limit of 0.1 nig/L and a reaction time of 7 mm), albumin, and gentamicin.

Many enzyme-catalysed reactions are exothermic, generating heat which may be used as a basis for measuring the rate of reaction and, hence, the analyte concentration. This represents the most generally applicable type of biosensor. The temperature changes are usually determined by means of thermistors at the entrance and exit of small, packed bed columns containing immobilized enzymes within a constant temperature environment. Under such closely controlled conditions, up to 80% of the heat generated in the reaction may be registered as a temperature change in the sample stream. This may be simply calculated from the enthalpy change and the amount reacted. If a 1 mM reactant is completely converted to product in a reaction generating 100 kJ $mole^{-1}$ then each ml of solution generates 0.1J of heat. At 80% efficiency, this will cause a change in temperature of the solution amounting to approximately 0.02°C. This is about the temperature change commonly encountered and necessitates a temperature resolution of 0.0001°C for the biosensor to be generally useful.

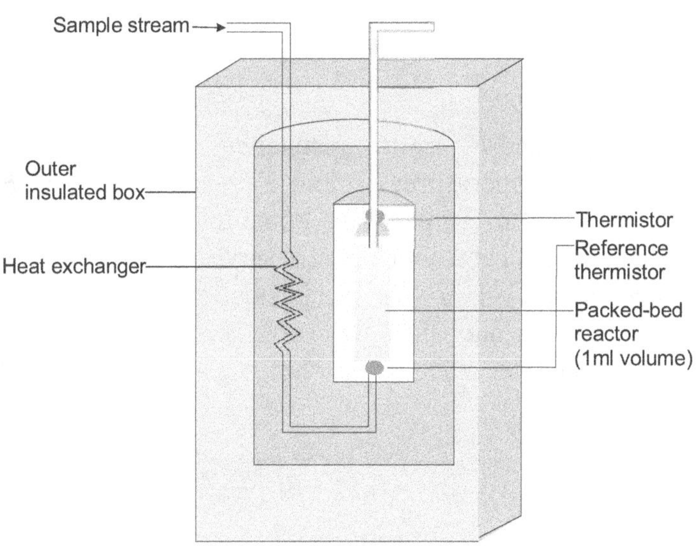

Figure 17.7 Schematic diagram of a calorimetric biosensor

Figure 17.7 shows a calorimetric biosensor in which the sample stream passes through the outer insulated box to the heat exchanger within an aluminium block. From there, it flows past the reference thermistor and into the packed bed bioreactor, containing the biocatalyst, where the reaction occurs. The change in temperature is determined by the thermistor and the solution passed to waste. External electronics determines the difference in the resistance, and hence temperature, between the thermistors.

The thermistors, used to detect the temperature change, function by changing their electrical resistance with the temperature, obeying the relationship

$$\ln\left(\frac{R_1}{R_2}\right) = B\left(\frac{1}{T_1} - \frac{1}{T_2}\right)$$

Therefore

$$\frac{R_1}{R_2} = e^{B\left(\frac{1}{T_1} - \frac{1}{T_2}\right)}$$

where, R_1 and R_2 are the resistances of the thermistors at absolute temperatures T_1 and T_2 respectively and B is a characteristic temperature constant for the thermistor. When the temperature change is very small, as in the present case, $B(1/T_1)-(1/T_2)$ is much smaller than one and this relationship may be substantially simplified using the approximation when $x<<1$ that $e^x>>1 + x$ (x here being $B(1/T_1)-(1/T_2)$,

$$R_1 = R_2\left\{1 + B\left(\frac{T_2 - T_1}{T_1 T_2}\right)\right\}$$

As $T_1 T_2$, they both may be replaced in the denominator by T_1.

$$\frac{\Delta R}{R} = -\left(\frac{B}{T_1^2}\right)\Delta T$$

The relative decrease in the electrical resistance (R/R) of the thermistor is proportional to the increase in temperature. A typical proportionality constant $\left(-B/T_1^2\right)$ is $-4\%°C^{-1}$. The resistance change is converted to a proportional voltage change, using a balanced Wheatstone bridge incorporating precision wire-wound resistors, before amplification. The expectation that there will be a linear correlation between the response and the enzyme activity has been found to be borne out in practice. A major problem with this biosensor is the difficulty encountered in closely matching the characteristic temperature constants of the measurement and reference thermistors.

An equal movement of only 1°C in the background temperature of both thermistors commonly causes an apparent change in the relative resistances of the thermistors equivalent to 0.01°C and equal to the full-scale change due to the reaction. The sensitivity of the glucose analysis using glucose oxidase can be more than doubled by the co-immobilization of catalase within the column reactor in order to disproportionate the hydrogen peroxide produced. An extreme case of this amplification is shown in the following recycle scheme for the detection of ADP:

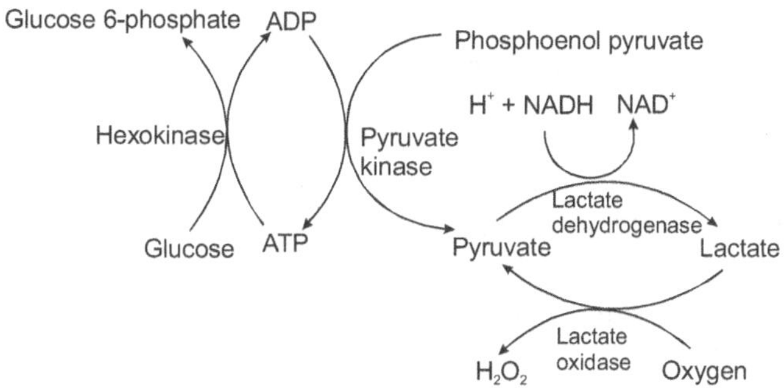

ADP is the added analyte and excess glucose, phosphoenol pyruvate, NADH and oxygen are present to ensure maximum reaction. Four enzymes (hexokinase, pyruvate kinase, lactate dehydrogenase and lactate oxidase) are co-immobilized within the packed-bed reactor. In spite of the positive enthalpy of the pyruvate kinase reaction, the overall process results in a

1000-fold increase in sensitivity, primarily due to the recycling between pyruvate and lactate. Reaction limitation due to low oxygen solubility may be overcome by replacing it with benzoquinone, which is reduced to hydroquinone by flavo-enzymes. Such reaction systems do, however, have the serious disadvantage in that they increase the probability of the occurrence of interference in the determination of the analyte of interest.

Reactions involving the generation of hydrogen ions can be made more sensitive by the inclusion of a base having a high heat of protonation. For example, the heat output by the penicillinase reaction may be almost doubled by the use of tris (tris-(hydroxymethyl)aminomethane) as the buffer. In conclusion, the main advantages of the thermistor biosensor are its general applicability and the possibility for its use on turbid or strongly coloured solutions. The most important disadvantage is the difficulty in ensuring that the temperature of the sample stream remains constant (± 0.01°C).

Conductimetric Biosensors

Conductimetric measurements are widely applicable to chemical systems, given that many chemical reactions produce or consume ionic species and therefore alter the overall electrical conductivity of a solution. Biosensors can be designed by immobilizing a suitable enzyme over a set of electrodes made of noble metal (e.g. gold, silver, copper, nickel, or chromium) and measuring the change in conductance of a solution containing the analyte of interest when an electric field is applied. For example, when urea is converted to its ionic product NH by the enzyme urease, the increase in solution conductance measured is proportional to the urea concentration. Variations in ionic strength and buffer capacity of measured samples have caused problems with this type of biosensor in the past, but these drawbacks have been overcome with more recent designs [SO]. Conductimetric sensors may also have problems with non-specificity of measurements, because the resistance of a solution is determined by the migration of all ions present. To date, the development of immunosensors based on conductance has hardly been exploited, and it will be interesting to see if this technology progresses in the future.

Optical Biosensors

There are two main areas of development in optical biosensors. These involve determining changes in light absorption between the reactants and products of a reaction, or measuring the light output by a luminescent process. The former usually involves the widely established, if rather low-technology, use of colorimetric test strips.

These are disposable, single-use cellulose pads impregnated with enzyme and reagents. The most common use of this technology is for whole-blood monitoring in diabetes control. In this case, the strips include glucose oxidase, horseradish peroxidase (EC 1.11.1.7) and a chromogen (e.g. *o*-toluidine or 3,3′,5,5′-tetramethylbenzidine). The hydrogen peroxide, produced by the aerobic oxidation of glucose, oxidizes the weakly coloured chromogen to a highly coloured dye.

$$\text{Chromogen(2H)} + H_2O_2 \xrightarrow{\text{Peroxidase}} \text{Dye} + 2H_2O$$

The evaluation of the dyed strips is best achieved by the use of portable reflectance meters, although direct visual comparison with a coloured chart is often used. A wide variety of test strips involving other enzymes are commercially available at present. The most promising biosensor involving luminescence uses firefly luciferase (*Photinus*-luciferin

4-monooxygenase (ATP-hydrolysing), EC 1.13.12.7) to detect the presence of bacteria in food or clinical samples. Bacteria are specifically lysed and the ATP released (roughly proportional to the number of bacteria present) reacted with D-luciferin and oxygen in a reaction which produces yellow light in a high-quantum yield.

$$ATP + \text{D-luciferin} + O_2 \xrightarrow{\text{Luciferase}} \text{Oxyluciferin} + AMP + \text{Pyrophosphate} + CO_2 + \text{light}$$

The light produced may be detected photometrically by use of high-voltage, and expensive, photomultiplier tubes or low-voltage, cheap photodiode systems. The sensitivity of the photomultiplier-containing systems is, at present, somewhat greater ($< 10^4$ cells ml^{-1}, $< 10^{-12}$ M ATP) than the simpler photon detectors which use photodiodes.

IMMUNOSENSORS

Biosensors concerned with monitoring solely antibody–antigen interactions can also be termed immunosensors. Similar to conventional immunoassays, these devices are based on the principles of solid-phase immunoassay, with either antibody or antigen immobilized at the sensor surface (Figure 17.8). An extensive range of analytes can be detected and measured by immunosensors, e.g. medical diagnostic markers such as hormones (steroids and pituitary hormones), drugs (therapeutic and abused), and bacteria, and environmental pollutants such as pesticides.

However, the advantages of an absence of labelling requirements and the ability to investigate the reaction dynamics of antibody–antigen binding have given these devices the potential to revolutionize conventional immunoassay techniques. The sensitivity and specificity of an immunosensor

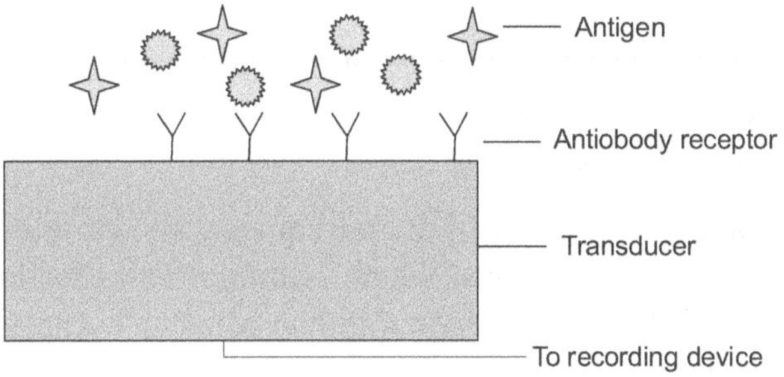

Figure 17.8 Schematic diagram of an immunosensor device

are determined by the same characteristics as in other solid-phase immunoassays, namely, the affinity and specificity of the binding agent and the background noise of the detection system (transducer).

Mass-detecting Immunosensors

An alternative property used in immunosensor design is mass change, which is measured with the use of piezoelectric crystals and acoustic wave techniques. The first generations of these devices were expensive to produce, lacked sensitivity, and had a high failure rate in manufacture. However, they have since improved dramatically and are now viable commercial products.

Piezoelectric crystals The phenomenon of piezoelectricity was discovered in 1880. Electric dipoles generated in anisotropic natural crystals (with no centre of symmetry) were subjected to mechanical stress, thus causing them to oscillate at frequencies between 9 and 14 MHz. There are 20 naturally abundant piezoelectric crystals, including quartz (SiO_2), lithium niobate (LiNbO), zinc oxide (ZnO), and Rochelle salt. Quartz is the most commonly used piezoelectric material because of its chemical stability in aqueous solutions and resistance to high temperatures without loss of piezoelectric properties (Figure 17.9). In piezoelectric biosensors, the crystals are coated with an adsorbent that selectively interacts with the analyte of interest; subsequent binding increases the mass of the coated crystal and alters its basic frequency of oscillation.

Monitoring the oscillation frequency allows determination of the change in mass, which is proportional to the analyte concentration. This technique has had much success in the analysis of gaseous environmental pollutants. Piezoelectric crystal sensors have also been investigated for monitoring antibody–antigen interactions; in this procedure, the crystals are coated with antibody and the amount of antigen bound is determined, or vice versa.

Piezoelectric immunosensors have also been used to detect a range of human cell types such as erythrocytes, granulocytes, and T lymphocytes and viruses such as human herpes and hepatitis. The measurement of HIV-specific antibodies in serum has also been reported, involving piezoelectric quartz crystals coated with peptides to HIV, although this was complicated by non-specific binding of serum proteins in the sample. In fact, use of piezoelectric immunosensors is greatly affected by non-specific binding to the adsorbed substrate on the crystal.

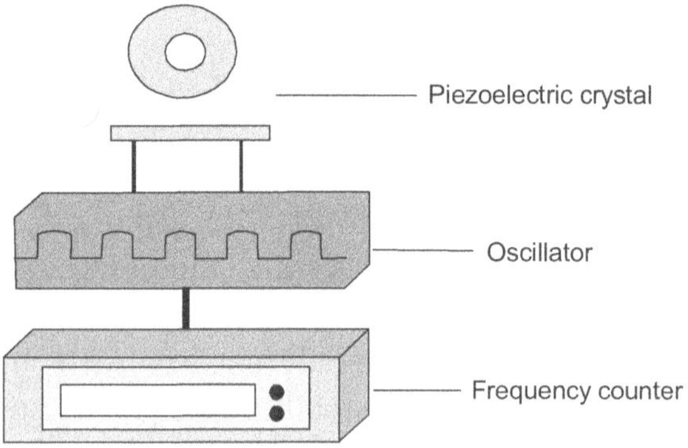

Figure 17.9 Experimental apparatus for a piezoelectric sensor

Acoustic waves Acoustic wave technology, e.g. surface acoustic wave measurements, has also attracted interest in biosensor design. In this method, the oscillation of the piezoelectric crystals is at a higher frequency (30–200 MHz), and an acoustic wave is generated by application of an alternating voltage across a pattern of interlaced metal electrodes (e.g. titanium or gold), known as an interdigital transducer. The acoustic signal produced is detected by a second interdigital transducer situated a few millimetres away. The adsorption of sample to the crystals slows the acoustic wave, and the recorded change in velocity is proportional to the analyte concentration.

Besides mass, however, factors such as temperature, pressure, and surface conductivity may also alter the properties of the acoustic wave; this has inhibited the successful design of such devices. Acoustic wave devices have been used in specific areas such as environmental gas monitoring and chemical detection; more recently, their incorporation in immunosensor design has been described, including rapid assays of antigen present in foodstuffs and human IgG measurements. Given the similarities between surface acoustic waves and piezoelectricity, the problem of non-specificity may also plague the development of acoustic wave immunosensors.

Types of Immunosensors

The two widely available immunosensors are both indirect and direct optical systems—the BIAcore and the IAsys—and both have surfaces of carboxylated dextran. These have proved to have very low non-specific binding in biological matrices and achieve good detection limits for a variety of molecules, but their major impact has been to revolutionize the kinetic rate analysis of biomolecular interactions.

Indirect optical sensing Optical transducers can be designed to respond to ultraviolet or visible radiation or to the production of bio- and chemiluminescence and can be adapted for fibre-optic-containing devices. Early optical systems were based on spectrophotometry, enzymes immobilized on a column, and the absorbance of the product measured. Later, the enzymes were immobilized on nylon coils and the system was linked to flow-injection or bubble analysers. In an effort to reduce the size of the device, fibre-optic technology was then introduced; the reagent phase was immobilized on a single optical fibre or a fibre bundle, so that changes in the optical properties of the reagent phase attributable to the analyte could be monitored. These devices, now known as optrodes, have been widely used and offer several advantages over electrodes, including the lack of a requirement for a reference electrode. Furthermore, fibre optics have increased versatility, being suitable for clinical applications, *in vivo* monitoring, and measurement of hazardous materials.

Direct optical sensing The development of optical techniques for direct monitoring of minuscule changes in adsorption, fluorescence, light scatter, or RI at a sensor surface is an extremely promising area in immunosensor technology. Optical transducers based on reflectance, ellipsometry, surface plasmon resonance (SPR), and waveguides have all been described for this purpose. In these systems, light entering the device is directed towards the sensing surface and then reflected out again. The light emerging from the device is then monitored, revealing information regarding the physical events occurring at the sensing surface.

The principles behind direct optical sensing lie in internal reflectance spectroscopy, This device consists of two materials of differing RI—a layer of high RI, often consisting of a glass prism, and a layer of lower RI that contains the sample.

A light beam is directed through the layer of higher RI to the interface between the two media. When the angle of the incident beam exceeds the critical angle, the light is totally internally reflected out of the device. In the material of lower RI, a high-frequency electromagnetic field is generated, known as the evanescent wave. This wave, a fraction of the wavelength of the incident light, advances along the x-axis in the plane of incidence, and penetrates into the sample medium (i.e., along the y-axis) for a short distance (approximately one wavelength) with exponentially decreasing amplitude.

Biomolecules present in the sample that have been adsorbed at, or are in close contact with, the interface, interact with the evanescent wave and cause a reduction in intensity of the reflected light beam. This distinctive decrease in light intensity therefore reflects any changes in RI occurring at the interface and is directly related to the mass (and concentration) of adsorbed biomolecules in the sample. As an alternative to light intensity, the optical interaction can also be monitored as a change in the phase of polarized light emerging from the sensing layer. This property has been successfully applied in several waveguide sensors.

Direct optical sensing has an advantage over indirect techniques, in that less-sophisticated instrumentation is required for detection of light, but problems have been encountered with non-specific binding and poor sensitivity to small molecules. In some cases, therefore, the best features of both techniques have been combined to improve sensitivity, e.g. the use of SPR with fluorescent labels and latex particles. Optical transducers may be used for sensing with or without the use of a label; thus, some sensors make use of labelled reagents, e.g. an enzyme or fluorophore, to provide the detected signal. These require sophisticated instrumentation because of the low-light levels detected. Direct optical sensors, which do not involve labelling and which constitute a substantial proportion of the immunosensors are currently available. These include attenuated total internal reflection, ellipsometry, SPR and monomode dielectric waveguides. In some cases labels have been incorporated into these immunosensors to increase sensitivity.

Optical sensing with a signal-generating label An enzyme can be used as a label to generate a range of products that absorb light, fluoresce, or luminesce—the last offering particularly high sensitivities. Biosensors have been designed to monitor luminescence spectroscopy, either bioluminescence (light emitted from a biological reaction) or chemiluminescence (light emitted from a chemical reaction). Unlike other optical biosensors, no light source is required for these systems, which greatly simplifies instrument design. Bioluminescence-based systems such as that of the luciferin–luciferase reaction, in which luciferin is oxidized by firefly luciferase with the production of light, have been used in assays for oxygen, ATP, and NADH/NAD(P)H in dehydrogenase systems and in environmental monitoring all demonstrating a high sensitivity.

Bioluminescent systems utilizing luminous bacteria have also been reported. Fibre-optic bioluminescence sensors have been developed by immobilizing luciferase onto an optical fibre to detect ATP and NAD(P)H. Chemiluminescence sensor systems include the luminol-peroxide system, which has been applied to many oxidase reactions involving hydrogen peroxide production. Chemiluminescence adapted for fibre-optic immunosensors has facilitated the determination of various antigens, e.g. estradiol, α-interferon, hCG, total IgG, and antibodies to influenza virus. Problems associated with luminescence sensors include reagent replenishment. For irreversible oxidation of substrate to take place, the luminescent compound must be available in excess. This can be arranged by immobilizing the enzyme onto an optical fibre, and loosely embedding co-reactants for their slow release.

Fluorescence The phenomenon of monitoring reactions occurring at the interface between the two layers of different RI has been adapted for the analysis of fluorescence. One example is the technique known as total internal reflection fluorescence, used to assess the fluorescence characteristics of a compound of interest. In total internal reflection fluorescence, the incident light excites molecules near the sensor surface, which, in turn, creates a fluorescent evanescent wave. This couples back (i.e., re-enters) into the waveguide and the emerging fluorescence is detected.

Reflectance Attenuated total reflection is a form of internal reflection spectroscopy (IRS) in which light energy absorbed from the evanescent wave is monitored as an attenuation of the internally reflected light beam. An optically absorbing film is present on the surface of an IRS sensor, resulting in a structure known as the Kretschmann configuration. Incident light directed towards the upper surface is absorbed by this film, and the attenuated light intensity can be measured as a function of the incident wavelength. This technique has been applied to the monitoring of biological interactions in the infrared, visible, and ultraviolet regions of the spectrum. Attenuated total reflection has been applied to the study of antibody–antigen interactions and a reflectance method has been reported for the direct detection of immunological reactions at high-RI surfaces.

Ellipsometry In ellipsometry, the change in the state of polarization of light caused by reflection from a planar-layered structure is monitored. This method generated a little interest for many years after the first observations, and miniaturization and low-cost fabrication were thought to be unfeasible for this technology. However, ellipsometry has become more popular, particularly for studying the binding of proteins to surfaces. Once reflected from a mirrored surface, the polarized light is interrogated so as to produce two readings; these are subsequently converted into changes in amplitude and in phase of light, respectively. These properties are altered when a molecule is adsorbed to the surface, and the new readings can be interpreted as changes in RI and coating thickness, from which the concentration of bound analyte can be calculated.

Surface plasmon resonance (SPR) The phenomenon of SPR was first observed in 1902 and is by now well documented. The optical arrangement for SPR is the Kretschmann configuration; it differs from that of the previously described IRS set-up by the incorporation of an additional layer of a thin metal film between the prism and sample. The metal film characteristic of SPR usually consists of gold or silver. Specific ligands can be immobilized at the upper surface of the device (the sensor surface), to interact directly with biomolecules in the sample. SPR technology is based on the excitation of surface plasmons present within the metal film of the sensor (Figure 17.10).

Light is directed towards a layer of low RI from one of higher RI; at an angle equal to or greater than the critical angle, the light undergoes total internal reflection (TIR). At the resonance angIe (O), surface plasmons in the metal film are excited by the light energy, causing them to oscillate and generate on evanescent wave. This condition of SPR results in a characteristic decrease in reflected light intensity.

Of the two types of plasmons—radiative and non-radiative—the latter is more commonly used for biosensor applications. Non-radiative plasmons are excited by light directed towards the metal film via a glass prism. When excited, the surface

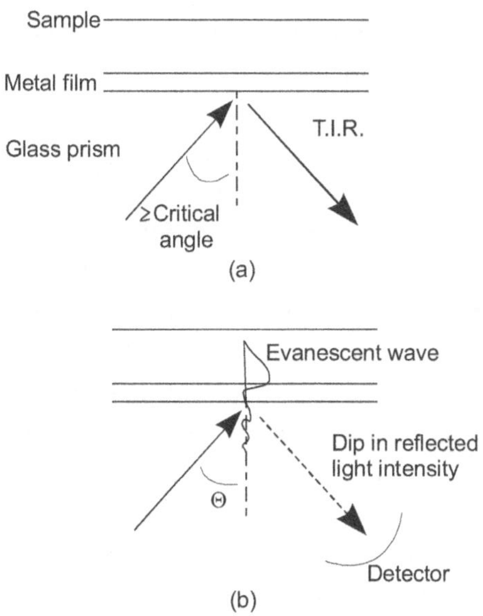

Figure 17.10 SPR technology (a) Principles (b) Resonance.

plasmons, which oscillate at a different frequency from that in the bulk of the metal film, absorb some of the light energy to generate an evanescent wave. This wave penetrates into the sample layer with exponentially decreasing amplitude, the condition known as SPR. In the SPR system, polarized light is directed in one plane into the prism, which is of higher RI than the metal layer. When the angle of reflection of light is greater than or equal to the critical angle, light is totally internally reflected back, out of the prism. At the resonance angle (θ), SPR is initiated; the absorption of light energy by the surface plasmons during resonance causes a sharp decrease in the intensity of the reflected light, which is monitored throughout the biomolecular interaction. The evanescent wave does not propagate parallel to the prism interface in SPR, but is redirected upwards into the metal layer. This allows it to sense the metal–sample interface (the sensor surface).

Applications of SPR to biosensing Applications of SPR to biosensing were followed by demonstrations that a wide range of molecules could be analysed in this manner, from small peptides to larger analytes such as virus particles and antibodies. SPR has been commercially developed as a biosensor known as the BIAcoreIM (Biosensor, Uppsala, Sweden), which is capable of label-free, real-time analysis of biological interactions; its use of integrated software facilitates kinetic evaluation of binding reactions, which has made this technique useful in many areas, e.g. biomolecular engineering, drug design, and monoclonal antibody characterization. SPR immunosensors for sex-hormone-binding globulin and syphilis screening have also been reported. An earlier example of the technology had a detection limit for thyroxine of 5×10^{-3} mol. A recent advance in SPR immunosensing is the incorporation of labels to increase sensitivity. The combination of fluorescent labelling with SPR to produce an SPR fluoro-immunoassay has successfully been used to assay hCG in serum.

Clinical Applications of Immunosensors

Not all of these sensing technologies developed for immunological quantification have achieved practical usage in biological fluids. The most sensitive (i.e., lowest detection limit) is of the order of 2×10^{-13} mols/L for hepatitis B surface antigen in serum, which compares well with many conventional immunoassay techniques. Hapten assays operate in the low nanomolar region for digoxin and estradiol. There is even a highly sensitive assay for prostate-specific antigen in whole blood by FCFD (3.3×10^{12} mol/L) technology, which is comparable with quoted detection limits for automated heterogeneous immunoassays.

Large sums of money have been invested by diagnostic companies to develop commercially viable immunosensors, as yet with little obvious success. Surface-effect measurement of kinetic rate constants by real-time, stable-automated immunosensors has received widespread usage in the diagnostic, biotechnology, and pharmaceutical industries as well as in basic clinical science. The potential of immunosensing technologies has been around for quite a while and new opportunities, e.g. microspot technology, are continually developing. However, the main impact of these technologies has been (a) the innovative detection systems that have furthered conventional immunoassay systems and (b) the feasibility studies in both clinical and environmental testing areas.

PRINCIPLES OF DETECTION

Optical Biosensors

Optical biosensors, based on the phenomenon of surface plasmon resonance, are evanescent wave techniques. This utilizes a property shown by gold and other materials, wherein a thin layer of gold on a high-refractive-index glass surface can absorb laser light, producing electron waves (surface plasmon) on the gold surface. This occurs only at a specific angle and wavelength of incident light and is highly dependent on the surface of the gold, such that binding of a target analyte to a receptor on the gold surface produces a measurable signal. Other optical biosensors are mainly based on changes in absorbance or fluorescence of an appropriate indicator compound.

Surface Plasmon Resonance Sensors

Surface plasmon resonance sensors operate using a sensor chip consisting of plastic cassette supporting a glass plate, one side of which is coated with a microscopic layer of gold. This side contacts the optical correction apparatus of the instrument. The opposite side is then contacted with a microfluidic flow system. The contact with the flow system creates channels across which reagents can be passed in solution. This side of the glass sensor chip can be modified in a number of ways, to allow easy attachment of molecules of interest. Normally, it is coated in carboxymethyl dextran or similar compound. Light at a fixed wavelength is reflected off the gold side of the chip, at the angle of total internal reflection and detected inside the instrument. This induces the evanescent wave to penetrate through the glass plate and, in some way, into the liquid flowing over the surface. A widely used research tool, the microarray is basically a biosensor.

Electrochemical Biosensors

Electrochemical biosensors are normally based on enzymatic catalysis of a reaction that produces or consumes electrons. The sensor substrate usually contains three electrodes, a reference electrode, an active electrode and a sink electrode. A counter electrode may also be present as an ion source. We can either measure the current at a fixed potential or the potential can be measured at zero current.

The Potentiometric Biosensor

The potentiometric biosensor works contrary to the current understanding of its ability. Such biosensors are screen-printed conducting-polymer-coated, open-circuit potential biosensors based on conjugated polymer immunoassays. They have only two electrodes and are extremely sensitive, robust and accurate. They enable the detection of analytes at levels previously only achievable by HPLC and LC/MS and without rigorous sample preparation. The signal is produced by electrochemical and physical changes in the conducting polymer layer due to changes occurring at the surface of the sensor. Such changes can be attributed to ionic strength, pH, hydration and redox reactions, and the latter due to the enzyme label turning over a substrate.

Piezoelectric Sensors

Piezoelectric sensors utilize crystals which undergo an elastic deformation when an electric potential is applied to them. This frequency is highly dependent on the elastic properties of the crystal, such that if a crystal is coated with a biological recognition element then the binding of a target analyte to a receptor will produce a change in the resonance frequency, which gives a binding signal. This is a special application of the quartz crystal microbalance in biosensor.

Thermometric and magnetic-based biosensors are rare.

BENEFICIAL FEATURES OF A BIOSENSOR

A successful biosensor must possess at least some of the following beneficial features:

1. The biocatalyst must be highly specific for the purpose of the analyses, be stable under normal storage conditions and except in the case of colorimetric enzyme strips and dipsticks, show good stability over a large number of assays.
2. The reaction should be as independent of physical parameters such as stirring, pH and temperature as it is manageable. If the reaction involves cofactors or coenzymes these should be co-immobilized with the enzyme.
3. The response should be accurate, precise, reproducible and linear over the useful analytical range, without dilution or concentration. It should be free from electrical noise.

4. If the biosensor is to be used for invasive monitoring in clinical situations, the probe must be tiny and biocompatible, having no toxic or antigenic effects. If it is to be used in fermenters it should be sterilizable.
5. The complete biosensor should be cheap, small, portable and capable of being used by semi-skilled operators.
6. Simple and low-cost instrumentation, fast response times, minimum sample pretreatment, and high-sample throughput are also essential.

The biological response of the biosensor is determined by the biocatalytic membrane which accomplishes the conversion of reactant to product. Immobilized enzymes possess a number of advantageous features which makes them particularly applicable for use in such systems. They may be reused, which ensures that the same catalytic activity is present for a series of analyses. When the reaction occurring at the immobilized enzyme membrane of a biosensor is limited by the rate of external diffusion, the reaction process will possess a number of valuable analytical assets. The biocatalyst gives a proportional change in reaction rate in response to the reactant (substrate) concentration over a substantial linear range, several times the intrinsic K_m.

APPLICATIONS OF BIOSENSORS

Biosensors have many potential applications in:

1. agricultural, horticultural and veterinary analysis
2. pollution, water and microbial contamination analysis
3. clinical diagnosis and biomedical applications
4. fermentation analysis and control
5. industrial gases and liquids analysis

6. mining and toxic gases analysis
7. explosives and military arena analysis
8. flavours, essences and pheromone analysis and identification of availability of a suitable biological recognition molecule
9. as disposable portable detection systems

Combining the exquisite specificity of biological recognition probes and the excellent sensitivity of laser-based optical detection; biosensors are capable of detecting and differentiating chemical constituents of complex systems in order to provide unambiguous identification and accurate quantification.

Example

1. Glucose monitoring in diabetic patients—historical market driver
2. Environmental applications—the detection of pesticides and river water contaminants
3. Detection and determining of organophosphate
4. Determination of drug residues in food, such as antibiotics and growth promoters
5. Detection of toxic metabolites such as mycotoxins

REVIEW QUESTIONS

1. Define a biosensor.
2. Discuss the components of biosensor.
3. Discuss the different types of biosensors giving suitable examples.

4. Write short notes on:
    i. Electrochemical affinity biosensors
    ii. Biocatalysis-based biosensors
    iii. Bioaffinity-based biosensors
    iv. Microorganism-based biosensors
    v. Potentiometric biosensors
    vi. Amperometric biosensors
    vii. Calorimetric biosensors
    viii. Conductimetric biosensors
    ix. Optical biosensors
    x. Immunosensors
5. Discuss the different types of immunosensors.
6. Discuss how fluorescence and reflectance are used in optical sensing.
7. What is surface plasmon resonance?
8. Discuss the clinical applications of immunosensors.
9. Write an account on:
    i. Optical biosensors
    ii. Surface plasmon resonance sensors
    iii. Electrochemical biosensors
    iv. The potentiometric biosensor
    v. Piezoelectric sensors
10. Discuss the beneficial features of a biosensor.
11. Write notes on the applications of biosensors.

18

INDUSTRIAL ENZYMES AND THEIR APPLICATIONS

INTRODUCTION

For many thousands of years, man has used naturally occurring microorganisms and the enzymes they produce to make foods such as bread, cheese, beer and wine. Today, enzymes are used for an increasing range of applications—bakery, cheese-making, starch processing and production of fruit juices and other drinks. Hence, they can improve texture, appearance and nutritional value and may generate desirable flavours and aromas. Currently used food enzymes sometimes originate in animals and plants, but, most come from a range of beneficial microorganisms.

NATURALLY OCCURRING ENZYMES

Enzymes are extracted from animal or plant tissues. Plant-derived commercial enzymes include proteolytic enzymes, papain, bromelain and ficin and some other special enzymes like lipoxygenase from soybeans. Animal-derived enzymes include proteinases like pepsin and rennin. Most of the enzymes are, however, produced by microorganisms in submerged cultures in large reactors called fermenters. Raw foods contain many enzymes that may help facilitate digestion. When they are eaten, mastication breaks down the cell membranes allowing the release and activation of the naturally occurring enzymes present in food. Enzymatic activity begins in the mouth where salivary amylase, lingual lipase and ptyalin initiate starch and fat digestion. In the stomach, hydrochloric acid activates pepsinogen to pepsin which breaks down protein, and gastric lipase begins the hydrolysis of fats.

Most of digestion and absorption takes place in the small intestine and is mediated by pancreatic amylase, protease, lipase and bile. These enzymes may play a beneficial role by initiating the breakdown of the food and giving the body a head start on the digestive process. Unfortunately, in this part of the world we tend to eat mostly cooked foods. When a food is cooked at 118°F, or greater the natural enzymes are destroyed. Without enzyme production, the body has a harder time digesting food which may lead to a variety of chronic disorders.

Inclusion of appropriate raw foods combined with enzyme replacement may be the most balanced and rational way to support those with digestive enzyme insufficiencies. Enzyme replacement entails providing enzyme supplements, whether they be animal or non-animal derived, in the quantity necessary to maintain adequate digestive capacity and facilitate absorption of essential nutrients.

PRIMARY FOOD PROCESSING

Some foods are prepared or predigested with enzymes. This makes it much more tasty, pleasant and longer life. Few such enzymes are:

Trypsin It is a serine protease (EC 3.4.21.4) from digestive system. In organisms trypsin disintegrate proteins from food. It is used for primary food processing for babies.

Papain It is a cystein protease (EC 3.4.22.2) from fruit. It is used for meat tendering.

Rennin It is an aspartic acid protease (EC 3.4.23.4) from stomach of young mammalian animals. It hydrolyses protein from milk in cheese production.

Glucose isomerase Glucose isomerase (EC 5.3.1.5) catalyses the reversible isomerization of D-glucose to D-fructose. It is used for syrup production.

Cellulase Cellulases (EC 3.2.1.4) are a group of enzymes from bacteria, fungi and protozoans which can hydrolyse cellulose to beta D-glucan. It is used for cellulose processing in coffee, beans, etc.

FOOD PRODUCTION

Many enzymes used in food processing (Table 18.1) are very similar to those from human body. Almost all enzymes are disintegrated at high temperatures and remain safe for first step of food processing. Furthermore many enzymes may be activated during food processing. In food production, enzymes have a number of advantages.

1. They provide alternatives to traditional chemical-based technology, and can replace synthetic chemicals in many processes. This allows advances in the environmental performance of production processes, through lower energy consumption and biodegradability.
2. They are more specific in their action than synthetic chemicals. Food processes which use enzymes have fewer side reactions and waste by-products generated, reducing the likelihood of pollution.
3. They allow some processes to be carried out which would otherwise be impossible. For example,

the production of clear apple juice concentrate relies on the use of the enzyme, pectinase.

Enzymes in Fruit and Vegetable Juice Technology

Enzymes are used in fruit juice manufacturing. Fruit cell wall needs to be broken down to improve juice liberation. Addition of enzymes pectinase, xylanase and cellulase improve the liberation of the juice from the pulp. Pectinases and amylases are used in juice clarification. Enzymes are used to maximize the production of clear or cloudy juice. This cloudiness arises from pectin which occurs in nearly all fruits. Pectins are structural polysaccharides which occur mainly in the middle lamella and the primary cell wall of higher plants; and are largely responsible for the integrity and coherence of plant tissues. Though they cause cloudiness in extracted juices, they interfere with rapid filtration.

Table 18.1 Enzymes in food production

Market	Enzyme	Function
Dairy	Rennet (protease)	Coagulant in cheese production
	Lactase	Hydrolysis of lactose to give lactose-free milk products
	Protease	Hydrolysis of whey proteins
	Catalases	Removal of hydrogen peroxide
Brewing	Cellulases, β -glucanases, α -amylases, proteases, maltogenic amylases	For liquefaction, clarification and to supplement malt enzymes

(Contd.)

Table 18.1 (Continued)

Market	Enzyme	Function
Alcohol production	Amyloglucosidase	Conversion of starch to sugar
Baking	α -amylases	Breakdown of starch, maltose production
	Amyloglycosidases	Saccharification
Baking	α -amylases	Breakdown of starch, maltose production
	Amyloglycosidases	Saccharification
	Maltogen amylase (Novamyl)	Delays process by which bread becomes stale
	Protease	Breakdown of proteins
	Pentosanase	Breakdown of pentosan, leading to reduced gluten production
	Glucose oxidase	Stability of dough
Wine and fruit juice	Pectinase	Increase of yield and juice clarification
	Glucose oxidase	Oxygen removal
Meat	Protease	
	Papain	Meat tenderizing
Protein	Proteases, trypsin, aminopeptidases	Breakdown of various components
Starch	Alpha amylase, glucoamylases, hemicellulases, maltogenic amylases, glucose isomerases	Modification and conversion (e.g. to dextrose or high-fructose syrups)
Inulin	Inulinases	Production of fructose syrups

The addition of pectin-degrading enzymes at the pressing stage increases the amount of juice produced and can reduce cloudiness. Two groups of such enzymes are distinguished, pectinesterases and pectin depolymerases. The latter group includes polygalacturonase, pectate lyases and pectin lyases. Pectinesterases appear in many fruit and vegetables, and are particularly abundant in citrus fruits and tomatoes. They are also produced by many fungi (*Aspergillus niger*).

The application of enzymes in these processes is a combination of economic and cosmetic factors. In contrast to the above, the desired flavour and colour of citrus juices especially orange depends on the insoluble, cloudy materials of the pressed juice. In these situations, the pectin component is manipulated requiring a balance between pectin methyl esterase, to promote cloudiness by increasing the pectin/calcium complex formation; and polygalacturonase, to break cloudiness by depolymerization of the pectin.

Enzymes in Baking Technology

Enzymes also affect the baking process. α-amylase is the most widely studied enzyme in connection to baking process particularly with improved bread quality. Both fungal and bacterial amylases are used. Over-dosage may lead to sticky dough so the added amount needs to be carefully controlled. In addition to starch, the flour typically contains minor amounts of cellulose, glucans and hemicelluloses like arabinoxylan, and arabinogalactan. There is evidence that the use of xylanases decreases the water absorption and thus reduces the amount of added water needed in baking. This leads to more stable dough. Proteinases can also be added to improve dough-handling properties; glucose oxidase has been used to replace chemical oxidants and lipases to strengthen gluten, which leads to more stable dough and better bread quality.

Bread Making

Throughout the world, about 85 million tons of wheat flour is used every year to make bread. In breadmaking, baker's yeast is added to a mixture of flour and dough and it is the carbon dioxide that is produced during fermentation which causes the dough to rise. The alcohol is driven off during the baking process. Flour, which gives bread its structure, is made by milling cereal grains such as wheat, barley, or rye. In this process, the grain seeds are crushed, releasing starch and proteins, some of which are enzymes. These enzymes modify the starch, protein and fibre of the flour when water is added. Enzymes, called carbohydrases, from the added yeast cells initially attack starch, breaking it down into a sugar called maltose. Other enzymes (maltases) convert maltose molecules into carbon dioxide and ethanol.

There are two proteins found in flour, gliadin and glutenin. When water is added to flour and kneaded, these proteins swell up like sponges and form a tough, sticky, elastic substance called gluten. Some maltose molecules are attacked by other enzymes present in the flour and are converted into alcohols, acids, and esters, all of which add to bread's flavour. To improve consistency and efficiency, enzymes such as xylanase, α-amylase, protease, glucose oxidase and lipase are added. Nowadays, inconsiderable quantity of wheat flour is thrown away due to its becoming stale. This occurs because the bread loses moisture, causing the bread to become hard within a few days. To slow this effect down an enzyme called maltogenic amylase, obtained from microorganisms, is added to the flour. This alters the structure of the starch enabling it to retain moisture better and thus stay fresher for longer.

Cheese Making

The process of cheese making is an ancient craft that dates back thousands of years, which may even pre-date beer and bread making. It is a complicated process. In cheese making, the basic raw material is milk, which may be obtained from a variety of animals, such as cows, sheep and goats. The milk is warmed and a mixture of two enzymes (chymosin and pepsin) known as rennet, which is obtained from the fourth stomach of the milk-fed calf, is added. This coagulates the milk to form "curds and whey". The whey is a cloudy liquid which contains some protein and sugars (including lactose), while the curds are precipitated protein which is pressed and subsequently packed in containers of various sizes for maturing. However, the necessary enzymes are also found in certain plants, which are used in some European countries and the Far East, since the use of calf stomach extract is a problem for vegetarians. Many modern cheese are produced using chymosin from fungi or bacteria, and sold as vegetarian cheese.

Enzymes for Animal Feed

Enzymes have become an important aspect of the animal feed industry. Intensive use of enzymes in animal feed started in the early 80s. The first commercial success was addition of β-glucanase into barley-based feed diets. Barley contains β-glucan, which causes high viscosity in the chicken gut. The net effect of enzyme usage in feed has been the increased animal weight gain with the same amount of barley resulting in increased feed conversion ratio. Enzymes were tested later also in wheat-based diets. Xylanase enzymes were found to be the most effective ones. Addition of xylanase to wheat-based broiler feed has increased the available metabolizable energy to 7–10%. Xylanases are routinely used in feed formulations.

For ruminants, alteration in milk fat content and improvement in the productivity of beef cattle has been demonstrated by adding the enzymes β-glucanase and xylanase to animal feed. Also, hemicellulase complexes in animal feed improve the digestion and nutritional value of various feed diets.

Usually a feed-enzyme preparation is a multienzyme cocktail containing glucanases, xylanases, proteinases and amylases. Enzyme addition reduces viscosity, which increases absorption of nutrients, liberates nutrients either by hydrolysis of non-degradable fibres or by liberating nutrients blocked by these fibres, and reduces the amount of faeces.

Another type of important feed enzyme is phytase. Phytase is a phosphoesterase which liberates phosphate from phytic acid which is a common compound in plant-based feed materials. The net effect is reduced phosphorus in faeces resulting in reduced environmental pollution. The use of phytase reduces the need to add phosphorus to the feed diet. In addition to poultry, enzymes are also used in pig feeds and turkey feeds. They are added as enzyme pre-mixes (enzyme–flour mixture) during the feed manufacturing process, which involves extrusion of wet feed mass in high temperature (80–90°C). Therefore the feed enzymes need to be thermotolerant during the feed manufacturing and operative at the animal body temperature.

ENZYMES IN BREWING INDUSTRY

The brewing industry relies on enzymes to create low-malt and malt-free brewing environments, to induce faster maturation of beer, to remove carbohydrates for light beer, and to induce chill proofing. Brewing of beer had its origins in ancient Mesopotamia from about 4000 BC. Here, bread was mashed, malted and fermented, and the resulting brew was flavoured with spices, dates or honey. Beer is made when yeast consumes the sugar derived from grain, particularly barley. The naturally occurring starch found in grain must first be converted into sugar before yeast can consume it. Thus, beer-making is a more complex art than wine-making.

Beer-Making

Beer can be prepared by several discrete processes. The first stage, malting, involves soaking the grain in water long enough to begin germination or sprouting. During malting, enzymes are produced which break down starches to sugars. Germination then has to be halted and this is done by **kilning**, a drying process, which involves heating in several stages. The malt is then ground, or lightly crushed between rollers, and at this stage is known as "grist", thus giving rise to the expression "grist to the mill". This process is known as **cracking**. The next stage, mashing, is the most important operation in brewing, for it is here that the principal enzymatic changes occur.

Mashing is the process of heating grains mixed with water at controlled temperatures for designated periods of time to activate various enzyme activities that convert starches to fermentable sugars. Converting starches to sugars is called saccharification. During mashing, the enzyme cytase dissolves the protective cellulose coating of the barley grains, giving access to the starch. The enzyme diastase then liquefies and converts starch to maltose and dextrins, which dissolve in the water to form sweet, malt-flavoured liquor known as **sweet wort**. Once mashing is completed, the brewer must separate the wort from the spent grain husks. This is done by **sparging** or rinsing the spent grains with hot water to extract as much sugar from the grains as possible. The grain husks act as a filter bed on the false bottom of the mash tun. After the wort is collected, it is boiled for one to two hours to prevent further enzyme activity. This is a critical step in the brewing process because it is at this step that hops, the aromatic flowers of hop vines, are added. Hops impart aroma, flavour, and bitterness to beer, which balances the sweetness of the wort. When the boil is completed, the temperature of the wort is dropped to about 15°C, the fermentation temperature, following which yeast is added.

The main genus of yeast used by brewers is *Saccharomyces*, and the species used to make ales is a top-fermenting strain called *S. cerevisiae*, while the species used to make lagers is a bottom-fermenting strain called *S. uvarum*. The yeast produces enzymes which convert malt sugar (maltose) to ethanol and carbon dioxide. This usually takes three to seven days and is referred to as primary fermentation. The traditional malting process, as described above, is really just an expensive and inefficient way of manufacturing enzymes. So nowadays industrially produced enzymes such as amylases, glucanases and proteases are added to unmalted barley to produce the same products that malting would produce by more controlled means. Use of industrial enzyme preparations in the brewing industry allows it to be more economical and have consistent quality. The enzymes used in brewing of beer include amylases and proteases.

Amylase Amylases are a group of enzymes belonging to EC 3.2.1 group. Amylase is used to split polysaccharides from malt during the mashing process. This enzyme is used in addition to those naturally present in malt.

Protease Proteases are a group of hydrolases acting on peptide bonds, and belongs to EC 3.4 group. These enzymes allow removing of proteins remaining from yeasts, which makes beer clearer and easily filtrated.

ENZYMES IN DETERGENTS

Detergents were the first large-scale application for microbial enzymes. Bacterial proteinases are still the most important detergent enzymes. Proteases are used for removing of protein contamination. Lipase allows removing of fatty stains. Some hostile agents including anionic detergents and oxidizing agents have been genetically engineered to be more stable in the hostile environment of washing machines with several different chemicals present. In the late 1980s lipid-degrading enzymes were introduced in powder and liquid detergents. Lipases decompose fats into more water-soluble compounds by hydrolysing the ester bonds between the glycerol backbone and fatty acid. The most important lipase in the market was originally obtained from *Humicola lanuginose.* It is produced on a large scale by *Aspergillus oryzae* host after cloning the *Humicola* gene into this organism.

Amylases are used in detergents to remove starch-based stains. Amylases hydrolyse gelatinized starch, which tends to stick on textile fibres and bind other stain components. Cellulases have been part of detergents since early 1990s. In textile washing, cellulases remove cellulose microfibrils, which are formed during washing and the use of cotton-based cloths. This can be seen as colour brightening and softening of the material. Alkaline cellulases are produced by *Bacillus* strains and neutral and acidic cellulases by *Trichoderma* and *Humicola* fungi.

Enzymes in Washing Powders

Rohm developed the first method for washing protein-stained cloth in detergents containing enzymes and manufactured the first detergent preparation containing enzymes. The main stains that need to be removed from clothing are organic in origin; for example, proteins, fats and carbohydrates which are all present in foodstuff. Hence biological washing powders may contain proteases, lipases and carbohydrases.

The carbohydrases may include amylases, which are effective in removing starchy food deposits. Some powders contain cellulase to brighten colours and soften fabrics. Proteases and amylases are also effective in dishwasher detergents, to remove food particles. These detergents are based on the enzymatic activities which help to remove many types of dirt from clothes. They are also used for cleaning of some sensitive equipment. These detergents are environmentally friendly with fewer bleaching agents and phosphates, allowing the enzymes to work more effectively whilst having minimal effects on public and environmental health. However, as biological washing powders contain proteases and the connective tissue in skin is largely made up of the proteins, collagen and elastin, biological washing powders can cause "irritant contact dermatitis".

ENZYMES IN TEXTILE INDUSTRY

The use of enzymes in textile industry is one of the most rapidly growing fields in industrial enzymology. Enzymes are used to treat and modify fibres, particularly during textile processing and in caring for textiles afterwards. Previously, textiles were treated with acid, alkali or oxidizing agents, or soaked in water for several days so that naturally occurring microorganisms could break down the starch. However, both of these methods damaged or discoloured the material. Enzymes are used in the textile industries in finishing processes. Proteases help in the de-hairing of animal hides, and lipases are used for de-greasing. The application of a cellulase enzyme can give a smoother, glossier, brighter fabric to cellulose fibres like cotton. This technique is known as bio-polishing.

In the denim industry, cloth was traditionally stonewashed with pumice stones to fade the fabric. This results in physical damage to the fibres in denim to allow the dye molecules to escape. A small application of cellulase minimizes damage to the garments and also to machinery. This technique is known as bio-stoning and can ensure greater fading without high abrasive damage to the fabric and accessories (buttons, rivets). Enzymes like catalases are used to treat cotton fibres and prepare them for the dyeing processes.

Some bacterial enzymes are used to separate the tough stem of the flax plant from the flax fibres used in textiles. By degrading surface fibres, many enzymes, including some cellulases and xylanases, are used to finish fabrics, give jeans a stonewashed effect, or help in the tanning of leathers.

A recombinant enzyme called laccase, made by certain fungi, may also be used to treat fabrics and even catalyse the synthesis of some synthetic fibres. There are even enzymes in regular laundry detergent to help break down dirt, to clean clothes more effectively, and to prevent the dulling of fabric colours. Enzymes are frequently used in laundry detergents. Without them, very high temperatures and mechanical shaking would be required to effectively clean clothes and other textiles. The main enzymes used in laundry detergents are acid cellulases, neutral cellulases and hybrid cellulases. The use of cellulases for stonewashing has allowed the cycle time to be cut in half when used in conjunction with pumice stones. Additionally, stonewashing with cellulase is superior to washing with pumice alone because it increases garment softness and results in higher levels of fabric abrasion. Cellulases are also used in biofinishing for the removal of surface hairs from garments for increased comfort and fashion.

Starch has for a long time been used as a protective glue of fibres in weaving of fabrics. This is called sizing. Enzymes are used to remove the starch in a process called desizing. Bacterial amylase derived from *Bacillus subtilis* was used for the first time in 1917 by Boidin and Effront for desizing. Amylases are used in this process since they do not harm the textile fibres. Recently, hydrogen peroxides have been tested as bleaching agents to replace chlorine-based chemicals. Catalase enzyme, which destroys hydrogen peroxide, may then be used to degrade excess peroxide.

Another recent approach is to use oxidative enzymes directly to bleach textiles. Laccases, a polyphenol oxidase from fungi is a new candidate in this field. Laccase is a copper-containing enzyme, which is oxidized by oxygen, and which in an oxidized state can oxidatively degrade many different types of molecules like dye pigments. Laccases are produced by white-rot fungi, which is used to degrade lignin, the aromatic polymer found in all plant materials.

ENZYMES IN LEATHER BATING

Leather industry uses proteolytic and lipolytic enzymes in leather processing. The use of these enzymes is associated with the structure of animal skin as a raw material. Alkaline bacterial proteases are added in the soaking phase. This improves water uptake by the dry skins and removal and degradation of protein, dirt and fats, and reduces the processing time.

In some cases pancreatic trypsins were originally used but they are being partly replaced by bacterial and fungal enzymes. In dehairing and dewooling phases enzymes are used to assist the alkaline chemical process. The bating phase aims at deliming and deswelling of collagen. In this phase the protein is partly degraded to make the leather soft and easier to dye. The use of lipases is a fairly new development in leather industry. Lipases used in the bating phase help specifically to remove grease.

ENZYMES IN PULP AND PAPER INDUSTRY

Enzymes have been used in the pulp and paper industry to soften wood fibres, improve drainage, and present alternatives to chemical bleaching. The use of enzymes in paper and pulp industry started with the discovery of lignin-degrading peroxidases in the early 80s.

Pulp Industry

The major application of enzymes in the pulp industry is the use of xylanases in pulp bleaching. Xylanases liberate lignin fragments by hydrolysing residual xylan. This reduces considerably the need for chlorine-based bleaching chemicals. Other minor enzyme applications in pulp production include the use of enzymes to remove fine particles from pulp. This facilitates water removal. In the use of secondary (recycled) cellulose fibre the removal of ink is important. The fibre is diluted to 1% concentration with water, flocculating surfactants and ink solvents added and the mixture is aerated. The ink particles float to the surface. There are reports that this process is facilitated by addition of cellulase enzymes.

Paper Industry

Paper is made from cellulose fibres, which must be separated from a tough wood fibre called lignin. The step-by-step process used to separate cellulose from lignin and other wood components is known as pulping. It is a time- and energy-consuming process, involving the mechanical processing of wood or the treatment of wood with harsh chemicals. In biopulping, cellulase and xylanase enzymes made by lignin-degrading fungi are used to pre-treat wood and break down the lignin fibres. Removing lignin prior to further wood pulping saves time and energy, and decreases the quantities of chemicals used.

In paper-making, enzymes are used especially in modification of starch, which is used as an important additive. Starch improves the strength, stiffness and erasability of paper. The starch suspension must have a certain viscosity, which is achieved by adding amylase enzymes in a controlled process. Pitch, composed of lipids, is a sticky substance present mainly in softwoods. It is a special problem when mechanical pulps of red pine are used as a raw material. Pitch causes problems in paper machines and can be removed by lipases.

Draining Enzymes can also improve water drainage during wood pulping, a process that often slows down paper production. When fine lignin fibres are degraded by enzymes, less water is absorbed, thereby reducing drainage times, lessening the energy required to dry the paper, and producing a clean water runoff.

Bleach boosting Lignin fibres that remain in wood pulp are coloured and must be bleached, usually by harsh chlorine compounds under high pressures. As an alternative, enzymes may be used to remove fine surface fibres, thereby reducing the bleaching process or eliminating it altogether.

ENZYMES IN PERSONAL-CARE PRODUCTS

Enzymes are used in personal-care products which is a relatively new area for enzymes. One application is contact lens cleaning. Enzyme solutions containing proteinase and lipase are used for this purpose. Hydrogen peroxide is used in disinfection of contact lenses. The residual hydrogen peroxide after disinfection can be removed by a haem-containing catalase enzyme, which degrades hydrogen peroxide.

Some toothpastes contain glucoamylase and glucose oxidase, because, glucoamylase liberates glucose from starch-based oligomers produced by α-amylase and glucose oxidase converts glucose to gluconic acid and hydrogen peroxide, both functioning as disinfectants. Dentures can be cleaned with protein-degrading enzyme solutions. Enzymes are also used in skin- and hair-care products.

ENZYMES IN FINE CHEMICAL PRODUCTION

Enzymes have been used in fine chemical production from the past. Usually the catalyst has been a living organism. Ethanol, acetic acid, antibiotics, vitamins, pigments, solvents are biotechnical products. One of the reasons to use whole cell catalysts lies in the need to combine chemical energy source (in the form of ATP) or reducing/oxidizing power (in the form of NAD(P)H) to the production process. *Candida* yeasts can reduce the 5-carbon sugar xylose to polyol called xylitol by a xylose reductase enzyme:

$$\text{Xylose} + \text{NADH} \longrightarrow \text{Xylitol} + \text{NAD}$$

Commercial production of chemicals by living cells using pathway engineering is still in many cases the best alternative to apply biocatalysis. Isolated enzymes have, however, been successfully used in fine chemical synthesis. The enzymes used for various industrial applications are listed in Table 18.2.

Table 18.2 Applications of enzymes in various fields

Industry	**Enzymes**	**Uses**
Detergent	Proteinase	Protein degradation
	Lipase	Fat removal
	Cellulase	Colour brightening
Textile	Cellulose	Microfibril removal
	Laccase	Colour brightening
Animal feed	Xylanase	Fibre solubility
	Phytase	Release of phosphate

(*Contd.*)

Table 18.2 (Continued)

Industry	Enzymes	Uses
Starch	Amylases	Glucose formation
	Glucose isomerase	Fructose formation
Pulp and paper	Xylanase	Biobleaching
Fruit juice	Pectinase Cellulase Xylanase	Juice clarification and extraction
Baking	Xylanase	Dough conditioning
	Alpha-amylase	Loaf volume
	Glucose oxidase	Shelf-life, dough quality
Dairy	Rennin	Protein coagulation
Brewing	Lactase	Lactose hydrolysis
	Glucanase	Filter aid
	Papain	Haze control
Photography	Protease	Dissolve gelatin off scrap film, allowing recovery of its silver content
Molecular Biology	Restriction enzymes, DNA ligase and polymerase	Used to manipulate DNA in genetic engineering essential for restriction digestion and polymerase chain reaction

CHIRALLY PURE AMINO ACIDS AND ASPARTAME

Natural as well as synthetic amino acids are widely used in the food, feed, agrochemical and pharmaceutical industries. Many proteinogenic amino acids are used in infusion solutions and essential amino acids as animal feed additives. Aspartic acid and phenylalanine methyl ester are combined to form the low-calorie sweetener aspartame. For example, several thousand tons of D-phenylglycine and D-*p*-hydroxyphenylglycine are produced annually for the synthesis of the broad-spectrum antibiotics ampicillin, amoxycillin, cefalexin and others. Natural amino acids are usually produced by microbial fermentation.

Novel enzymatic resolution methods have been developed for the production of L- as well as for D-amino acids. Racemic mixture of amino acid amides is synthesized by Strecker synthesis. Permeabilized cells of *Pseudomonas putida* containing amino acid amidase enzyme that are used to specifically hydrolyse the natural L-form of the amino acid are produced and separated. The D-form can then be chemically formed or recycled after racemization. Aspartame, the intensive non-calorie sweetener, is synthesized in non-aqueous conditions by thermolysin, a proteolytic enzyme, from N-protected aspartic acid and phenylalanine methyl ester.

Rare Sugars

Non-natural monosaccharides are needed as starting materials for new chemicals and pharmaceuticals. Examples are L-ribose, D-psicose, L-xylose, D-tagatose and others. Some of the sugars are produced by chemical isomerization or epimerization. Recently, enzymatic methods have been developed to manufacture practically all D- and L-forms of simple sugars. Glucose isomerase is one of the important industrial enzymes used in fructose manufacturing. It can catalyse previously unknown conversions. For example L-arabinose is isomerized to L-ribulose and slowly also to L-ribose. D-xylose is isomerized to D-xylulose and slowly to D-lyxose. 4-carbon sugars are also good substrates. Enzymatic methods are an important tool in production of rare sugars.

Semi-synthetic Penicillins

Penicillin is produced by genetically modified strains of *Penicillium*. Most of the penicillin is converted by immobilized acylase enzyme to 6-aminopenicillanic acid, which serves as a backbone for many semi-synthetic penicillins. These can be synthesized by chemical or enzymatic methods.

Lipase-based Reactions

In addition to detergent applications, lipases can be used in versatile chemical reactions since they are active in organic solvents. The transferase activity of lipases is used to convert low-value fats into more valuable ones in transesterification reactions. Lipases have also been used to form aromatic and aliphatic polymers. The enzyme can be used for enantiomeric separation of alcohols. In place of alcohols, amines can also be used as the nucleophile. This makes it possible to separate racemic amine mixtures. Chirally pure amines can be used as building blocks for bioactive molecules.

Asymmetric Synthesis

Chiral compounds can be produced in biocatalytic asymmetric syntheses in which a prochiral precursor is converted into a chiral molecule by enantioselective addition reaction. Lyases catalyse the addition of a substance to a double bond or the elimination of a group resulting in an unsaturated bond. A chiral compound is formed in such a reaction. Ammonia lyases are used to produce amino acids from α-keto acid precursors. An example is L-aspartate ammonia lyase in the production of L-aspartic acid.

A novel lyase application involves hydroxynitrile lyase, which catalyses the addition of HCN to aldehydes and ketones. The enzyme from rubber tree has been cloned and over expressed in microorganisms. This enzyme produces valuable chemical intermediates. A third important biocatalytic enzyme group is nitrile hydratases. They catalyse the addition of water to nitriles resulting in the formation of amides. They are used in the production of acrylamide from acrylonitrile and nicotine amide.

ENZYMATIC OLIGOSACCHARIDE SYNTHESIS

Oligosaccharides have found applications in cosmetics, medicines and as functional foods. The chemical synthesis of oligosaccharides is a complicated multi-step effort. Biocatalytic synthesis with isolated enzymes like glycosyltransferases and glycosidases or engineered whole cells are powerful alternatives to chemical methods. Glycosyltransferases catalyse the transfer of monosaccharides from a donor (nucleotide) to acceptors saccharide. These enzymes are extracellular in nature. *Leuconostoc* lactic acid bacteria produce an enzyme called dextran sucrase. It converts sucrose into fructose and a glucose polymer called dextran. Dextran is used in biomedical applications and as a matrix in separation processes. The enzyme can use molecules other than glucose as acceptor and thus novel oligomers can be produced, e.g. antibacterial properties. Glycosidases are hydrolytic enzymes, which can be used for synthetic reactions in a manner similar to thermolysin, used for aspartame synthesis.

High-Fructose Corn Syrup

The process for making the sweetener, known as "high-fructose corn syrup" (HFCS) from sweet corn was developed in the 1970s. High-fructose corn syrup is produced by processing corn starch to yield glucose, and then processing the glucose to produce a high percentage of fructose. The process involves the use of three different enzymes. Firstly, corn starch is treated with α-amylase (bacterial origin), to produce shorter chains of sugars called polysaccharides. Next, an enzyme called glucoamylase, obtained from the fungus *Aspergillus niger*, breaks the sugar chains down even further to yield the simple sugar glucose. The third enzyme, glucose isomerase, converts glucose to a mixture of about 42% fructose and 50–52% glucose.

Two of the enzymes α-amylase and glucose isomerase, are genetically modified to make them more thermostable. HFCS has the same sweetness and taste as sucrose from cane or beet sugar. HFCS is cheaper and easy to transport. This involves exchanging specific amino acids in the primary sequence so that the enzyme is resistant to unfolding or denaturing. This allows the industry to use the enzymes at higher temperatures without loss of activity.

Starch hydrolysis and fructose production The use of starch-degrading enzymes was the first large-scale application of microbial enzymes in food industry. Two enzymes carry out the conversion of starch to glucose: α-amylase and glucoamylase. α-amylase cuts rapidly the large $(\alpha 1 \rightarrow 4)$-linked glucose polymers into shorter oligomers at high temperature. This phase is called liquefaction and is carried out by bacterial enzymes. In the next phase called saccharification, glucoamylase hydrolyses the oligomers into glucose. This is done by fungal enzymes, which operate at a pH and temperature lower than that of α-amylase. Sometimes additional debranching enzymes like pullulanase are added to improve the glucose yield. Fructose is separated from glucose by large-scale chromatographic separation, and crystallized. Alternatively, fructose is concentrated to 55% and used as a high-fructose corn syrup in the soft drink industry. An alternative method to produce fructose is to use sucrose as a starting material. Sucrose is split by invertase into glucose and fructose, fructose is separated and crystallized and then glucose is circulated back to the process.

MEDICAL APPLICATIONS OF ENZYMES

Enzymes make the most desirable therapeutic agents as they reflect the magnitude of the potential rewards: for example, pancreatic enzymes have been in use since the nineteenth century for the treatment of digestive disorders. At present, the most successful applications are extracellular: the removal of toxic substances and the treatment of life-threatening disorders within the blood circulation. Therapeutically useful enzymes (Table 18.3) are required in relatively tiny amounts but at a very high degree of purity and specificity. The kinetic properties of these enzymes are low K_m and high V_{max} in order to be maximally efficient even at very low enzyme and substrate concentrations.

Table 18.3 Therapeutic enzymes

Enzyme	EC number	Reaction	Use
Asparaginase	3.5.1.1	L-Asparagine + $H_2O \rightarrow$ L-aspartate + NH_3	Leukaemia
Collagenase	3.4.24.3	Collagen hydrolysis	Skin ulcers
Glutaminase	3.5.1.2	L-Glutamine + $H_2O \rightarrow$ L-glutamate + NH_3	Leukaemia
Hyaluronidase	3.2.1.35	Hyaluronate hydrolysis	Heart attack
Lysozyme	3.2.1.17	Bacterial cell wall hydrolysis	Antibiotic
Rhodanase	2.8.1.1	$S_2O_3^{2-} + CN^- \rightarrow SO_3^{2-} + SCN^-$	Cyanide poisoning
Ribonuclease	3.1.26.4	RNA hydrolysis	Antiviral
β-Lactamase	3.5.2.6	Penicillin → Penicilloate	Penicillin allergy
Streptokinase	3.4.22.10	Plasminogen → Plasmin	Blood clots
Trypsin	3.4.21.4	Protein hydrolysis	Inflammation
Uricase	1.7.3.3	Urate + $O_2 \rightarrow$ Allantoin	Gout
Urokinase	3.4.21.31	Plasminogen → Plasmin	Blood clots

A number of factors severely reduce this potential utility of enzymes. They are too large to be distributed simply within the body's cells. This is the major reason why enzymes have not yet been successfully applied to the large number of human genetic diseases. A number of methods are being developed in order to overcome this by targeting enzymes, e.g. enzymes with covalently attached β-galactose residues are targeted at hepatocytes and enzymes covalently coupled to target-specific monoclonal antibodies are being used to avoid non-specific side reactions.

Cancer Treatment

A major potential therapeutic application of enzymes is the use of asparaginase that has been proved to be promising for the treatment of acute lymphocytic leukaemia. Its action depends upon the fact that tumour cells are deficient in aspartate-ammonia ligase activity, which restricts their ability to synthesize normally the non-essential amino acid, L-asparagine. Therefore, they are forced to extract it from body fluids. The action of asparaginase does not affect the functioning of normal cells which are able to synthesize enough for their own requirements, but reduces the free exogenous concentration and so induces a state of fatal starvation in the susceptible tumour cells.

Diabetes

In normal individuals the pancreas produces the hormone insulin, enabling them to absorb glucose from the blood, required for energy. Some individuals are unable to produce sufficient insulin and must therefore inject themselves with prescribed insulin to measure their blood sugar level.

Enzyme Deficiency Diseases

About 2000 different enzymes are required for the human body to function. A variety of metabolic diseases are now known to be caused by deficiencies or malfunctions of enzymes. Albinism, for example, is often caused by the absence of tyrosinase, an enzyme essential for the production of cellular pigments. The hereditary lack of phenylalanine hydroxylase results in the disease phenylketonuria (PKU) which, if untreated, leads to severe mental retardation in children.

"Enzyme replacement therapy" can sometimes be employed in some enzyme deficiency diseases. One such example is Gaucher's disease type 1, where the body lacks sufficient levels of the enzyme glucocerebrosidase. This is needed for breakdown of fatty materials, or lipids. This deficiency causes lipids to accumulate, swelling the spleen and liver, crowding out the marrow in the bones, and triggering anaemia and low blood-platelet counts. This can be treated using intravenous enzyme replacement therapy.

Heart Attacks

The enzyme streptokinase is administered intravenously to patients as soon as possible after the onset of a heart attack to dissolve clots in the arteries of the heart wall. This minimizes the amount of damage to the heart muscle. Streptokinase works by stimulating extra production of a naturally produced protease called plasmin, a major constituent of blood clots, therefore dissolving clots once they have fulfilled their purpose of stopping bleeding.

Drug Manufacture

Enzymes are particularly useful when it comes to pharmaceutical chemicals. This is because they are stereospecific and are thus able to make single-isomer (chiral) compounds, whereas ordinary chemical methods normally yield mixtures of stereo-isomers. Many organic molecules exist in left-handed and right-handed forms. Enantiomers have virtually identical physical properties and identical chemical properties except towards other chiral compounds like enzymes found in the body. Drug manufacturers may either use the purified enzymes or enzyme-containing microorganisms to biocatalyse a given chemical reaction. A typical example is the enzyme-based production of β-lactam, an intermediate for carboxylic nucleosides such as the AIDS drug **abacavir sulphate**, produced by ChiroTech.

IMMOBILIZED ENZYMES

An alternative way to use enzymes is to immobilize them so that they can be reused. The largest application of an immobilized enzyme is the conversion of glucose syrup to high-fructose syrup for food applications.

Another way to immobilize enzymes is by using ultrafiltration membranes in the reactor system. The large enzyme molecules cannot pass through the membrane but the small molecular reaction products can. Therefore enzymes are retained in a reaction system and the products leave the system continuously. This method has been used in production of chirally pure amino acids from racemic mixtures of amino acid derivatives. Enzymes have also been immobilized on membranes for analytical purposes. The best known example is glucose oxidase, which is used to measure glucose concentrations in biological samples.

Other uses of immobilized enzyme technology include the production of whey syrup, which is used in confectionery in place of sweetened, condensed milk. The whey is a by-product of cheese-making, which contains lactose and protein.

The lactose is hydrolysed to glucose and galactose, which taste sweeter and dissolve well, compared to the rather tasteless, poorly soluble lactose.

ENZYMES IN BIOFUELS

Enzymes may be used to produce fuels from renewable sources of biomass. Such enzymes include cellulases, which convert cellulose fibres from feedstock like corn into sugars. These sugars are subsequently fermented into ethanol by microorganisms. A new process called simultaneous saccharification and fermentation has greatly improved ethanol production efficiency. In this new process, cellulase enzymes and fermentation microorganisms are combined in a single reaction mixture to produce ethanol in one step, rather than producing sugars from cellulose and then fermenting them into ethanol separately.

ENZYMES IN RECOMBINANT DNA TECHNOLOGY

By using recombinant DNA technology, microorganisms may be genetically modified to produce a desired enzyme under specific conditions. This is accomplished by insertion through small circular pieces of DNA, known as plasmids to produce genes into the genomes of organisms that possess another desirable trait, such as the ability to thrive on inexpensive nutrients. Therefore, both the enzyme and the original trait will be expressed in a single recombinant microorganism.

PROTEIN ENGINEERING

All enzymes are made of proteins, which are large molecules formed from basic units, called amino acids, strung together like beads on a chain. To form functional enzymes, long chains of amino acids must be folded properly. Some enzymes consist of only one chain, whereas others are made of several chains that fit together. Different ways to improve and engineer enzymes can be known by studying the relation between the structure of a protein and its functions. Proteins may be modified by changing one or more amino acids, and/or changing the way the amino acid chains fold and fit together.

Therefore, the effective catalytic properties of enzymes have promoted their introduction into several industrial products and processes. Recent developments in biotechnology, particularly in areas such as protein engineering and directed evolution, have provided important tools for the efficient development of new enzymes. This has resulted in the development of enzymes with improved properties for established technical applications and in the production of new enzymes tailor-made for entirely new areas of application where enzymes have not previously been used.

19

BIOINFORMATICS IN ENZYMOLOGY

CATALYTIC SITE ATLAS (CSA)

The Catalytic Site Atlas (CSA) is a database documenting enzyme active sites and catalytic residues in enzymes of 3D structure. A classification of catalytic residues which includes only those residues is thought to be directly involved in some aspect of the reaction catalysed by an enzyme. The CSA contains two types of entries:

1. Original hand-annotated entries, derived from the primary literature.
2. Homologous entries, found by PSI-BLAST alignment to one of the original entries.

Access to the CSA is through a PDB code, SWISS-PROT entry or E.C. number. Each CSA entry lists the catalytic residues found in that entry, using PDB residue numbering. Each site is also marked with an evidence tag, which is either "Literature reference" or "PSI-BLAST hit".

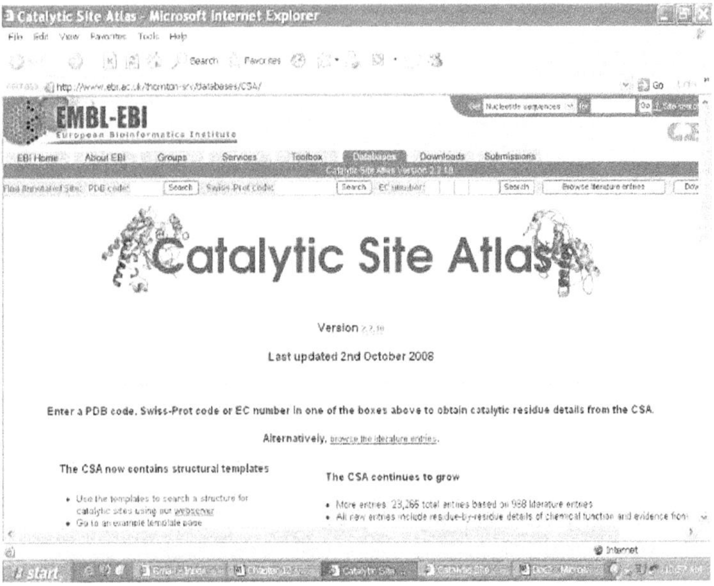
Catalytic Site Atlas - Microsoft Internet Explorer
EMBL-EBI
European Bioinformatics Institute
EBI Home About EBI Groups Services Toolbox Databases Downloads Submissions
Catalytic Site Atlas
Version
Last updated 2nd October 2008
Enter a PDB code, Swiss-Prot code or EC number in one of the boxes above to obtain catalytic residue details from the CSA.
Alternatively, browse the literature entries.
The CSA now contains structural templates
The CSA continues to grow
start

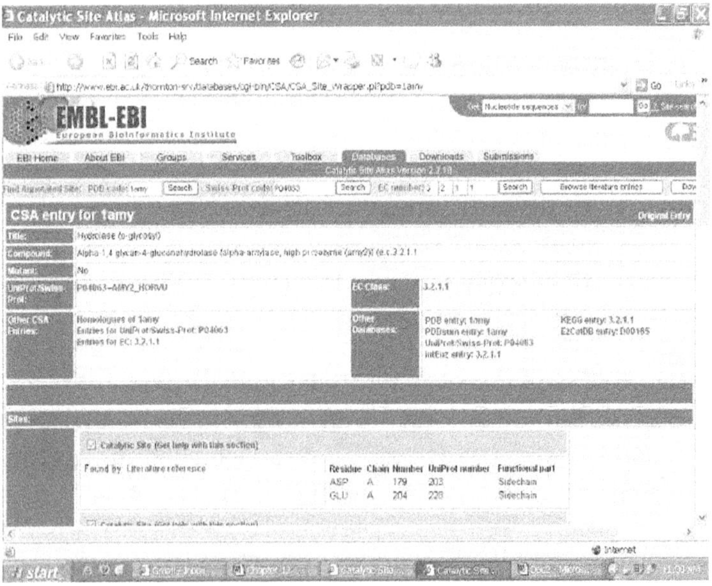
Catalytic Site Atlas - Microsoft Internet Explorer
EMBL-EBI
European Bioinformatics Institute
EBI Home About EBI Groups Services Toolbox Databases Downloads Submissions
CSA entry for 1amy
Original Entry
Title:
Compound:
Mutant:
No
EC Class:
3.2.1.1
Other CSA Entries:
Homologues of 1amy
Other Databases:
PDB entry: 1amy
KEGG entry: 3.2.1.1
Sites:
Found by: Literature reference
Residue Chain Number UniProt number Functional part
ASP A 179 203 Sidechain
GLU A 204 228 Sidechain
start

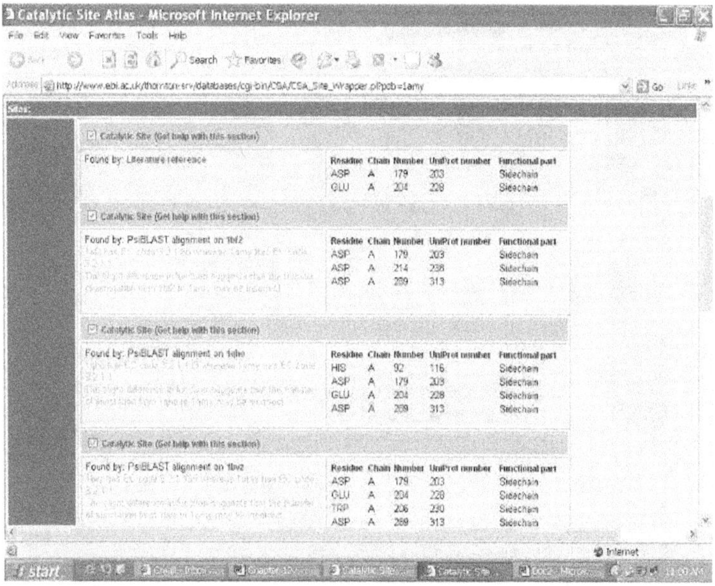
Catalytic Site Atlas - Microsoft Internet Explorer
http://www.ebi.ac.uk/thornton-srv/databases/cgi-bin/CSA/CSA_Site_Wrapper.pl?pdb=1amy
Found by: Literature reference
Residue Chain Number UniProt number Functional part
ASP A 179 203 Sidechain
GLU A 204 228 Sidechain
Found by: PsiBLAST alignment on 1bf2
ASP A 179 203 Sidechain
ASP A 214 238 Sidechain
ASP A 289 313 Sidechain
Found by: PsiBLAST alignment on 1qho
HIS A 92 116 Sidechain
ASP A 179 203 Sidechain
GLU A 204 228 Sidechain
ASP A 289 313 Sidechain
Found by: PsiBLAST alignment on 1bvz
ASP A 179 203 Sidechain
GLU A 204 228 Sidechain
TRP A 206 230 Sidechain
ASP A 289 313 Sidechain

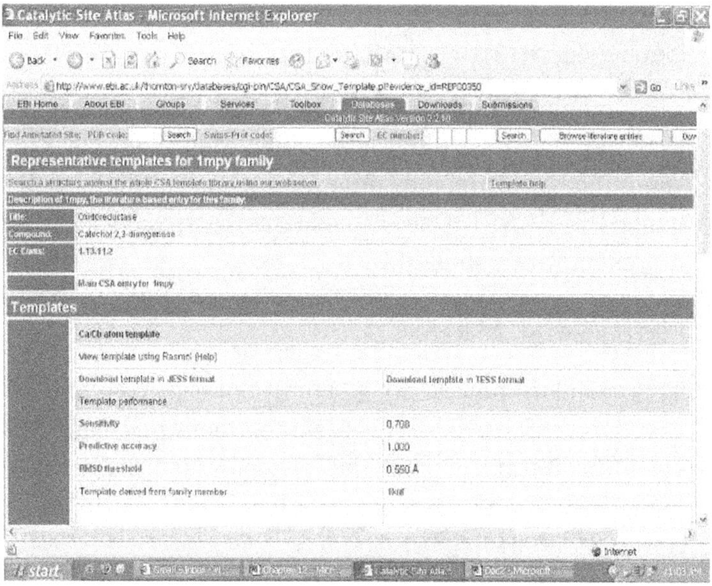
Catalytic Site Atlas - Microsoft Internet Explorer
Representative templates for 1mpy family
Templates
CaCb atom template
View template using Rasmol (Help)
Download template in JESS format
Download template in TESS format
Template performance
Sensitivity 0.708
Predictive accuracy 1.000
RMSD threshold 0.560 A

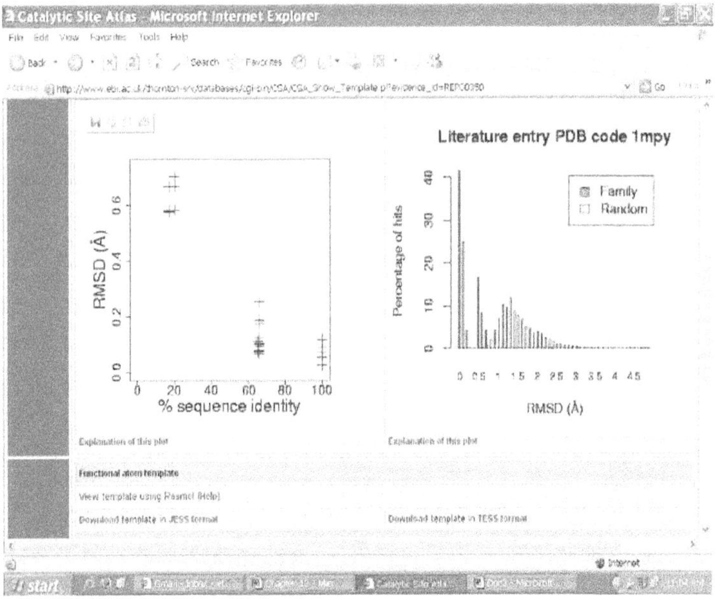

Sequence-Structure Plot

This template represents a "family" of enzymes: one enzyme that has had its catalytic residues identified based on the scientific literature, plus other enzymes identified as relatives using PSI-BLAST. One of these has been used to derive the representative template.

Histogram of Matches

This template represents a "family" of enzymes: one enzyme that has had its catalytic residues identified based on the scientific literature, plus other enzymes identified as relatives using PSI-BLAST. One of these has been used to derive the representative template.

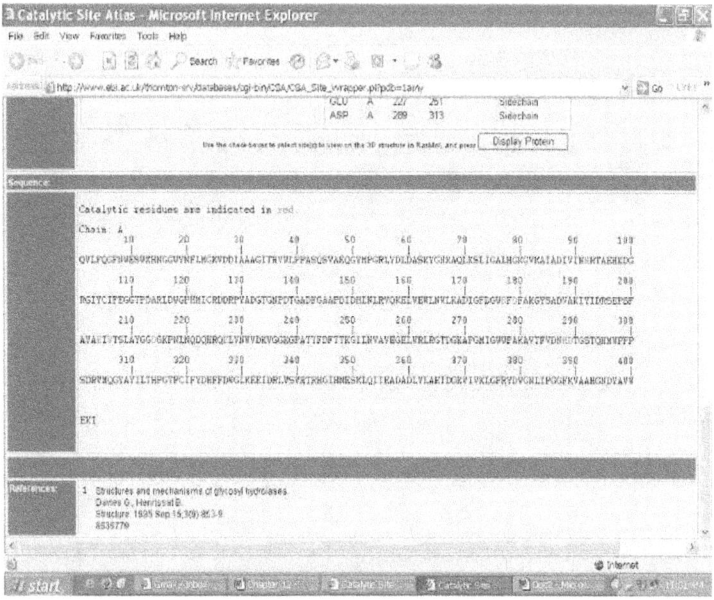

INTEGRATED RELATIONAL ENZYME DATABASE INTENZ

IntEnz is a freely available resource focused on enzyme nomenclature. IntEnz is created in collaboration with the Swiss Institute of Bioinformatics (SIB). This collaboration is responsible for the production of the enzyme resource. The goal of IntEnz is to create a single relational database containing enzyme data from three different sources: the official version of the enzyme nomenclature comprising recommendations of the Nomenclature Committee of the International Union of Biochemistry and Molecular Biology (NC-IUBMB) on the nomenclature and classification of enzyme-catalysed reactions; Swiss Institute of Bioinformatics (SIB) produces an Enzyme Nomenclature database (ENZYME); BRENDA, the enzyme function database which contains information on substrates, products, and inhibitors.

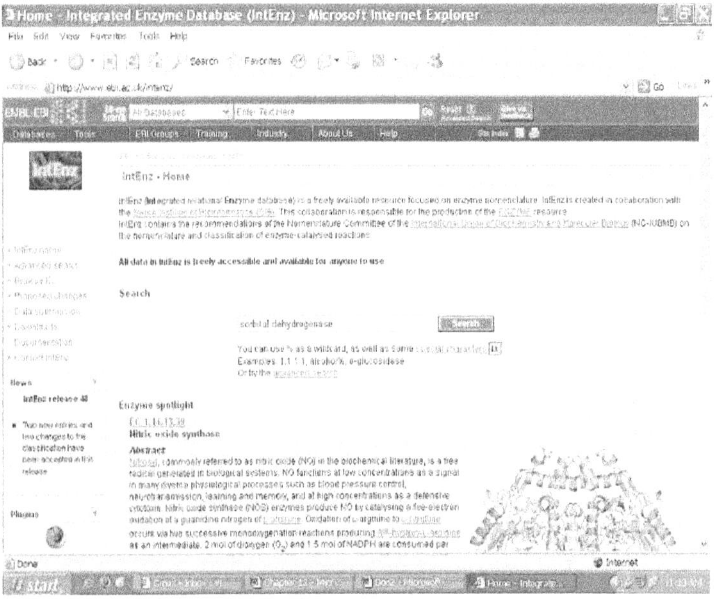
Home - Integrated Enzyme Database (IntEnz) - Microsoft Internet Explorer
IntEnz - Home
All data in IntEnz is freely accessible and available for anyone to use
Search
sorbitol dehydrogenase
Enzyme spotlight
Nitric oxide synthase
Abstract
Done
Internet

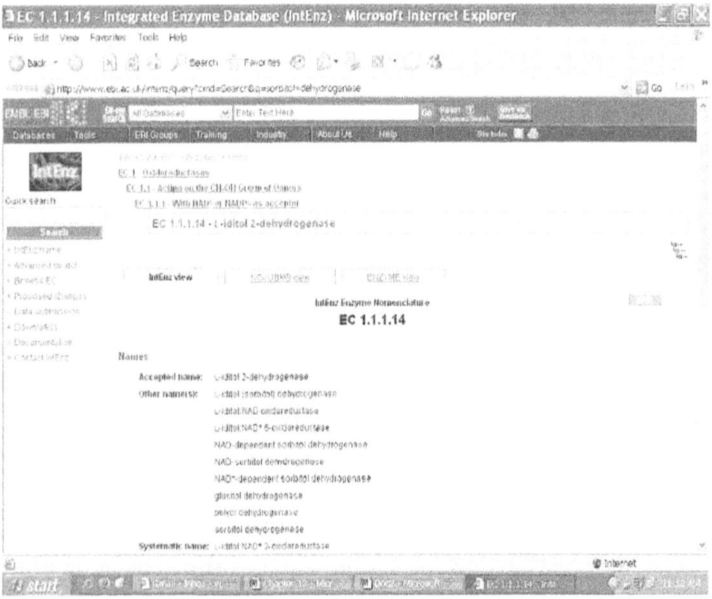
EC 1.1.1.14 - Integrated Enzyme Database (IntEnz) - Microsoft Internet Explorer
EC 1.1.1.14 - L-iditol 2-dehydrogenase
IntEnz view
IntEnz Enzyme Nomenclature
EC 1.1.1.14
Names
Accepted name: L-iditol 2-dehydrogenase
Other name(s):
glucitol dehydrogenase
sorbitol dehydrogenase
Systematic name:
Internet

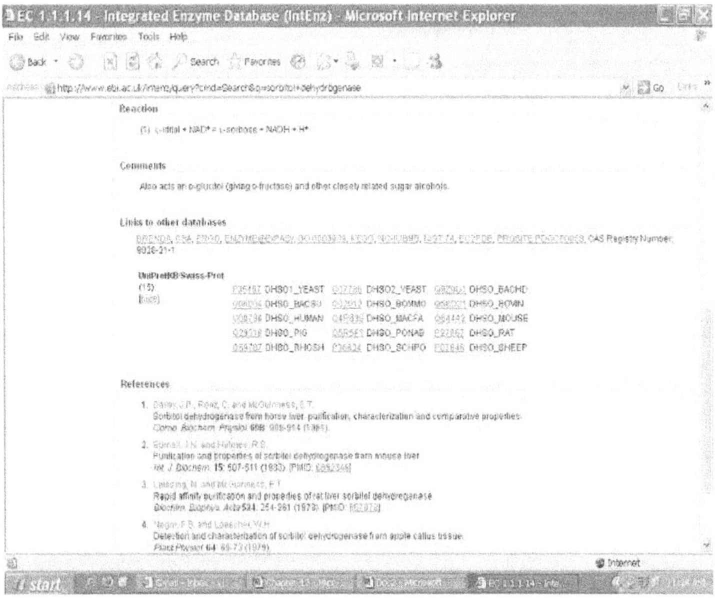

REBASE

The Restriction Enzyme Database (REBASE), is a collection of information about restriction enzymes and related proteins. Restriction enzymes are a group of enzymes that can cut double-stranded DNA molecules into smaller restriction fragments. The enzymes work by cleaving the chemical bonds in the phosphate backbone of the DNA molecule. These bonds can be reformed by ligase enzymes and in this manner restriction fragments with different origin can be spliced together in novel ways. This process forms the basis of many procedures in molecular biology and genetic engineering. Bioinformatical algorithms can detect sites in a DNA sequence where a given enzyme will cut. They do so by searching the input sequence for regions that match the specific recognition sequence of the enzyme in question.

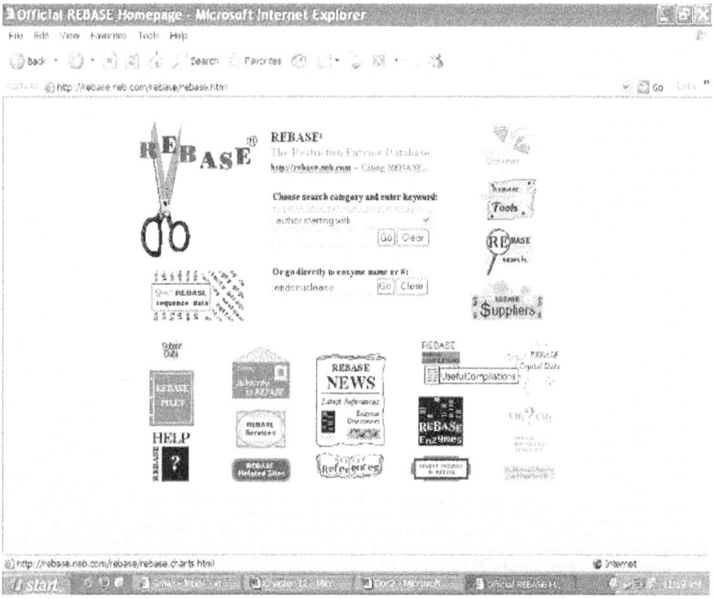
Official REBASE Homepage - Microsoft Internet Explorer
http://rebase.neb.com/rebase/rebase.html
REBASE
The Restriction Enzyme Database
Choose search category and enter keyword:
Or go directly to enzyme name or #:
REBASE NEWS
HELP
Suppliers
References
http://rebase.neb.com/rebase/rebase.charts.html

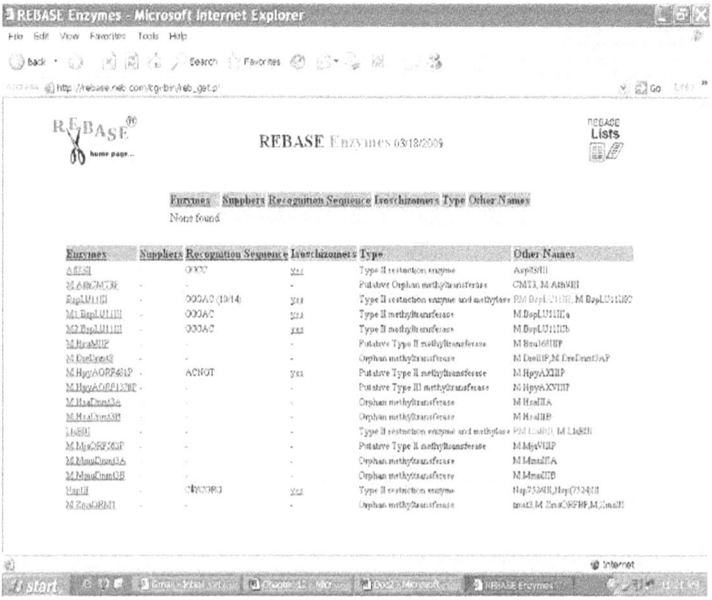
REBASE Enzymes - Microsoft Internet Explorer
http://rebase.neb.com/cgi-bin/reb_get.pl
REBASE Enzymes 03/18/2009
REBASE Lists
Enzymes Suppliers Recognition Sequence Isoschizomers Type Other Names
None found

The enzymes that have palindromic recognition sequences will match the same region no matter which DNA strand is examined. Therefore, only one strand is searched for in this group of enzymes. Enzymes with non-palindromic recognition sequences can match the regions on one strand that do not correspond to a match on the opposite strand, and both strands must therefore be searched.

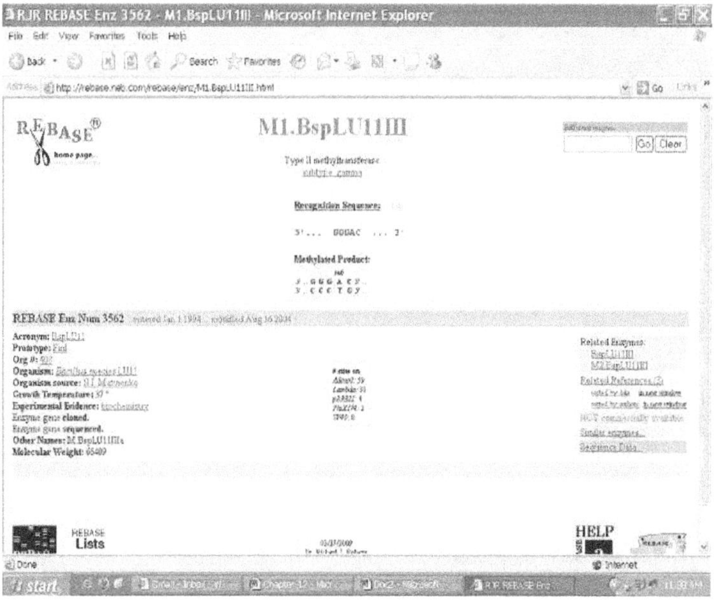

ENZYME STRUCTURES DATABASE

This database contains the known enzyme structures that have been deposited in the Protein Data Bank (PDB). There are currently 24,840 PDB enzyme entries in the PDB involving 22,103 separate PDB files. The enzyme structures can also be accessed via the PDBsum database. The enzyme structures are classified by their EC number.

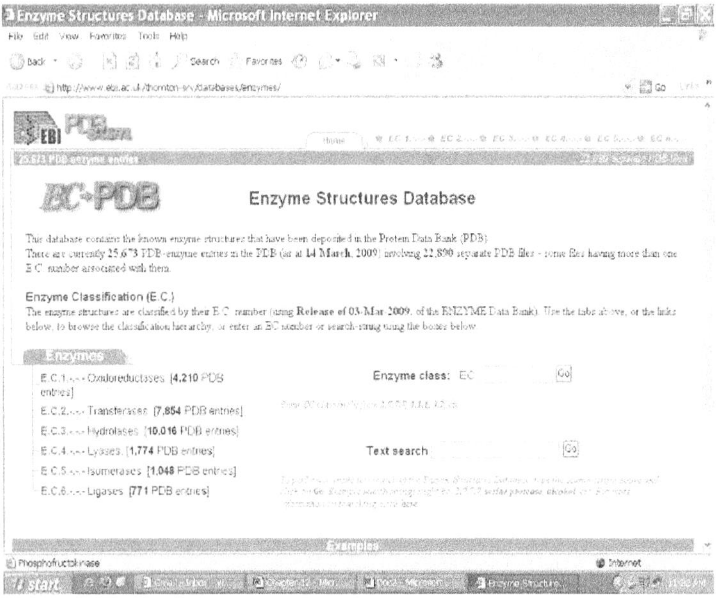
Enzyme Structures Database - Microsoft Internet Explorer
EC-PDB
Enzyme Structures Database
This database contains the known enzyme structures that have been deposited in the Protein Data Bank (PDB).
Enzyme Classification (E.C.)
Enzymes
E.C.1.-.-.- Oxidoreductases [4,210 PDB entries]
E.C.2.-.-.- Transferases [7,854 PDB entries]
E.C.3.-.-.- Hydrolases [10,016 PDB entries]
E.C.4.-.-.- Lyases [1,774 PDB entries]
E.C.5.-.-.- Isomerases [1,048 PDB entries]
E.C.6.-.-.- Ligases [771 PDB entries]
Enzyme class: EC
Text search
Examples
Phosphofructokinase

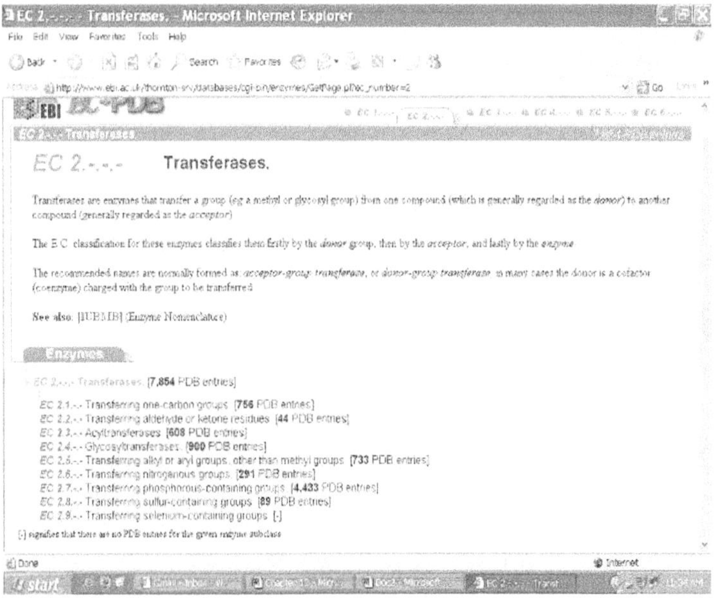
EC 2.-.-.- Transferases. - Microsoft Internet Explorer
EC 2.-.-.- Transferases.
Transferases are enzymes that transfer a group (eg a methyl or glycosyl group) from one compound (which is generally regarded as the donor) to another compound (generally regarded as the acceptor).
See also: [IUBMB] (Enzyme Nomenclature)
Enzymes
EC 2.-.-.- Transferases [7,854 PDB entries]
EC 2.1.-.- Transferring one-carbon groups [756 PDB entries]
EC 2.2.-.- Transferring aldehyde or ketone residues [44 PDB entries]
EC 2.3.-.- Acyltransferases [608 PDB entries]
EC 2.4.-.- Glycosyltransferases [900 PDB entries]
EC 2.5.-.- Transferring alkyl or aryl groups, other than methyl groups [733 PDB entries]
EC 2.6.-.- Transferring nitrogenous groups [291 PDB entries]
EC 2.7.-.- Transferring phosphorous-containing groups [4,433 PDB entries]
EC 2.8.-.- Transferring sulfur-containing groups [89 PDB entries]
EC 2.9.-.- Transferring selenium-containing groups [-]

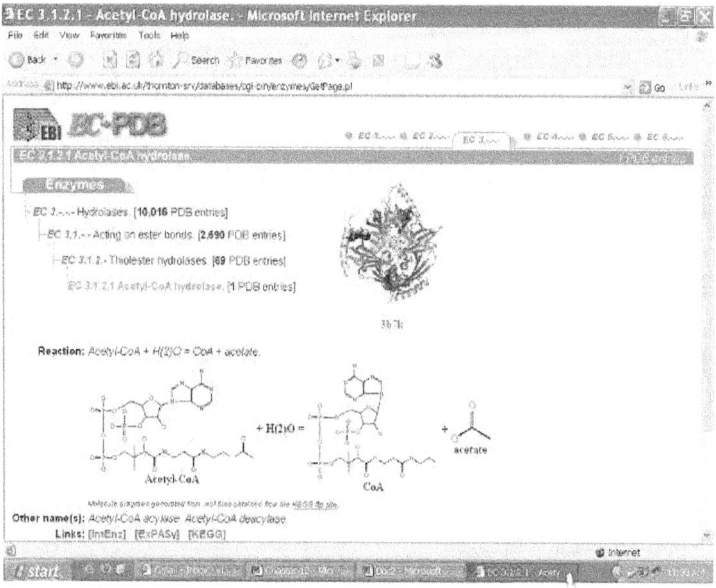
EC 3.1.2.1 - Acetyl-CoA hydrolase. - Microsoft Internet Explorer
EBI EC-PDB
Enzymes
EC 3.-.-.- Hydrolases [10,016 PDB entries]
EC 3.1.-.- Acting on ester bonds [2,690 PDB entries]
EC 3.1.2.- Thiolester hydrolases [69 PDB entries]
EC 3.1.2.1 Acetyl-CoA hydrolase [1 PDB entries]
3b7k
Reaction: Acetyl-CoA + H(2)O = CoA + acetate.
+ H(2)O =
acetate
Acetyl-CoA
CoA
Other name(s): Acetyl-CoA acylase Acetyl-CoA deacylase
Links: [IntEnz] [ExPASy] [KEGG]

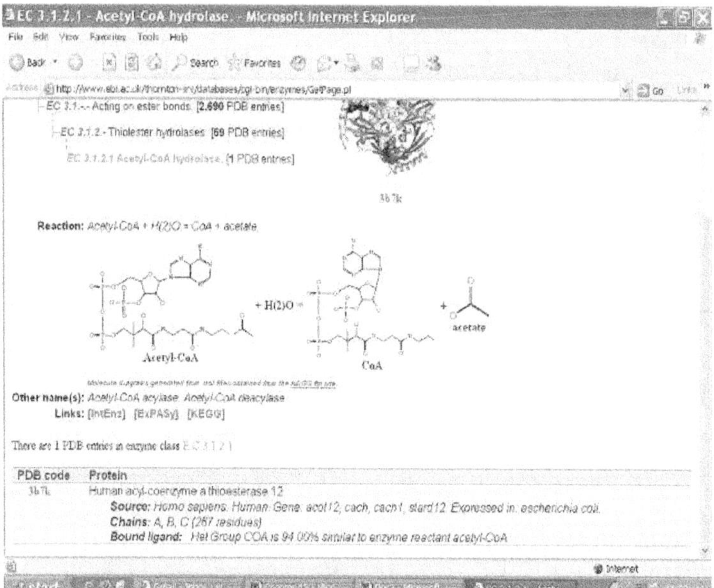
EC 3.1.2.1 - Acetyl-CoA hydrolase. - Microsoft Internet Explorer
EC 3.1.-.- Acting on ester bonds [2,690 PDB entries]
EC 3.1.2.- Thiolester hydrolases [69 PDB entries]
EC 3.1.2.1 Acetyl-CoA hydrolase [1 PDB entries]
3b7k
Reaction: Acetyl-CoA + H(2)O = CoA + acetate.
+ H(2)O =
acetate
Acetyl-CoA
CoA
Other name(s): Acetyl-CoA acylase Acetyl-CoA deacylase
Links: [IntEnz] [ExPASy] [KEGG]
There are 1 PDB entries in enzyme class E.C.3.1.2.1
PDB code
Protein
3b7k
Human acyl-coenzyme a thioesterase 12
Source: Homo sapiens. Human. Gene: acot12, cach, cach1, stard12. Expressed in: escherichia coli.
Chains: A, B, C (287 residues)
Bound ligand: Het Group COA is 94.00% similar to enzyme reactant acetyl-CoA

BioCyc OPEN CHEMICAL DATABASE (BOCD)

The BioCyc Open Chemical Database (BOCD) is a collection of chemical compound data from the BioCyc databases. Most of the compounds act as substrates in enzyme-catalysed metabolic reactions, but some compounds serve as enzyme activators, inhibitors, or cofactors. Chemical structures are provided for the majority of compounds. The compounds include extensive lists of synonyms, and have undergone several types of quality control, including validation by a reaction balance checker, which is applied to the BioCyc databases.

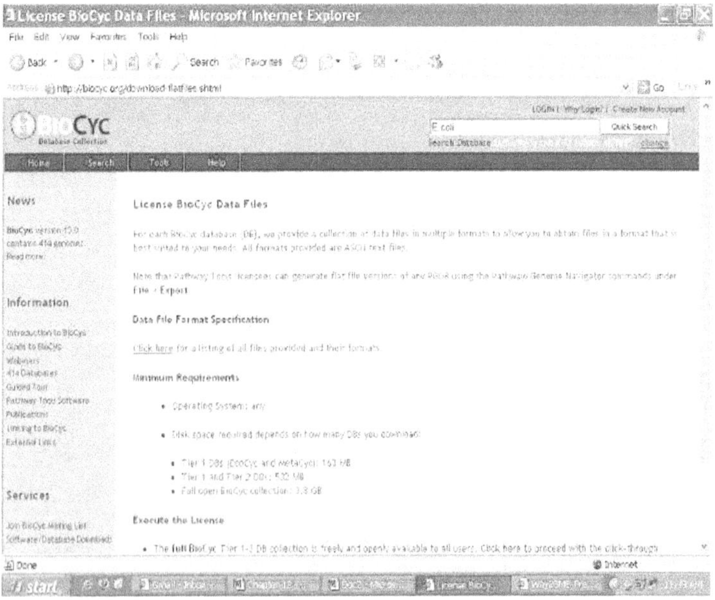

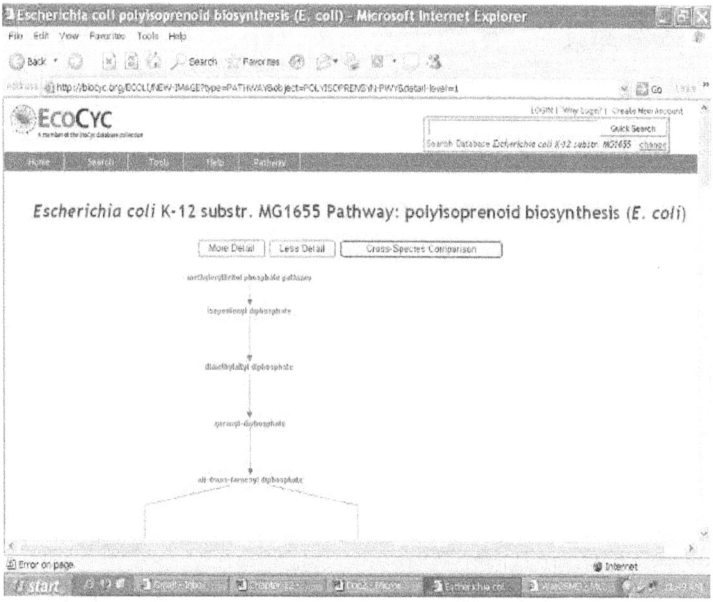
Escherichia coli polyisoprenoid biosynthesis (E. coli) - Microsoft Internet Explorer
File Edit View Favorites Tools Help
Back Search Favorites
http://biocyc.org/ECOLI/NEW-IMAGE?type=PATHWAY&object=POLYISOPRENSYN-PWY&detail-level=1
Go
EcoCyc
Quick Search
Escherichia coli K-12 substr. MG1655 Pathway: polyisoprenoid biosynthesis (E. coli)
More Detail
Less Detail
Cross-Species Comparison
Error on page
Internet
start

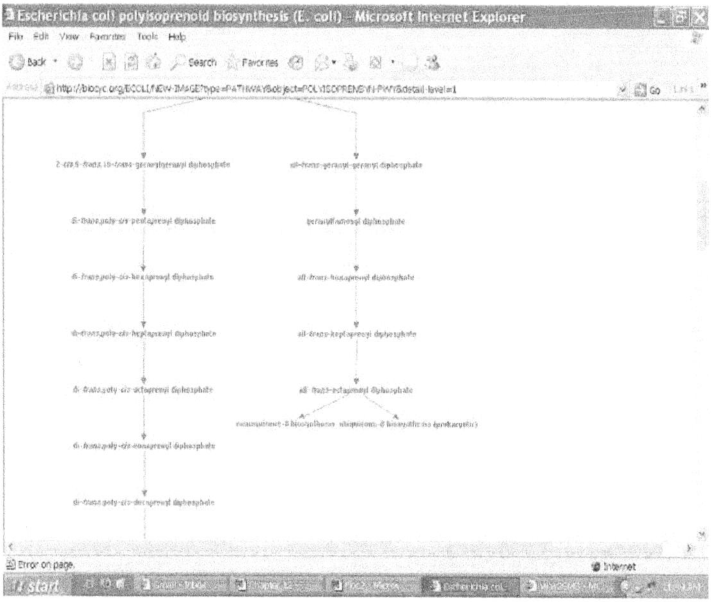
Escherichia coli polyisoprenoid biosynthesis (E. coli) - Microsoft Internet Explorer
File Edit View Favorites Tools Help
Back Search Favorites
http://biocyc.org/ECOLI/NEW-IMAGE?type=PATHWAY&object=POLYISOPRENSYN-PWY&detail-level=1
Go
Error on page.
Internet
start

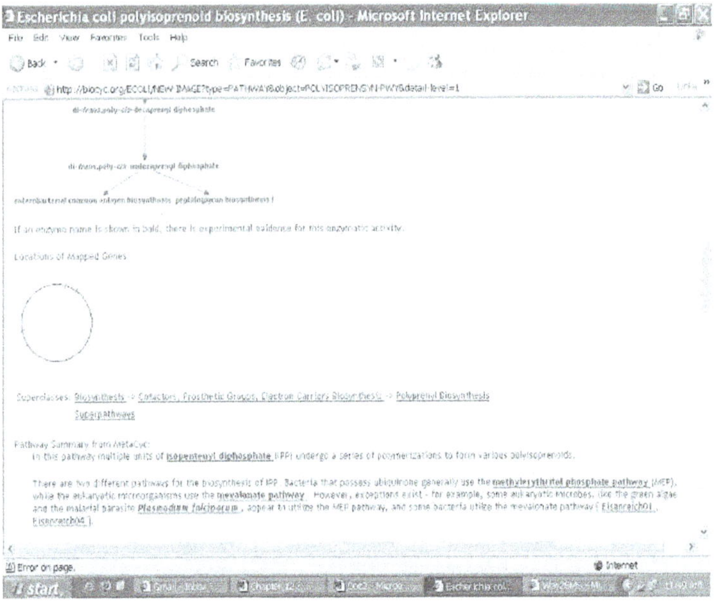
Escherichia coli polyisoprenoid biosynthesis (E. coli) - Microsoft Internet Explorer
File Edit View Favorites Tools Help
Back Search Favorites
http://biocyc.org/ECOLI/NEW-IMAGE?type=PATHWAY&object=POLYISOPRENSYN-PWY&detail-level=1
Go
di-trans,poly-cis-undecaprenyl diphosphate
enterobacterial common antigen biosynthesis
If an enzyme name is shown in bold, there is experimental evidence for this enzymatic activity.
Locations of Mapped Genes
Superclasses: Biosynthesis -> Cofactors, Prosthetic Groups, Electron Carriers Biosynthesis -> Polyprenyl Biosynthesis
Superpathways
Pathway Summary from MetaCyc:
In this pathway multiple units of isopentenyl diphosphate (IPP) undergo a series of polymerizations to form various polyisoprenoids.
There are two different pathways for the biosynthesis of IPP. Bacteria that possess ubiquinone generally use the methylerythritol phosphate pathway (MEP), while the eukaryotic microorganisms use the mevalonate pathway. However, exceptions exist - for example, some eukaryotic microbes, like the green algae and the malarial parasite Plasmodium falciparum, appear to utilize the MEP pathway, and some bacteria utilize the mevalonate pathway [Eisenreich01, Eisenreich04].
Error on page.
Internet
start

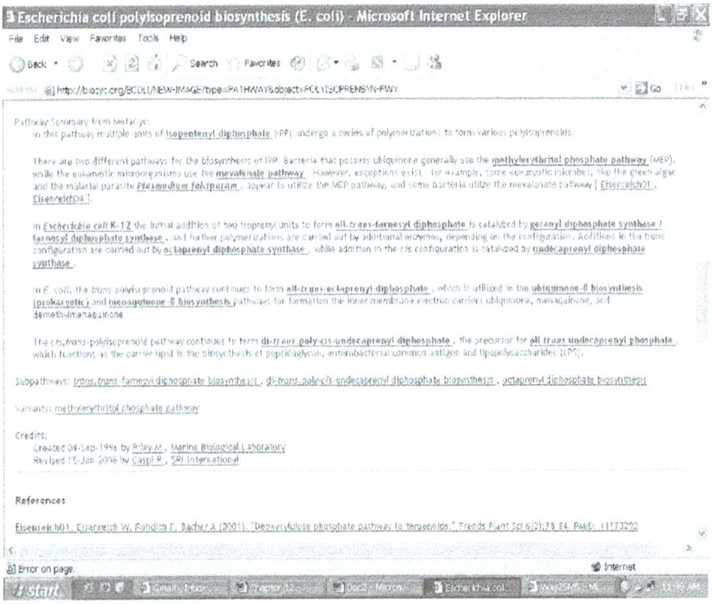
Escherichia coli polyisoprenoid biosynthesis (E. coli) - Microsoft Internet Explorer
File Edit View Favorites Tools Help
Back Search Favorites
http://biocyc.org/ECOLI/NEW-IMAGE?type=PATHWAY&object=POLYISOPRENSYN-PWY
Go
Pathway Summary from MetaCyc:
In this pathway multiple units of isopentenyl diphosphate (IPP) undergo a series of polymerizations to form various polyisoprenoids.
There are two different pathways for the biosynthesis of IPP. Bacteria that possess ubiquinone generally use the methylerythritol phosphate pathway (MEP), while the eukaryotic microorganisms use the mevalonate pathway. However, exceptions exist - for example, some eukaryotic microbes, like the green algae and the malarial parasite Plasmodium falciparum, appear to utilize the MEP pathway, and some bacteria utilize the mevalonate pathway [Eisenreich01, Eisenreich04].
In Escherichia coli K-12 the initial addition of two isoprenyl units to form all-trans-farnesyl diphosphate is catalyzed by geranyl diphosphate synthase / farnesyl diphosphate synthase, and further polymerizations are carried out by additional enzymes, depending on the configuration. Additions in the trans configuration are carried out by octaprenyl diphosphate synthase, while addition in the cis configuration is catalyzed by undecaprenyl diphosphate synthase.
In E. coli, the trans polyisoprenoid pathway continues to form all-trans-octaprenyl diphosphate, which is utilized in the ubiquinone-8 biosynthesis (prokaryotic) and menaquinone-8 biosynthesis pathways for formation the inner membrane electron carriers ubiquinone, menaquinone, and demethylmenaquinone
The cis,trans-polyisoprenoid pathway continues to form di-trans,poly-cis-undecaprenyl diphosphate, the precursor for all trans undecaprenyl phosphate, which functions as the carrier lipid in the biosynthesis of peptidoglycan, enterobacterial common antigen and lipopolysaccharides (LPS).
Subpathways: trans,trans-farnesyl diphosphate biosynthesis, di-trans,poly-cis-undecaprenyl diphosphate biosynthesis, octaprenyl diphosphate biosynthesis
Variants: methylerythritol phosphate pathway
Credits:
Created 04-Sep-1996 by Riley M., Marine Biological Laboratory
Revised 15-Jan-2006 by Caspi R., SRI International
References
Eisenreich01: Eisenreich W, Rohdich F, Bacher A (2001). "Deoxyxylulose phosphate pathway to terpenoids." Trends Plant Sci 6(2);78-84. PMID: 11173292
Error on page.
Internet
start

ENZYME NOMENCLATURE DATABASE

The ENZYME database is a repository of information related to the nomenclature of enzymes. In recent years it has became an indispensable resource for the development of metabolic databases. The current version contains information on 3705 enzymes. It is available through the ExPASy WWW server. (http://www.expasy.ch/enzyme/)

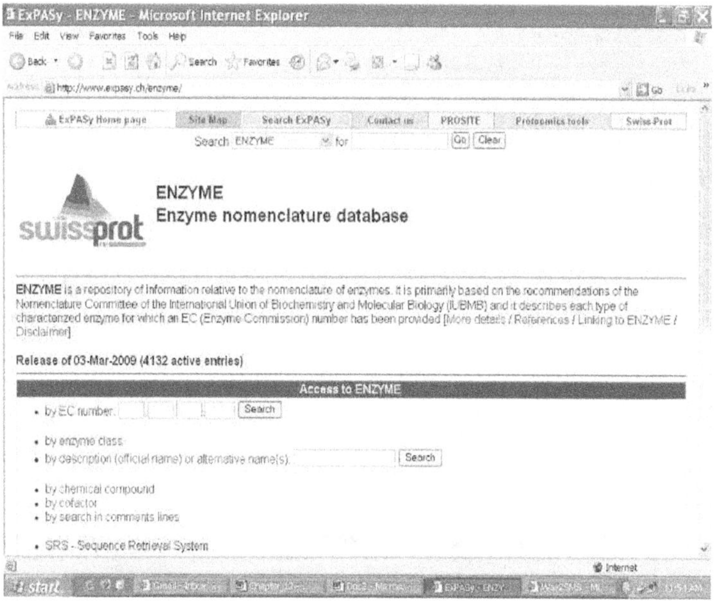

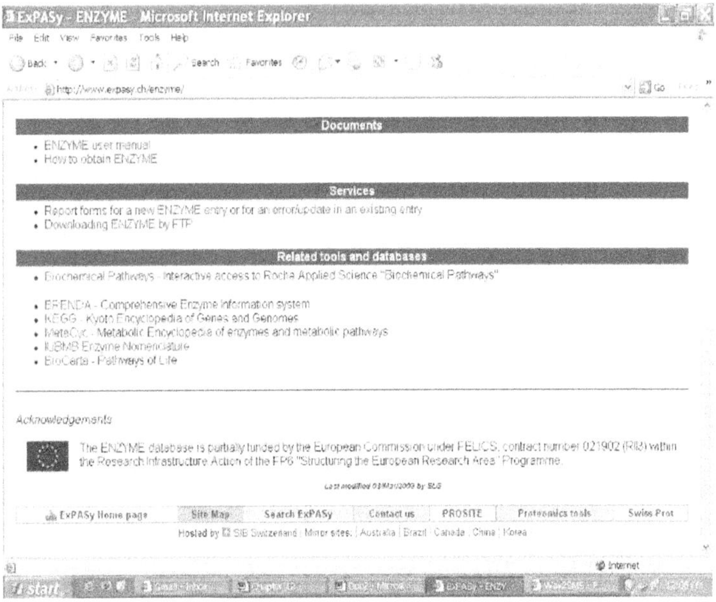
ExPASy - ENZYME - Microsoft Internet Explorer
File Edit View Favorites Tools Help
http://www.expasy.ch/enzyme/
Documents
ENZYME user manual
How to obtain ENZYME
Services
Report forms for a new ENZYME entry or for an error/update in an existing entry
Downloading ENZYME by FTP
Related tools and databases
Biochemical Pathways - Interactive access to Roche Applied Science "Biochemical Pathways"
BRENDA - Comprehensive Enzyme Information system
KEGG - Kyoto Encyclopedia of Genes and Genomes
MetaCyc - Metabolic Encyclopedia of enzymes and metabolic pathways
IUBMB Enzyme Nomenclature
BioCarta - Pathways of Life
Acknowledgements
The ENZYME database is partially funded by the European Commission under FELICS, contract number 021902 (RII3) within the Research Infrastructure Action of the FP6 "Structuring the European Research Area" Programme.
ExPASy Home page
Site Map
Search ExPASy
Contact us
PROSITE
Proteomics tools
Swiss-Prot
Hosted by SIB Switzerland | Mirror sites: Australia | Brazil | Canada | China | Korea
Internet
start

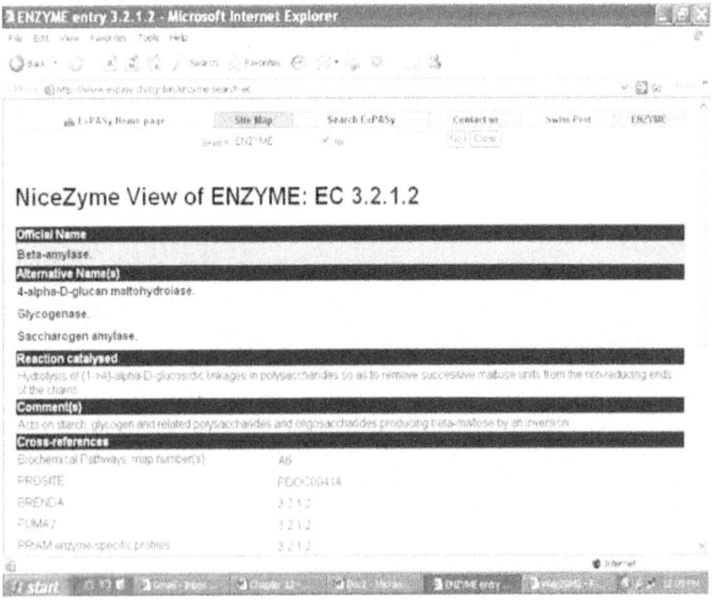
ENZYME entry 3.2.1.2 - Microsoft Internet Explorer
ExPASy Home page
Site Map
Search ExPASy
Contact us
Swiss-Prot
ENZYME
NiceZyme View of ENZYME: EC 3.2.1.2
Official Name
Beta-amylase.
Alternative Name(s)
4-alpha-D-glucan maltohydrolase.
Glycogenase.
Saccharogen amylase.
Reaction catalysed
Hydrolysis of (1->4)-alpha-D-glucosidic linkages in polysaccharides so as to remove successive maltose units from the non-reducing ends of the chains
Comment(s)
Acts on starch, glycogen and related polysaccharides and oligosaccharides producing beta-maltose by an inversion
Cross-references
Biochemical Pathways; map number(s)
A6
PROSITE
PDOC00414
BRENDA
3.2.1.2
PUMA2
3.2.1.2
PRIAM enzyme-specific profiles
3.2.1.2
Internet
start

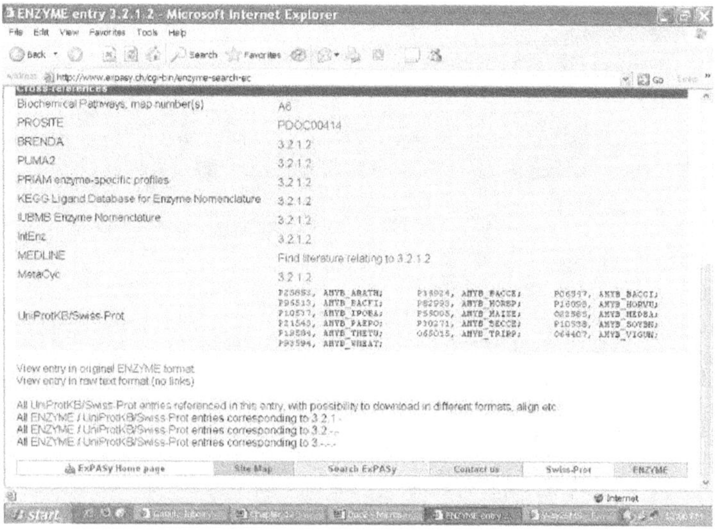

KINASE PATHWAY DATABASE

Protein kinases play a crucial role in the regulation of cellular functions. These molecules are important for understanding signalling pathways and organism characteristics. The Kinase Pathway Database is an integrated database involving major completely sequenced eukaryotes. It contains the classification of protein kinases and their functional conservation, ortholog tables among species, protein–protein, protein–gene, and protein–compound interaction data, domain information, and structural information. It also provides an automatic pathway graphic image interface. The protein, gene, and compound interactions are automatically extracted from abstracts for all genes and proteins by Natural Language Processing (NLP). The method of automatic extraction uses phrase patterns and the GENA protein, gene, and compound name dictionary. With this database, pathways are easily compared among species using data with more than 47,000 protein interactions and protein kinase ortholog tables. The database is available for querying and browsing at http://inasedb.ontology.ims.u-tokyo.ac.jp/.

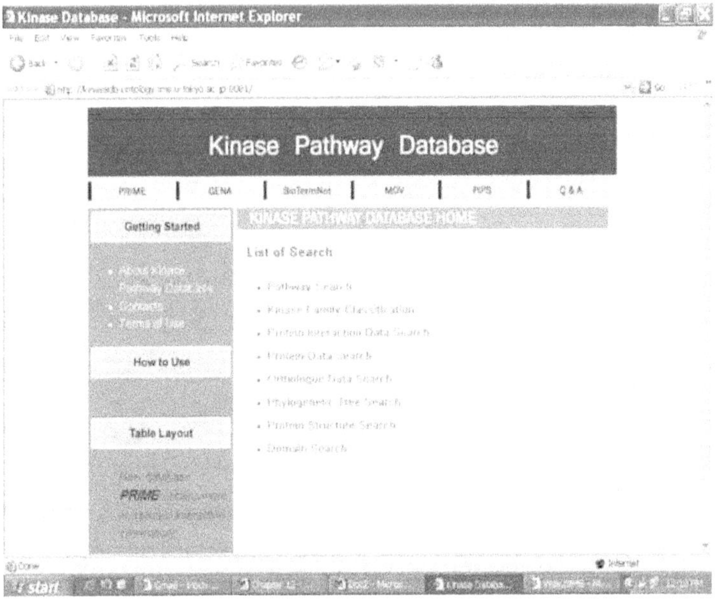
Kinase Database - Microsoft Internet Explorer
Kinase Pathway Database
PRIME
GENA
BioTermNet
MGV
PIPS
Q & A
KINASE PATHWAY DATABASE HOME
Getting Started
List of Search
How to Use
Table Layout
PRIME

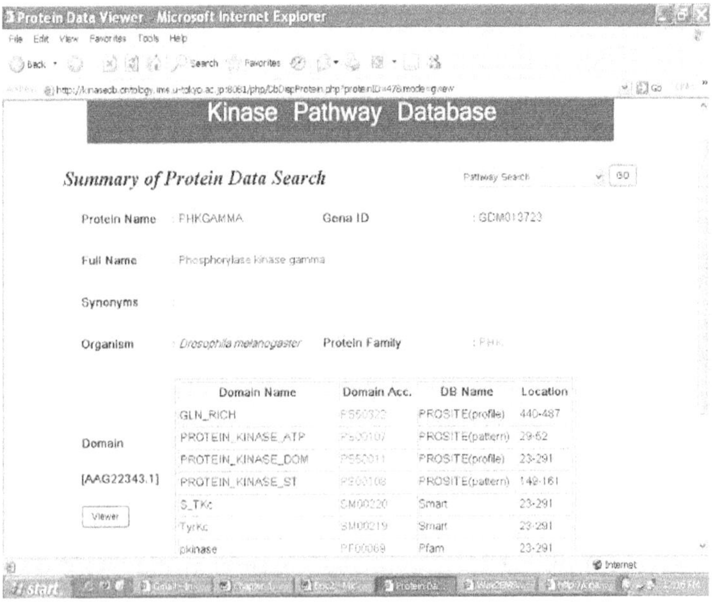
Protein Data Viewer - Microsoft Internet Explorer
Kinase Pathway Database
Summary of Protein Data Search
Pathway Search
GO
Protein Name
: PHKGAMMA
Gena ID
: GDM013723
Full Name
Phosphorylase kinase gamma
Synonyms
Organism
Drosophila melanogaster
Protein Family
Domain
[AAG22343.1]
Viewer
Domain Name
Domain Acc.
DB Name
Location
GLN_RICH
PROSITE(profile)
440-487
PROTEIN_KINASE_ATP
PROSITE(pattern)
29-52
PROTEIN_KINASE_DOM
PROSITE(profile)
23-291
PROTEIN_KINASE_ST
PROSITE(pattern)
149-161
S_TKc
SM00220
Smart
23-291
TyrKc
SM00219
Smart
23-291
pkinase
PF00069
Pfam
23-291

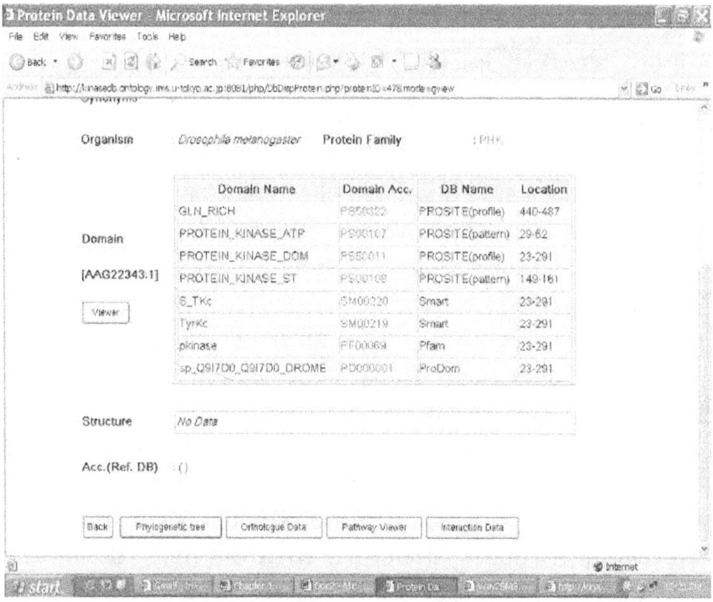
Protein Data Viewer - Microsoft Internet Explorer
Organism
Drosophila melanogaster
Protein Family
Domain Name
Domain Acc.
DB Name
Location
GLN_RICH
PROSITE(profile)
440-487
Domain
PROTEIN_KINASE_ATP
PROSITE(pattern)
PROTEIN_KINASE_DOM
PROSITE(profile)
23-291
[AAG22343.1]
PROTEIN_KINASE_ST
PROSITE(pattern)
Viewer
S_TKc
Smart
23-291
TyrKc
Smart
23-291
pkinase
Pfam
23-291
sp_Q9I7D0_Q9I7D0_DROME
ProDom
23-291
Structure
No Data
Acc.(Ref. DB)
Back
Phylogenetic tree
Orthologue Data
Pathway Viewer
Interaction Data

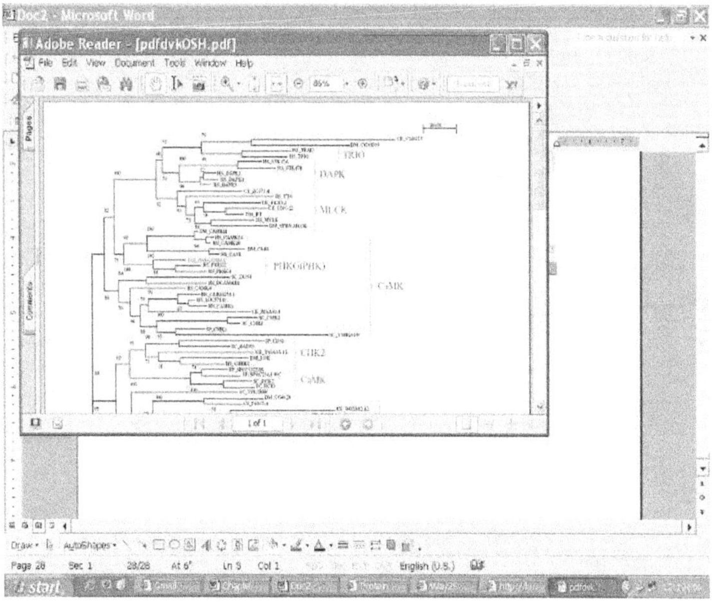
Doc2 - Microsoft Word
Adobe Reader - [pdfdvkOSH.pdf]
TRIO
DAPK
MLCK
PHKG(PHK)
CaMK
CHK2
1 of 1
English (U.S.)

A PROTEIN PHOSPHATASE INFORMATION RESOURCE (PHOSPHOBASE)

PhosphoBase is an ontology-driven database resource containing information on the protein phosphatase family. It is the first public resource dedicated to protein phosphatases, which are enzymes that perform dephosphorylation reactions. In conjunction with the phosphorylation action of protein kinases, phosphatases are involved in important control and communication mechanisms in the cell. They have also been implicated in many human diseases, including diabetes, obesity, cancers, and neurodegenerative conditions. The resource is built around a formal, domain-specific DAML+OIL ontology, and the data are collected from heterogeneous biological sources using Gene Ontology terms as a means of data extraction. The overall ontology-driven architecture provides a robust structure with distinct advantages for sustainability and provides the potential for the development of diagnostic tools, as well as a data repository.

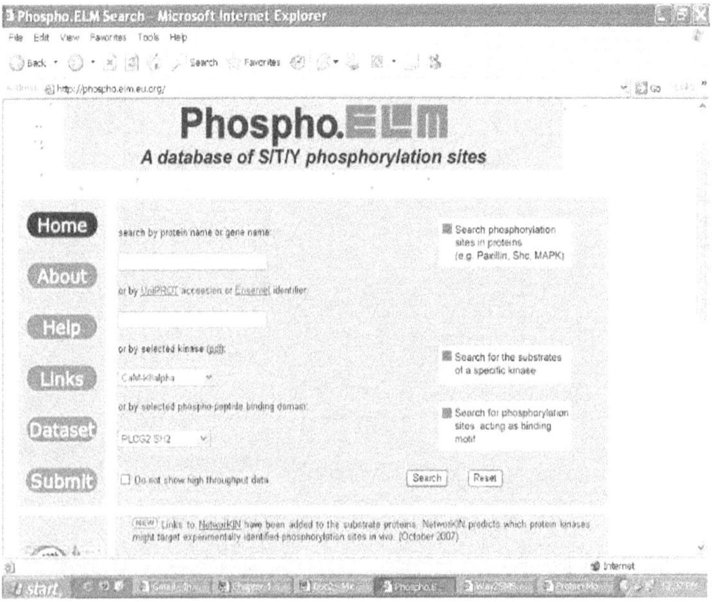

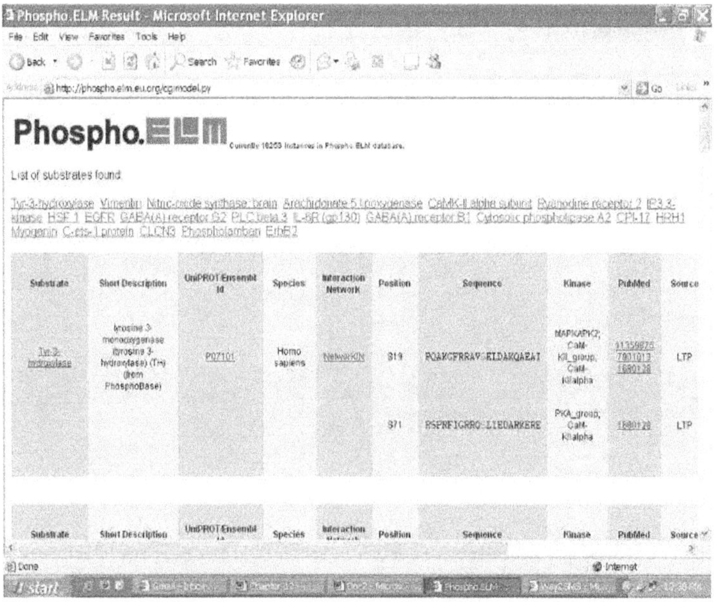
Phospho.ELM Result - Microsoft Internet Explorer
http://phospho.elm.eu.org/cgimodel.py
Phospho.ELM
List of substrates found
Substrate
Short Description
UniPROT/Ensembl Id
Species
Interaction Network
Position
Sequence
Kinase
PubMed
Source
Homo sapiens
NetworKIN
S19
PQAKGFRRAVSELDAKQAEAI
LTP
S71
RSPRFIGRRQSLIEDARKERE
LTP
P07101
Done
Internet

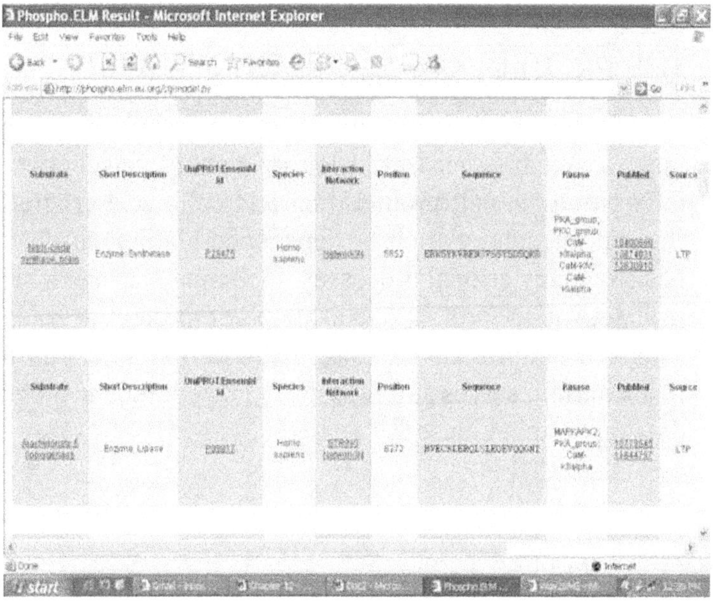
Phospho.ELM Result - Microsoft Internet Explorer
Substrate
Short Description
UniPROT/Ensembl Id
Species
Interaction Network
Position
Sequence
Kinase
PubMed
Source
LTP
LTP
Internet

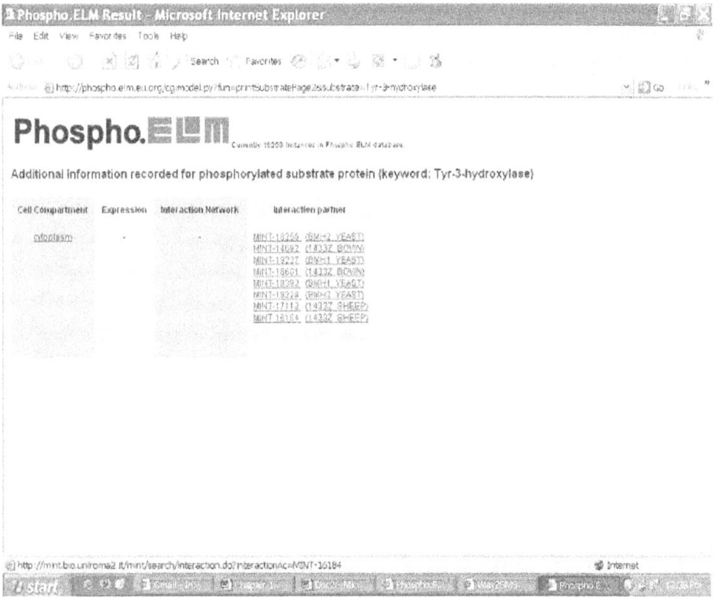

THE PROTEIN KINASE RESOURCE (PKR)

Protein kinases and phosphatases play crucial roles in all the major cellular processes, such as signal transduction, cell differentiation, cell proliferation and cell cycle progression. Protein phosphorylation or dephosphorylation can form the basis of many critical processes, including enzyme activation or inactivation, protein localization and protein degradation. The Protein Kinase Resource (http://pkr.sdsc.edu/html/index.shtml), serves as a repository for cellular and molecular data on protein kinases.

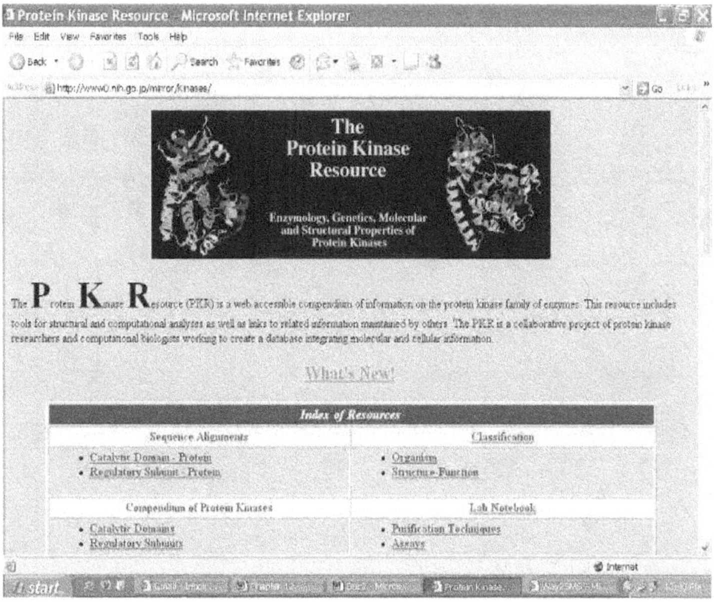
Protein Kinase Resource - Microsoft Internet Explorer
http://www0.nih.go.jp/mirror/Kinases/
The Protein Kinase Resource
Enzymology, Genetics, Molecular and Structural Properties of Protein Kinases
The Protein Kinase Resource (PKR) is a web accessible compendium of information on the protein kinase family of enzymes. This resource includes tools for structural and computational analyses as well as links to related information maintained by others. The PKR is a collaborative project of protein kinase researchers and computational biologists working to create a database integrating molecular and cellular information.
What's New!
Index of Resources
Sequence Alignments
Catalytic Domain - Protein
Regulatory Subunit - Protein
Classification
Organism
Structure-Function
Compendium of Protein Kinases
Catalytic Domains
Regulatory Subunits
Lab Notebook
Purification Techniques
Assays

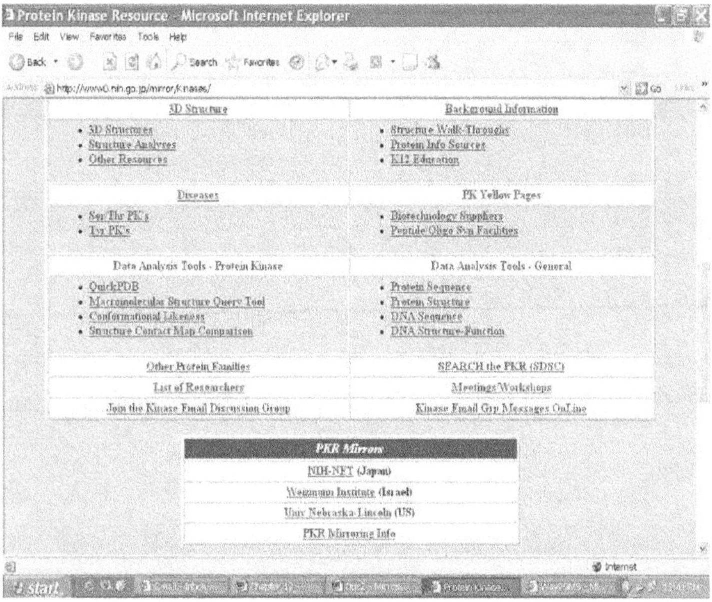
Protein Kinase Resource - Microsoft Internet Explorer
http://www0.nih.go.jp/mirror/Kinases/
3D Structure
3D Structures
Structure Analyses
Other Resources
Background Information
Structure Walk-Throughs
Protein Info Sources
K12 Education
Diseases
Ser/Thr PK's
Tyr PK's
PK Yellow Pages
Biotechnology Suppliers
Peptide/Oligo Syn Facilities
Data Analysis Tools - Protein Kinase
QuickPDB
Macromolecular Structure Query Tool
Conformational Likeness
Structure Contact Map Comparison
Data Analysis Tools - General
Protein Sequence
Protein Structure
DNA Sequence
DNA Structure-Function
Other Protein Families
SEARCH the PKR (SDSC)
List of Researchers
Meetings/Workshops
Join the Kinase Email Discussion Group
Kinase Email Grp Messages OnLine
PKR Mirrors
NIH-NET (Japan)
Weizmann Institute (Israel)
Univ Nebraska-Lincoln (US)
PKR Mirroring Info

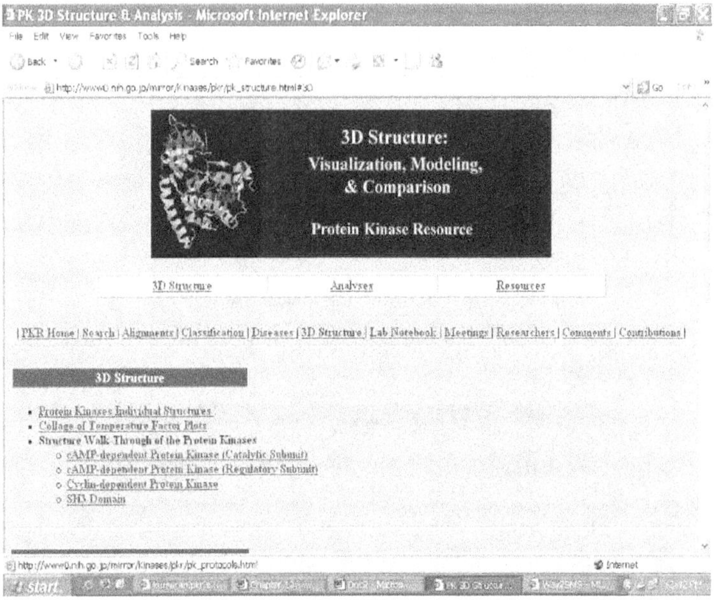
PK 3D Structure & Analysis - Microsoft Internet Explorer
http://www0.nih.go.jp/mirror/Kinases/pkr/pk_structure.html#3D
3D Structure:
Visualization, Modeling,
& Comparison
Protein Kinase Resource
3D Structure
Analyses
Resources
| PKR Home | Search | Alignments | Classification | Diseases | 3D Structure | Lab Notebook | Meetings | Researchers | Comments | Contributions |
3D Structure
Protein Kinases Individual Structures
Collage of Temperature Factor Plots
Structure Walk-Through of the Protein Kinases
cAMP-dependent Protein Kinase (Catalytic Subunit)
cAMP-dependent Protein Kinase (Regulatory Subunit)
Cyclin-dependent Protein Kinase
SH3 Domain
http://www0.nih.go.jp/mirror/Kinases/pkr/pk_protocols.html
Internet

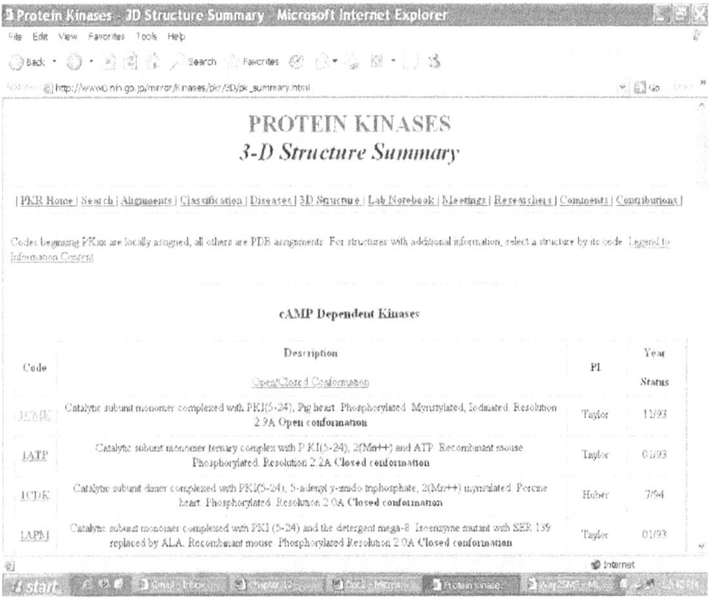
Protein Kinases - 3D Structure Summary - Microsoft Internet Explorer
http://www0.nih.go.jp/mirror/Kinases/pkr/3D/pk_summary.html
PROTEIN KINASES
3-D Structure Summary
| PKR Home | Search | Alignments | Classification | Diseases | 3D Structure | Lab Notebook | Meetings | Researchers | Comments | Contributions |
Codes beginning PKxxx are locally assigned, all others are PDB assignments. For structures with additional information, select a structure by its code. Legend to Information Content
cAMP Dependent Kinases
Code
Description
Open/Closed Conformation
PI
Year
Status
1CMK
Catalytic subunit monomer complexed with PKI(5-24), Pig heart. Phosphorylated. Myristylated, Iodinated. Resolution 2.9A Open conformation
Taylor
11/93
1ATP
Catalytic subunit monomer ternary complex with PKI(5-24), 2(Mn++) and ATP. Recombinant mouse. Phosphorylated. Resolution 2.2A Closed conformation
Taylor
01/93
1CDK
Catalytic subunit dimer complexed with PKI(5-24), 5-adenyl y-imido triphosphate, 2(Mn++) myristylated. Porcine heart. Phosphorylated. Resolution 2.0A Closed conformation
Huber
7/94
1APM
Catalytic subunit monomer complexed with PKI (5-24) and the detergent mega-8. Isoenzyme mutant with SER 139 replaced by ALA. Recombinant mouse. Phosphorylated Resolution 2.0A Closed conformation
Taylor
01/93
Internet

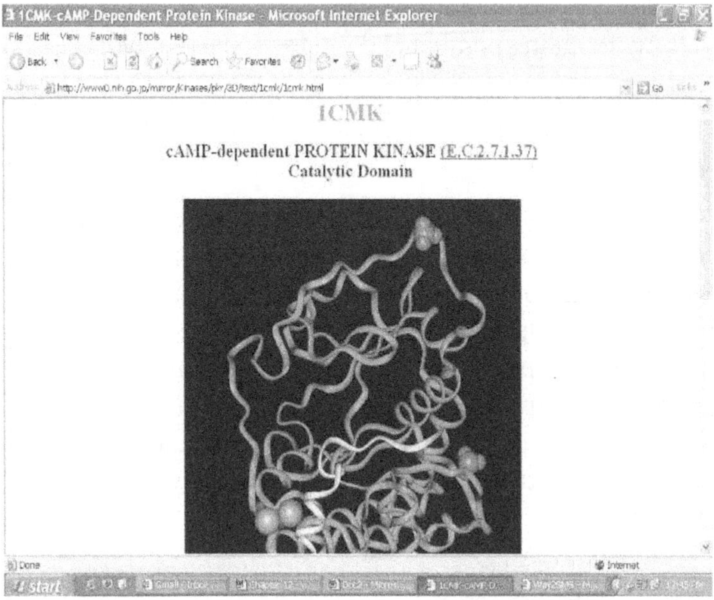

PREDICTED AND CONSENSUS INTERACTION SITES IN ENZYMES (PRECISE)

Predicted and Consensus Interaction Sites in Enzymes (PRECISE) is a database of interactions between the amino acid residues of an enzyme and its ligands (substrate and transition state analogues, cofactors, inhibitors and products). It is available online at http://precise.bu.edu/.

In the current version, all information on interactions are extracted from the enzyme–ligand complexes in the Protein Data Bank (PDB) by performing the following steps:

i. clustering homologous enzyme chains such that, in each cluster, the proteins have the same EC number and all sequences are similar;
ii. selecting a representative chain for each cluster;

iii. selecting ligand types;

iv. finding non-bonded interactions and hydrogen bonds and

v. summing the interactions for all chains within the cluster.

The output of the search is the colour-coded sequence of the representative. The colours indicate the total number of interactions found at each amino acid position in all chains of the cluster. Optional filters allow restricting the output to selected chains in the cluster, to non-bonded or hydrogen bonding interactions, and to selected ligand types. The binding site information is essential for understanding and altering substrate specificity and for the design of enzyme inhibitors.

A DATABASE OF PROTEIN CATALYTIC DOMAINS (SCOPEC)

Domains are the units of protein structure, function, and evolution. For this a database of catalytic domains, SCOPEC, was developed by combining structural domain information from SCOP, full-length sequence information from Swiss-Prot, and verified functional information from the Enzyme Classification (EC) database. Two major problems need to be overcome to create a database of domain–function relationships; 1) for sequences, EC numbers are typically assigned to whole sequences rather than the functional unit, and 2) the Protein Data Bank (PDB) structures elucidated from a larger multi-domain protein will often have EC annotation although the relevant catalytic domain may lie elsewhere. SCOPEC entries have high-quality enzyme assignments, having passed both computational and manual checks. SCOPEC currently contains entries for 75% of all EC annotations in the PDB. Overall,

EC number is fairly well-conserved within a superfamily, even when the proteins are distantly related. The SCOPEC database is a valuable resource in the analysis and prediction of protein structure and function. It can be obtained or queried at the website http://www.enzome.com.

THERMODYNAMICS OF ENZYME-CATALYZED REACTIONS DATABASE (TECRDB)

The Thermodynamics of Enzyme-catalyzed Reactions Database (TECRDB) is a comprehensive collection of thermodynamic data on enzyme-catalysed reactions. The data, which consist of apparent equilibrium constants and calorimetrically determined molar enthalpies of reaction, are the primary experimental results obtained from thermodynamic studies of the biochemical reactions.

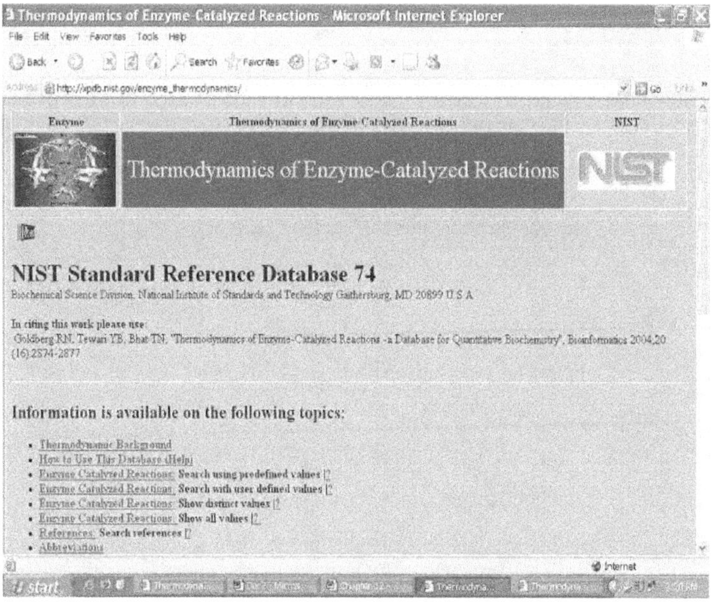

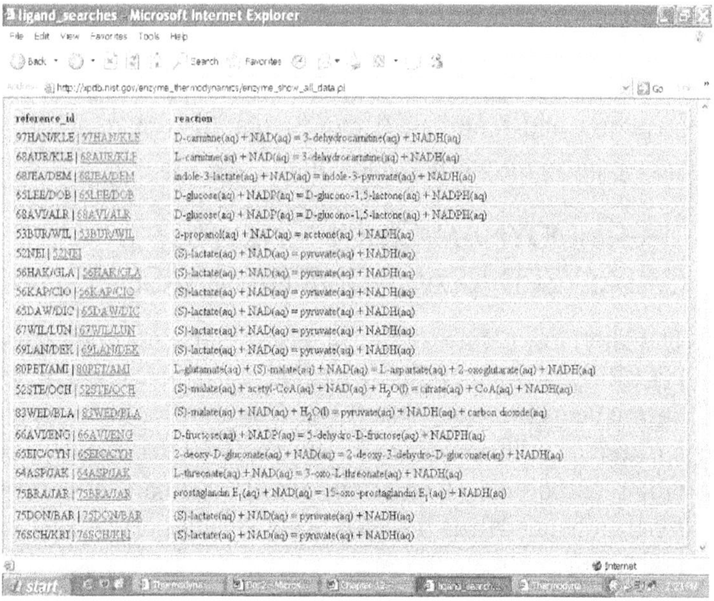

ligand_searches - Microsoft Internet Explorer

http://xpdb.nist.gov/enzyme_thermodynamics/enzyme_show_all_data.pl

reference_id	reaction
97HAN/KLE | 97HAN/KLE	D-carnitine(aq) + NAD(aq) = 3-dehydrocarnitine(aq) + NADH(aq)
68AUR/KLE | 68AUR/KLE	L-carnitine(aq) + NAD(aq) = 3-dehydrocarnitine(aq) + NADH(aq)
68JEA/DEM | 68JEA/DEM	indole-3-lactate(aq) + NAD(aq) = indole-3-pyruvate(aq) + NADH(aq)
65LEE/DOB | 65LEE/DOB	D-glucose(aq) + NADP(aq) = D-glucono-1,5-lactone(aq) + NADPH(aq)
68AVI/ALR | 68AVI/ALR	D-glucose(aq) + NADP(aq) = D-glucono-1,5-lactone(aq) + NADPH(aq)
53BUR/WIL | 53BUR/WIL	2-propanol(aq) + NAD(aq) = acetone(aq) + NADH(aq)
52NEI | 52NEI	(S)-lactate(aq) + NAD(aq) = pyruvate(aq) + NADH(aq)
56HAK/GLA | 56HAK/GLA	(S)-lactate(aq) + NAD(aq) = pyruvate(aq) + NADH(aq)
56KAP/CIO | 56KAP/CIO	(S)-lactate(aq) + NAD(aq) = pyruvate(aq) + NADH(aq)
65DAW/DIC | 65DAW/DIC	(S)-lactate(aq) + NAD(aq) = pyruvate(aq) + NADH(aq)
67WIL/LUN | 67WIL/LUN	(S)-lactate(aq) + NAD(aq) = pyruvate(aq) + NADH(aq)
69LAN/DEK | 69LAN/DEK	(S)-lactate(aq) + NAD(aq) = pyruvate(aq) + NADH(aq)
80PET/AMI | 80PET/AMI	L-glutamate(aq) + (S)-malate(aq) + NAD(aq) = L-aspartate(aq) + 2-oxoglutarate(aq) + NADH(aq)
52STE/OCH | 52STE/OCH	(S)-malate(aq) + acetyl-CoA(aq) + NAD(aq) + H_2O(l) = citrate(aq) + CoA(aq) + NADH(aq)
83WED/BLA | 83WED/BLA	(S)-malate(aq) + NAD(aq) + H_2O(l) = pyruvate(aq) + NADH(aq) + carbon dioxide(aq)
66AVI/ENG | 66AVI/ENG	D-fructose(aq) + NADP(aq) = 5-dehydro-D-fructose(aq) + NADPH(aq)
65EIC/CYN | 65EIC/CYN	2-deoxy-D-gluconate(aq) + NAD(aq) = 2-deoxy-3-dehydro-D-gluconate(aq) + NADH(aq)
64ASP/JAK | 64ASP/JAK	L-threonate(aq) + NAD(aq) = 3-oxo-L-threonate(aq) + NADH(aq)
75BRA/JAR | 75BRA/JAR	prostaglandin E_1(aq) + NAD(aq) = 15-oxo-prostaglandin E_1(aq) + NADH(aq)
75DON/BAR | 75DON/BAR	(S)-lactate(aq) + NAD(aq) = pyruvate(aq) + NADH(aq)
76SCH/KRI | 76SCH/KRI	(S)-lactate(aq) + NAD(aq) = pyruvate(aq) + NADH(aq)

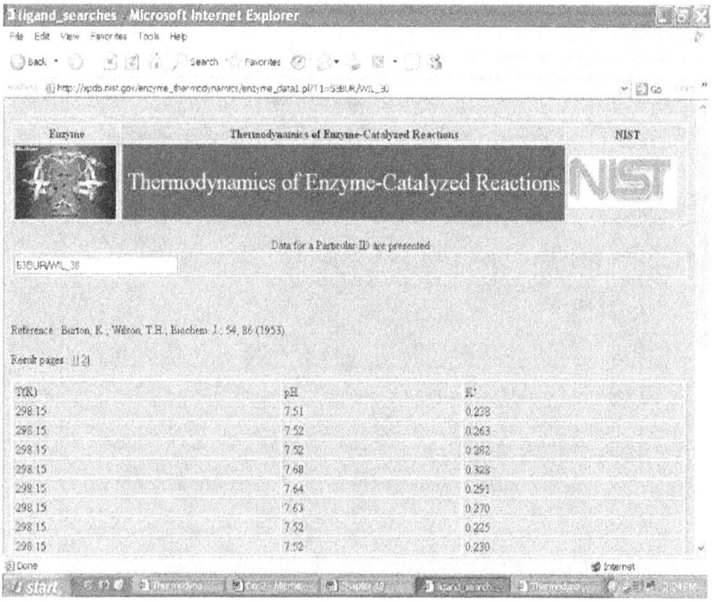

ligand_searches - Microsoft Internet Explorer

http://xpdb.nist.gov/enzyme_thermodynamics/enzyme_data1.pl?T1=53BUR/WIL_30

Enzyme | Thermodynamics of Enzyme-Catalyzed Reactions | NIST

Thermodynamics of Enzyme-Catalyzed Reactions

Data for a Particular ID are presented

53BUR/WIL_30

Reference: Burton, K.; Wilson, T.H., Biochem. J., 54, 86 (1953)

Result pages: [1] [2]

T(K)	pH	K'
298.15	7.51	0.238
298.15	7.52	0.263
298.15	7.52	0.282
298.15	7.68	0.328
298.15	7.64	0.291
298.15	7.63	0.270
298.15	7.52	0.225
298.15	7.52	0.230

The results from 1000 published papers containing data on 400 different enzyme-catalysed reactions constitute the essential information in the database. The information is managed using Oracle and is available on the web.

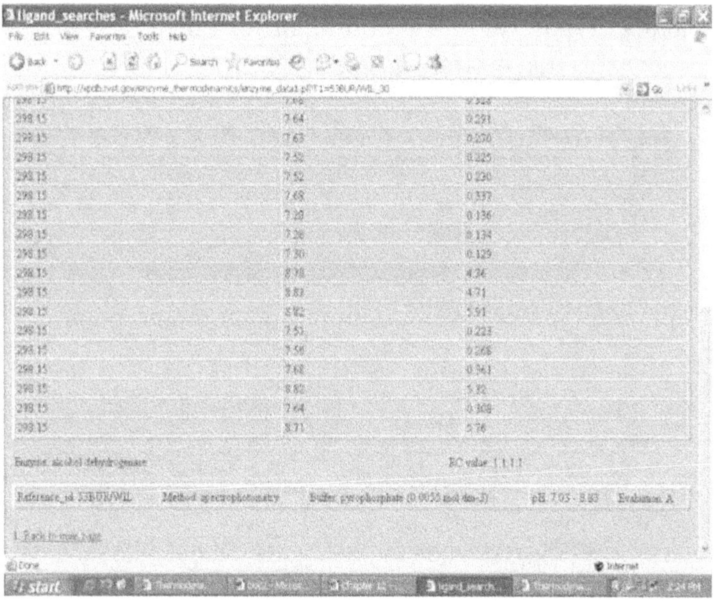

ERGO

ERGO is a third-generation bioinformatics suite offered exclusively from Integrated Genomics at: http://ergo.integratedgenomics.com/ERGO/. The ERGO system represents the development of a genome analysis strategy into a multi-dimensional environment, which supports both automatic and manual genome-wide curation. Rather than just repackaging known information, ERGO integrates genomic information with biochemical data, literature, and high-throughput analysis into a comprehensive user-friendly network of metabolic and non-metabolic pathways. In contrast to conventional systems, the ERGO user can take into account sequence similarity, protein and gene context clustering, occurrence profiles, regulatory and expression data, as well as functional hierarchies in order to achieve a set of the best possible functional predictions. In fact, using the ERGO system, a major part of the metabolism of an organism, can be reconstructed entirely *in silico*.

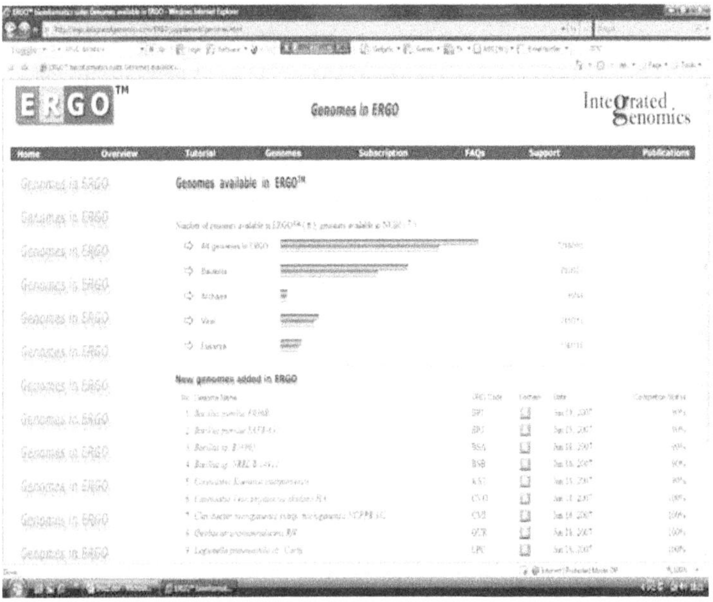

The current version of the ERGO™ database contains 618 complete or nearly complete genomes, of which 319 are bacteria, 116 eukarya, 34 archaea and 149 are viruses. In total, these genomes contain over 1,300,000 Open Reading Frames (ORFs), more than 60% of which have a functional annotation. This percentage of annotated genes is actually much higher for the bacterial genomes, reaching an average of 70%. Every genome that goes into the ERGO system, is annotated from scratch whether it has been sequenced at Integrated Genomics, or at another sequencing centre. More than 450 of the genomes are available for subscription or as part of a stand-alone ERGO server package from Integrated Genomics.

Data Types in ERGO

The data types in ERGO include the following

- Genomic data
- Pathway data
- Regulatory data
- Essentiality data
- Expression data

Genomic data includes

- DNA sequence data into contigs (from over 400 genomes)
- ORFs and their location (graphical visualization of ORFs on a contig)
- Translation of ORFs
- Pre-computed sequence similarities for each ORF (against the entire database)
- Functional assignments of proteins (with their history records)
- RNA assignments
- Identification and localization of insertion elements (ISs)
- Ortholog clusters
- Paralog clusters
- Protein family clusters
- Chromosomal clusters
- Fusion clusters

Pathway data includes

- Chemical structures
- Enzyme records
- Metabolic pathways
- Non-metabolic pathways
- Cellular overviews (networks of metabolic and non-metabolic pathways)
- Functional hierarchies (functional roles organized into Gene Ontologies)

The front page of ERGO is designed to provide the user with direct access into the system, as well as with general information regarding the company, and the system is as follows:

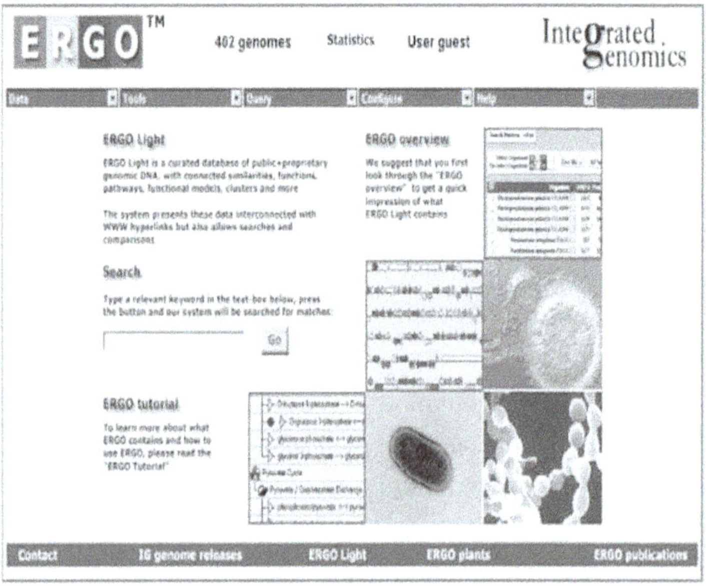

The ERGO™ system supports the creation of a user-specific environment throughout several of the provided tools/pages. From the green menu bar at the top of the page, the user can click on the arrow next to the **Configure** menu bar (see the menu inside the selected red box) in order to: (a) change the User Name; (b) select a preferred organism; or (c) select a preferred group of organisms, as shown in the following screenshot:

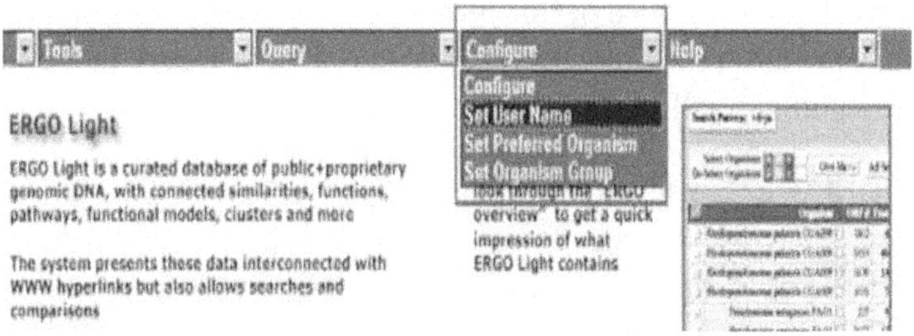

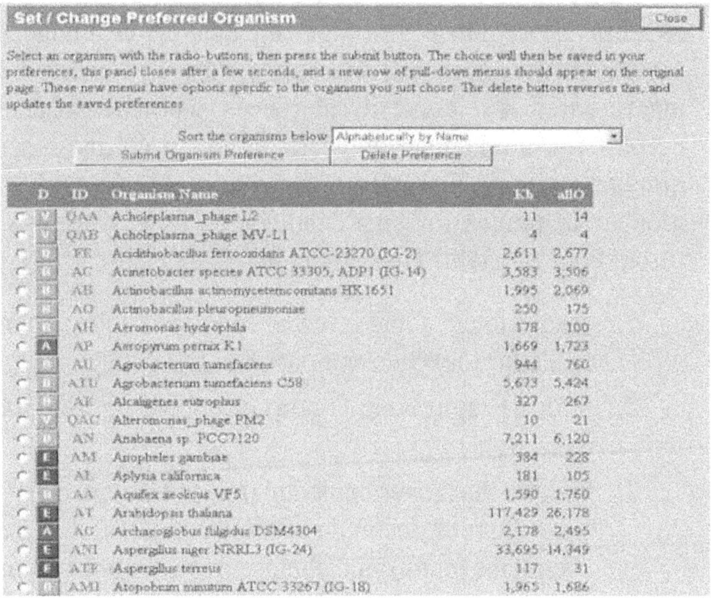

The coloured buttons in front of the organism names denote the different domains:

B —Bacterial A —Archaea V —Viral E —Eukarya

By selecting a preferred organism, a new window appears with the list of the genomes that are available on the subscribed ERGO system. The user can select any one of those genomes (only one at a time), by clicking on the radio button in front of the preferred organism. The selected organism is then saved in the preferences of this particular User Name. When the user logs in again into the system, using this name, the selected organism will automatically appear.

ENZYME NOMENCLATURE DATABASE (END)

END is a database built from the Enzyme Nomenclature. The Nomenclature is maintained by the IUBMB/IUPAC Joint Committees on Biochemical Nomenclature (JCBN). Several aspects of version 2.0 of END all accession numbers are permanently assigned; all minor data types have been incorporated or referenced; references to other databases are current, and distinguish between search strings and accession numbers; simple histories of entries are presented; and a number of names for small molecules have been changed throughout the data. These include:

1. resolution of some internal synonyms (more than one name for the same compound in the database);
2. making syntactic inconsistencies in the names consistent;
3. making the names of nicotinamide cofactors consistent with current recommendations in all fields except synonyms for enzyme names which are not systematic names and
4. correction of miscellaneous errors, including typographical ones.

CAZy (CARBOHYDRATE-ACTIVE ENZYMES)

The CAZy database describes the families of structurally related catalytic and carbohydrate-binding modules (or functional domains) of enzymes that degrade, modify, or create glycosidic bonds.

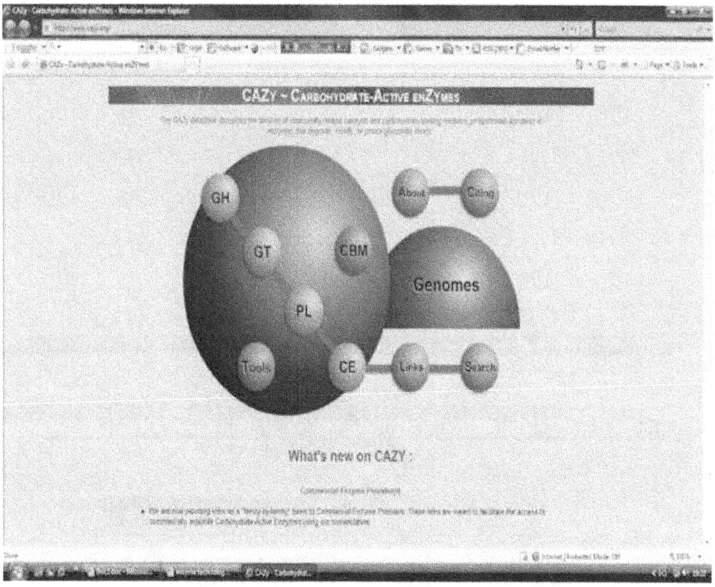

The content in Carbohydrate-Active Enzymes ("CAZome") of a genome provides an insight into the nature and relative importance of carbohydrate and glycoconjugate breakdown and biosynthesis in the metabolism of a species. The CAZomes of a free-living organism typically correspond to 1–5% of the predicted coding sequences. The above effort is the simple assignment of a protein sequence to a CAZy family and therefore it does not constitute a refined functional prediction tool for genomic annotation.

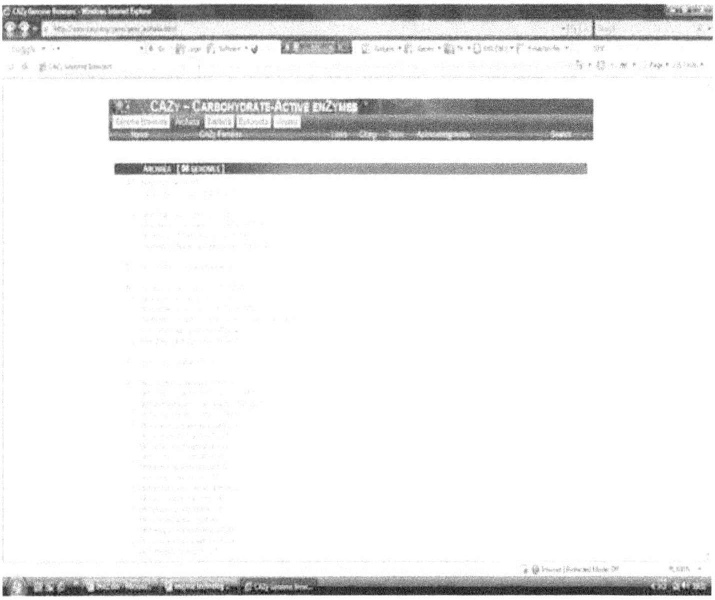
CAZy - CARBOHYDRATE-ACTIVE ENZYMES

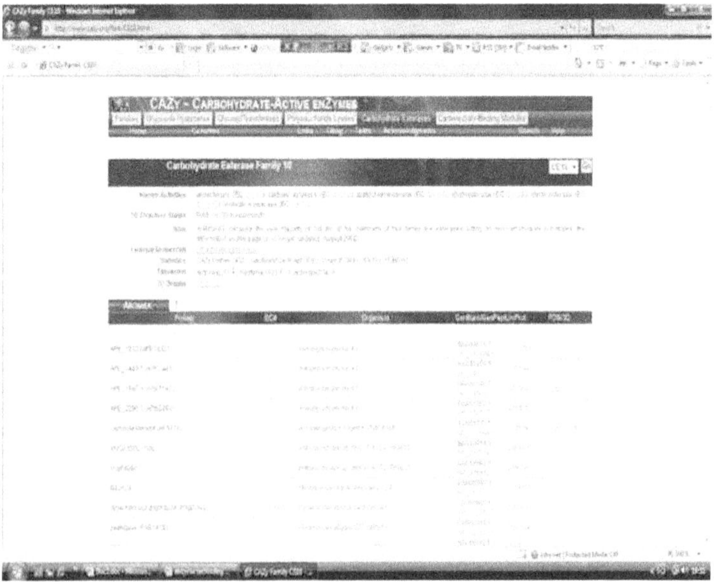
CAZy - CARBOHYDRATE-ACTIVE ENZYMES
Carbohydrate Esterase Family 10

THE PEPTIDASE DATABASE (MEROPS)

The MEROPS database is an information resource for peptidases (also termed proteases, proteinases and proteolytic enzymes) and the proteins that inhibit them. The MEROPS database uses a hierarchical, structure-based classification of the peptidases. In this, each peptidase is assigned to a Family on the basis of statistically significant similarities in amino acid sequence, and families that are thought to be homologous are grouped together in a Clan. There is a Summary page for each Family and Clan, and these again have indexes. Each of the Summary pages offers links to supplementary pages.

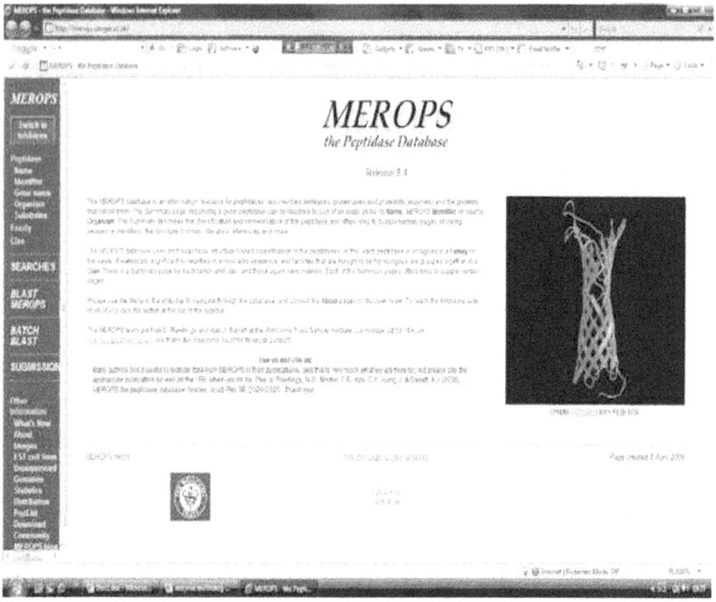

The following shows the index of peptidase names:

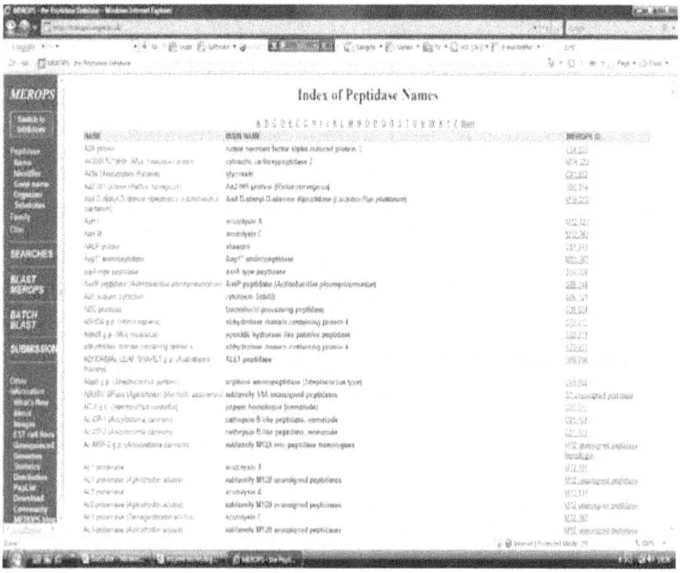

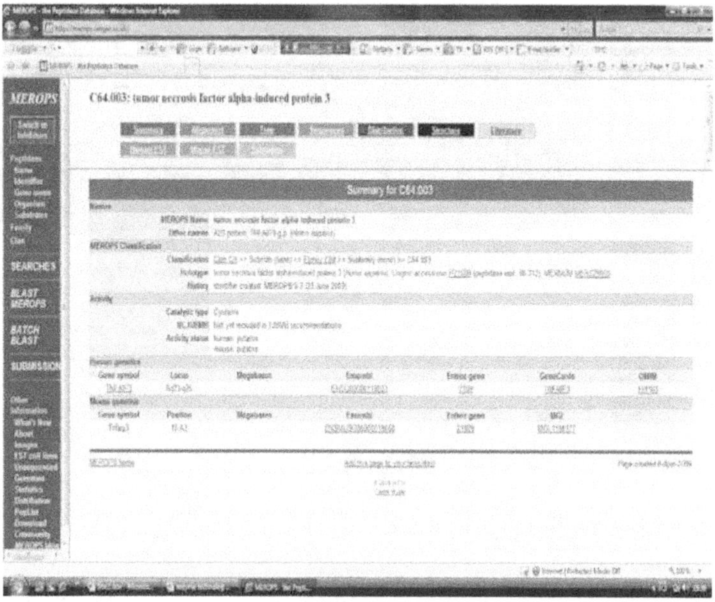

The structure of a peptidase is shown in the following screenshots.

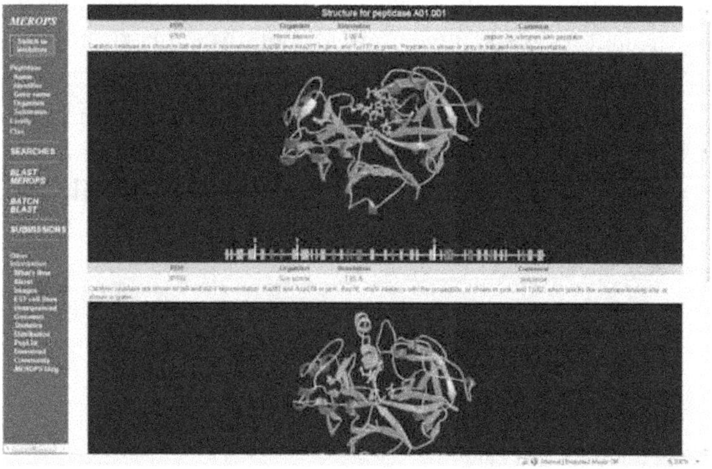

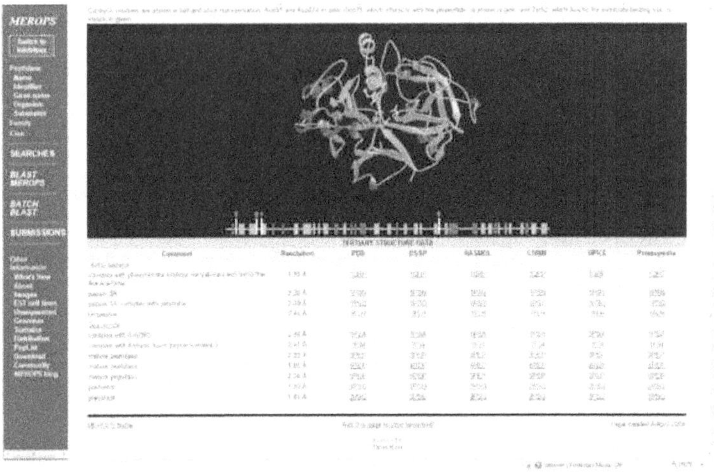

SABIO-REACTION KINETICS DATABASE

The System for the Analysis of Biochemical Pathways-Reaction Kinetics (SABIO-RK) is a web-based application based on the SABIO relational database that contains information about biochemical reactions, their kinetic equations with their parameters, and the experimental conditions under which these parameters were measured. It aims to support modellers in the setting up of models of biochemical networks, but it is also useful for the experimentalists or researchers with interest in biochemical reactions and their kinetics. Information about reactions and their kinetics can be exported in SBML (Systems Biology Mark-Up Language) format.

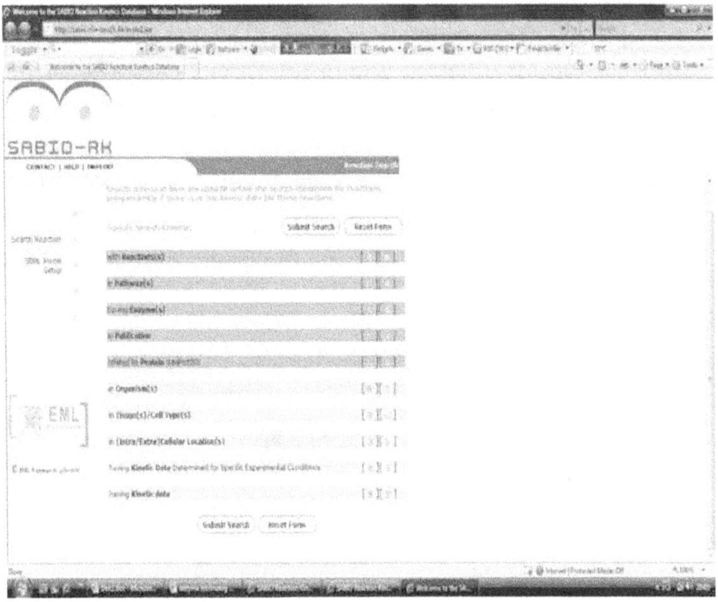

BRENDA—THE ENZYME DATABASE

BRENDA is the most comprehensive, expert-updated, relational database of molecular and metabolic information on more than 83,000 enzymes from 9,800 organisms according to the EC system of the IUBMB. Data on enzyme function are extracted directly from the primary literature. The database is developed at the Institute of Biochemistry at the University of Cologne. The curation is carried out by enzymeta GmbH. BRENDA provides researchers in the domain of biochemistry and medicine a user-friendly navigation system offering time-saving access to a comprehensive summary of enzyme-specific data from more than 46,000 publications.

The database includes biochemical and molecular information on classification, nomenclature, reaction, specificity, functional parameters, occurrence, enzyme structure, application, engineering, stability, disease, isolation, and preparation. The database also provides additional information on ligands, which function as natural or *in vitro* substrates/products, inhibitors, activating compounds, cofactors, bound metals, etc. BRENDA ligand provides a search for all included synonyms of a given compound and thus facilitates to find all enzyme- and ligand-related information. This is based on the generation of unique and chiral SMILES strings for ligand structures in the database.

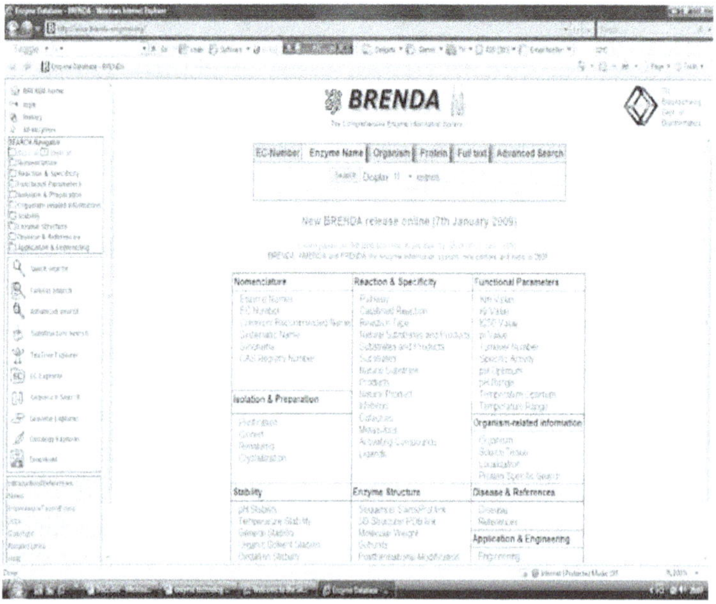

PROTEOLYTIC ENZYMES DATABASE

Proteolytic enzymes also referred to as proteases and proteinases degrade proteins by hydrolysing the peptide bond of the protein. Proteases can either break specific peptide bonds called limited proteolysis, or break down a complete polypeptide chain to amino acid residues known as unlimited proteolysis.

Four mechanistic classes of the proteolytic enzymes have been identified, they are serine proteinase, cysteine proteinase, aspartic proteinase and metalloproteinase.

Proteolytic enzymes play regulatory roles in a great variety of physiological processes. Proteases also play a key role in cellular processes like separation of the chromosome during mitosis, cell cycle and apoptosis. Proteolytic enzymes also serve as an analytical reagents for sequencing the polypeptide chain, in the identification and isolation of domain of multi-domain protein. Proteolytic enzymes database was developed using Microsoft Access. This is a web-based interface with the aim of helping users to search for specific information. This user interface has been built with ASP Server Side Scripts on a Microsoft Windows-based web server. The database can be regularly updated.

The bibliographic data for the database was collected from Pubmed (http://www.ncbi.nlm.nih.gov/). The database can be accessed through website http://www.proteolyticenzymes. info. There are 5144 records in the database. The database contains bibliographic information and information on physico-chemical properties of a total 30 proteolytic enzymes. The database can be searched by keywords (in which user can type enzyme name) and by simultaneously selecting "Search Type" option. Alternatively, the database can be searched by selecting enzyme along with by selecting "Search Type". Since proteolytic enzymes hydrolyse food proteins and play a regulatory role in a great variety of physiological processes. Therefore, database of proteolytic enzymes has the potential of becoming major hub of resource for the biological community. This database provides bibliography and information on physico-chemical properties of proteolytic enzymes to the researchers and students working in this area.

BRENDA, AMENDA AND FRENDA— THE ENZYME INFORMATION SYSTEM

BRENDA

The BRaunschweig ENzyme DAtabase (BRENDA) (http://www.brendaenzymes.org) represents the largest freely available information system containing a huge amount of biochemical and molecular information on all classified enzymes as well as software tools for querying the database and calculating molecular properties.

The database covers information on classification and nomenclature, reaction and specificity, functional parameters, occurrence, enzyme structure and stability, mutants and enzyme engineering, preparation and isolation, the application of enzymes, and ligand-related data. The development of the BRENDA enzyme information system started in 1987 at the former German National Research Centre for Biotechnology (now Helmholtz Centre for Infection Research) in Braunschweig. BRENDA is maintained and curated at the Technische University in Braunschweig, Institute of Bioinformatics and Systems Biology.

The Taxonomy Tree Explorer provides a search for enzymes or organisms in the taxonomic tree. The EC Explorer can be used to browse or search the hierarchical tree of enzymes. The Sequence Search is useful for enzymes with a known protein sequence. It is also possible to specifically search membrane proteins using the program TMHMM. BRENDA contains functional data for all enzyme classes that have been classified according to the EC scheme of the International Union of Biochemistry and Molecular Biology (IUBMB) irrespective of the enzyme's source. The range of data in BRENDA is not restricted to specific aspects but includes a wide area of biochemical and molecular properties of enzymes such as the following.

- Classification and nomenclature
- Reaction and specificity
- Functional parameters
- Organism-related information
- Enzyme structure
- Isolation and preparation
- Literature references
- Application and engineering
- Enzyme–disease relationships

All data and information are manually extracted from primary literature and are connected to the biological source of the enzyme, i.e., the organism, the tissue, the subcellular localization and/or the protein sequence.

The new feature is based on the UniProtKB/Swiss-Prot accession codes. Currently, BRENDA contains different isoforms for more than 3300 enzymes in nearly 500 different organisms.

AMENDA (Automatic Mining of ENzyme DAta) and FRENDA (Full Reference ENzyme DAta) have been developed as additional databases based on text-mining procedures and are completely reprogrammed to improve the quality and reduce false-positive entries.

FRENDA

FRENDA provides links to all literature references indexed by PubMed that cover enzyme-specific information in combination with the name of the organism or one of its synonyms. Another enhancement of FRENDA is the consequent exclusion of ambiguous enzyme and organism names by manually compiled exclusion lists.

AMENDA

AMENDA is a subset of FRENDA comprising information on enzyme occurrence in organism, localization, and source tissue. It includes the most reliable organism-specific enzyme information from FRENDA. In addition, it comprises data on the subcellular localization of enzymes and the source tissues in which the enzymes are active. Whereas the information in FRENDA is completely based on co-occurrence of enzyme names and organism names in title and abstract of a paper, for AMENDA, a refined and more rigorous text-mining procedure is used. BRENDA offers a SOAP-based web service, API (http://www.brendaenzymes.org/soap). It covers more than 50 different data fields and provides access to them via almost 150 different remote methods. For all data fields in the record sets, the corresponding literature references can be obtained. The methods for querying the individual data fields accept either the organism, the EC number or the ligand identifier or combinations of all three as input parameters. The data types of the return values of the new SOAP API (http:// www.brenda-enzymes.org/soap2) use exclusively strings and which are supported by almost every programming language.

ORENZA—A WEB RESOURCE FOR ORPHAN ENZYME ACTIVITIES

The presence of so many EC numbers that do not have an associated sequence appears rather extraordinary at a time when inundated by high genomic data. Such a situation is encroaching research at different levels. Alleviating this problem would be very helpful for the difficult task of annotating and/or reannotating genomes.

Thus, there was an urgent need to bridge this unwanted gap between biochemical knowledge and massive identification of coding sequences and others. It is also shown that orphans are present at about the same proportion in every class and subclass of enzyme activities. Likewise, no correlation between orphan distribution and main functional categories was found. 25.3% of the enzyme activities involved in well-studied metabolic pathways is sequence-less while 49.5% orphans were found among non-metabolic enzyme activities. Thus, it appears that there is an important gap between function and sequence, which implies that its progressive bridging would require a concerted effort.

ORENZA, a database of ORphan ENZyme Activities, was built to offer such a tool to the research community. ORENZA database, an efficient relational database, will help to identify the encoding gene for the maximum number of sequence-less enzyme activities (the so-called **orphan enzymes**). A Perl script screened the occurrence of EC numbers in UniProt Knowledgebase. Any EC number assigned by the NC-IUBMB that is not referenced in UniProt is defined as an **orphan enzyme activity**. Accordingly, PostgreSQL 8.1, one of the most advanced open source databases, was installed on a Linux platform. PHP language was used to structure the web service and to better exploit the queries from the relational database. One can browse and/or search ORENZA using three main avenues which are discussed in the following pages.

1. Browsing the Whole Set of EC Numbers

The complete list of EC numbers (recent version of NC-IUBMB) is available. The entire list, which can be easily downloaded as a text file, is completely dynamic. A click on a line opens a new view delivering a wealth of information about the selected EC number that is structured in three successive levels.

The first level consists of characteristics of the enzymatic activity and its history. The description section contains information taken from the NC-IUBMB data such as the different names (common, systematic, and others) of the enzyme, a scheme of the reaction(s) it catalyses and other data about the cofactors and NC-IUBMB comments about the reactions that are extracted from the ENZYME database. The second level presents information about the position of the enzyme in the cell metabolism with the corresponding number of a KEGG map. The third level exhibits information about the peptide molecule such as motifs (from PROSITE), the lists of amino acid sequences found in Swiss-Prot and TrEMBL, respectively. If there is no sequence, label "orphan", is mentioned.

2. Browsing the Orphan EC Numbers

The second main avenue offered by ORENZA to explore the enzyme universe is the entire list, periodically updated, of the orphan enzyme activities. There are several ways to retrieve these orphans besides browsing the list in its entirety.

- First, one can browse the different levels (class, subclass, etc.) of the EC hierarchy.
- A second approach is to explore the metabolism hierarchy proposed by KEGG.
- A third way to browse the orphan EC numbers is to sort them by their year of creation.

It is possible to query ORENZA for a specific enzyme activity by entering either the EC number or the enzyme name. For example, entering the word "aspartate" recovers 41 EC numbers, 13 being presently not assigned to a sequence.

ORENZA resource is freely available via the Internet at http://www.orenza.u-psud.fr. The web accessibility has been tested to work with the Mozilla 1.7.12, Mozilla Firefox 1.5, and

Internet Explorer 6.0 web browsers. Complete lists of all EC numbers and of orphan EC numbers are available and will be periodically updated. All data can be easily downloaded as text files.

MACiE—A DATABASE OF ENZYME REACTION MECHANISMS

Mechanism, Annotation and Classification in Enzymes (MACiE) is a publicly available web-based database, held in CMLReact (an XML application), that aims to help our understanding of the evolution of enzyme catalytic mechanisms and also to create a classification system which reflects the actual chemical mechanism (catalytic steps) of an enzyme reaction, not only the overall reaction. MACiE is available at http://www-mitchell.ch.cam.ac.uk/macie/. The MACiE dataset evolved from Catalytic Site Atlas (CSA) fulfils the following criteria:

1. There is a 3-dimensional crystal structure of the enzyme deposited in the Protein Data Bank (PDB).
2. There is a relatively well-understood mechanism, or at least a chemically meaningful suggestion of a mechanism. These were taken from the literature and cover a variety of methodologies, including chemical and biochemical studies, quantum mechanical calculations and structural biology reports.
3. The enzyme is unique at the H level of the CATH classification, a hierarchical classification system of protein domain structures, unless there is a homologue with a significantly different chemical mechanism.
4. Where there are a number of possible PDB codes available, the entry should be, if possible, a wild type enzyme, i.e., not engineered or mutated.

All MACiE enzymes are also contained in the Enzyme Commission (EC) classification system, that is they all have four number codes describing their overall reaction. The first level (class) describes the basic reaction type, of which there are six defined: oxidoreductases 1) transferases 2) hydrolases 3) lyases 4) isomerases 5) and ligases 6) The second and third levels (subclass and sub-subclass, respectively) describe the reaction in further detail and the final level (serial number) describes the substrate specificity. Each entry includes:

- Enzyme name and EC number
- PDB code and CATH codes of all domains in the enzyme
- Species or gene (where applicable)
- Diagram and annotation of the overall reaction
- Primary literature references

MACiE currently covers 56 of the 174 EC sub-subclasses present in the PDB. It is anticipated that all 158 subclasses for which both structures and reliable mechanisms are available will be represented in the forthcoming MACiE version 2. The data are initially entered in MDL's ISIS/Base, a database package for chemical reactions, validated by at least two people, and then converted into CMLReact using the Jumbo toolkit to create an information and semantically rich database. Jumbo is a set of Java-based software which converts the MDL file format produced from ISIS/Base into CMLReact.

The MACiE Converter section of Jumbo performs the following functions:

- Integration of the files in the ISIS/Base version of MACiE
- Identification of the reactant, product and spectator molecules
- Splitting of groups of molecules

- Automatic mapping of atoms within the reaction
- Checking for mass and charge conservation throughout the reaction (stoichiometry)
- Integration and checking of MACiE dictionary entries.

LIGAND—CHEMICAL DATABASE FOR ENZYME REACTIONS

LIGAND is a chemical database for enzyme reactions consists of two sections: the expanded ENZYME section and the new COMPOUND section. The ENZYME section is a collection of all known enzymatic reactions classified according to the nomenclature of the International Union of Biochemistry and Molecular Biology. The COMPOUND section is a collection of metabolic compounds, including substrates, products, inhibitors, cofactors and effectors, and other chemical compounds that play important functional roles in living cells. Both sections are tightly coupled with the KEGG metabolic pathway database: the ENZYME section for the network of genes and the COMPOUND section for the network of chemical compounds. The LIGAND database thus provides fundamental data on both biological and chemical aspects of life.

ENZYME Section

The ENZYME section of the LIGAND database accumulates information on all known enzymes and reactions classified according to the EC numbers given by the IUBMB. An entry of the ENZYME section consists of three parts:

i. the description of the enzyme and the reaction it catalyses;
ii. the collection of chemical compounds that are related to the enzyme;

iii. links to other databases for biological aspects of the enzyme.

The first description part is mandatory and is based on the IUBMB nomenclature. It is stored in five fields:

1. the EC number in the ENTRY field,
2. the recommended and alternative names of the enzyme in the NAME field,
3. the systematic name of the enzyme in the SYSNAME field,
4. the class of reaction in the CLASS field, and
5. the reaction formula in the REACTION field.

Additional information from the IUBMB nomenclature on the enzymatic reaction mechanisms may also be given in the COMMENT field. Non-enzymatic reactions are also included under the ENZYME entries that contain reactions adjacent to the non-enzymatic reactions.

The chemical substances that are related to the enzyme are described in the second part including SUBSTRATE, PRODUCT, INHIBITOR, COFACTOR and EFFECTOR fields. The metabolites in the SUBSTRATE and PRODUCT fields are extracted from the REACTION field, and the others are extracted from the COMMENT field, the textbooks and journal articles. The chemical compounds described in these fields are given unique accession numbers and form entries of the COMPOUND section of the LIGAND database.

The last part of the ENZYME entry contains link information to other databases that is automatically generated by the DBGET/LinkDB system.

The PATHWAY field describes the metabolic pathway in which the enzyme appears and it is a link to the KEGG/

PATHWAY database. The GENES field contains gene names that encode the enzyme for a number of organisms and they are linked to the KEGG/GENES database. The DISEASE field describes human genetic disorders caused by a lack of or mutation of the enzyme, which is linked to the OMIM database. The MOTIF field describes the protein sequence motifs that are linked to PROSITE, and the STRUCTURE field contains the code names of the protein is three-dimensional structures in the Protein Data Bank.

COMPOUND Section

The COMPOUND section collects all chemical compounds that appear in the ENZYME section, including substrates, products, inhibitors, cofactors and effectors. Each compound is given an accession number in the ENTRY field, which is followed by the compound name and its synonyms in the NAME field, and the molecular formula in the FORMULA field. The GIF image of the chemical structure is also attached when the entry is retrieved by DBGET/LinkDB under the WWW. The last portion of the COMPOUND entry contains link information to other databases.

The PATHWAY field describes the information on where in the known metabolic pathway the compound appears. The DBLINKS field includes the CAS (Chemical Abstract Services) registry number and the EC number link to the ENZYME section. The COMPOUND section is constructed manually, except for the link information to ENZYME and KEGG/PATHWAY.

The flat-file format of LIGAND database provides an easy way to integrate various databases within the DBGET/LinkDB system, which is the backbone system of the Japanese GenomeNet database service. Here, a flat file includes other types of files, including the GIF and MOL files for the COMPOUND section of LIGAND.

The DBGET/LinkDB system is especially suited to search information on related entries in other databases. During the update procedure of LIGAND, the link fields of ENZYME and COMPOUND entries are created automatically by DBGET/LinkDB. The LIGAND database is updated daily, as well as many other databases in the Japanese GenomeNet, so that the links are updated right after the linked database is updated. The other fields of ENZYME and COMPOUND are updated manually.

THE MetaCyc—DATABASE OF METABOLIC PATHWAYS AND ENZYMES

MetaCyc (MetaCyc.org) is a universal database of metabolic pathways and enzymes from all domains of life. The pathways in MetaCyc are curated from the primary scientific literature, and are experimentally determined small-molecule metabolic pathways. MetaCyc is a non-redundant reference database of a small-molecule metabolism that contains experimentally verified metabolic pathway and enzyme information curated from the scientific literature. Because MetaCyc contains only experimentally elucidated data, it is a unique and valuable resource.

MetaCyc is used in conjunction with the PathoLogic component of the pathway tools software to predict computationally the metabolic network of any organism whose genome has been sequenced and annotated. This automated process creates the predicted network in the form of a Pathway/ Genome Database (PGDB). A detailed description of Pathway Tools is available at http://bioinformatics.ai.sri.com/ptools/.

The pathways are curated from all kingdoms of life, with an emphasis on microbial (bacteria, archaea and fungi), higher plant (Viridiplantae) and, to a lesser extent, metazoan (animal) pathways. MetaCyc covers important, microbially catalysed environmental processes such as the recycling of nitrogen (10 pathways), sulphur compounds (37 pathways), the metabolism of single-carbon (C1) compounds (20 pathways), methanogenesis (13 pathways) and aromatic compound degradation (94 pathways).

The pathways in MetaCyc fall into four broad categories (or classes), namely, biosynthesis, degradation/utilization/assimilation, generation of precursor metabolites and energy, and detoxification. The largest category is currently biosynthesis, with 530 base pathways. The second largest category is degradation/utilization/assimilation, with 485 base pathways. The third category, generation of precursor metabolites and energy, contains 100 pathways. MetaCyc is used as a reference database for pathway predictions in new organisms, as part of the creation of a new PGDB by the PathoLogic software.

Recently the schema and user interface of MetaCyc and other PGDBs was enhanced by adding the ability to represent electron transport reactions (ETRs). These reactions are used to represent the electron transport processes that occur in membrane-associated enzyme complexes, involving membrane-bound electron carriers. ETRs are defined as a combination of two or more redox half-reactions that include such information as redox potential and the compartmental location of electrons. The system supports both scalar and vectoral ETRs. The MetaCyc query page contains a section for performing a text-based search of the PGDB. A "Search All" box is present at the bottom of every MetaCyc webpage to allow users to perform a new search without having to go back to the query page.

THE BioCyc DATABASE

BioCyc (BioCyc.org) is a collection of more than 350 organism-specific Pathway/Genome Databases (PGDBs) created by pathologic. Each BioCyc PGDB contains the predicted metabolic network of one organism, including metabolic pathways enzymes, metabolites and reactions predicted by the Pathway BioCyc PGDBs also contain predicted operons and predicted pathway hole fillers—predictions of enzymes which may catalyse pathway reactions that have not been assigned to an enzyme. The BioCyc website offers many tools for computational analysis of PGDBs, including comparative analysis and analysis of -omics data in a pathway context.

The BioCyc PGDBs generated by SRI are offered for adoption by any interested party for the ongoing integration of metabolic and genome-related information about an organism. But some were contributed to BioCyc from Pathway Tools users outside of SRI. The BioCyc PGDBs were created through an automated computational pipeline. Based on the amount of subsequent manual review and updating they received, the BioCyc databases are organized into three tiers.

Tier 1 PGDBs were created through intensive manual efforts, and receive continuous updating.

Tier 2 PGDBs have undergone moderate amounts of review and updating.

Tier 3 PGDBs received no subsequent manual review or updating.

The Pathway Tools software provides query and visualization services for MetaCyc and BioCyc. The software can run as a web server—in this mode it powers the BioCyc website. It can also run as a desktop application. Platforms supported for both modes are PC/Windows, PC/Linux (32-bit and 64-bit) and Sun/Solaris. SRI has developed several analysis tools that can be used with the PGDBs in the BioCyc collection. These tools are available on the BioCyc website and/or through a desktop installation of Pathway Tools, and include different viewers, comparative analysis, metabolite tracing and a set of Omics Viewers—tools for painting omics datasets (e.g. gene and protein expression and metabolomics data) onto different diagrams that can show the full genome, metabolic network or regulatory network of an organism.

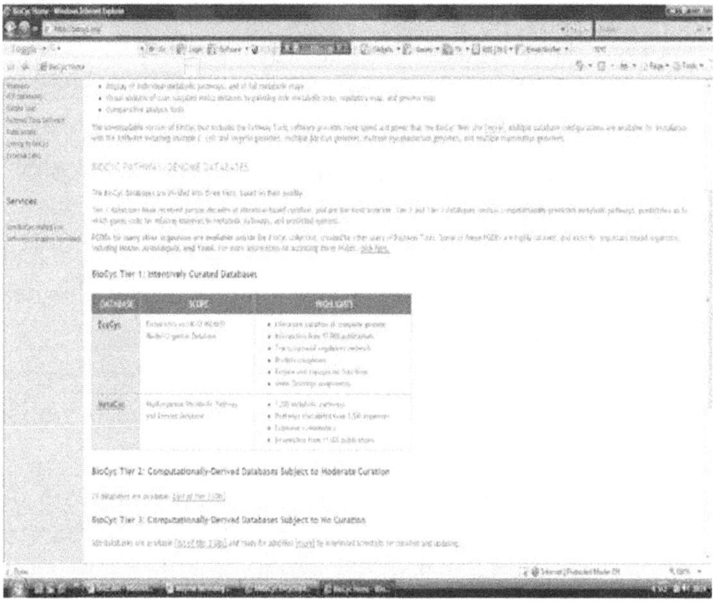

A BAYESIAN METHOD FOR IDENTIFYING MISSING ENZYMES

The PathoLogic program constructs Pathway/Genome Databases (PGDBs) by using a genome's annotation to predict the set of metabolic pathways present in an organism. PathoLogic determines the set of reactions catalysed by an organism from the enzymes annotated in its genome. For each pathway in a set of reference pathways, if one or more reactions is present in the organism, PathoLogic adds that pathway to the set of pathways present in that organism.

PathoLogic has assigned genes to three of the six reactions. When PathoLogic includes a pathway in a PGDB, the pathway may be missing one or more enzymes; we refer to these as "missing enzymes" or "pathway holes". Therefore, a fully computational approach for finding missing enzymes (i.e., filling pathway holes), was found, thus improving both the completeness and accuracy of the PGDB and the annotation of its associated genome. The pathway hole filler is implemented as part of the Pathway Tools software, a software environment for creating, editing, and querying PGDBs, such as EcoCyc and MetaCyc (http://biocyc.org). The steps of the algorithm applied to each reaction lacking an enzyme are:

1. Sequence retrieval—Retrieve from Swiss-Prot and PIR sequences for enzymes that catalyze the desired reaction in other organisms.
2. Homology search—BLAST each query isozyme sequence against the genome of the organism of interest.
3. Data consolidation—Congruence analysis of the resulting BLAST hits to consolidate data reported for sequences that align with one or more query isozymes.
4. Candidate evaluation—Determine the probability that each candidate protein has the activity required by the missing reaction.

The activity of the missing reaction is known from the inferred pathway method which uses multiple isozymes from other organisms to search for similar sequences in the genome which is used to build the PGDB. Searching the smaller genome database greatly increases search sensitivity by reducing the probability of finding a match by chance. Also, if a sequence with the desired function exists in the genome, its alignment with multiple isozymes will be more credible than an alignment between a single sequence queried against a large sequence database.

EzCatDB—THE ENZYME CATALYTIC-MECHANISM DATABASE

The EzCatDB is a novel enzyme catalytic-mechanism database, which specifically focuses on the catalytic mechanisms of enzymes. It is intended to classify them based on structural information of enzymes and the ligand molecules and proposed mechanisms. The EzCatDB is available at http://mbs.cbrc.jp/ezCatDB. It includes a search system that can retrieve some enzyme groups specifically by their respective types of catalytic residues, names or ligand molecule types that interact with enzymes as cofactors, substrates, and products. It allows querying of literature information, other database entry codes and a specific research stage.

The EzCatDB is updated on a weekly basis at the rate of roughly 10 entries a week. Each EzCatDB entry has been assigned EC numbers and CATH numbers that represent hierarchical classification of domain structures. Moreover, types of active site residues can be specified along with ligand names or types that are related to enzymes as cofactors, substrates or products. An author name and keywords in related literature can also be specified in addition to the code for literature in the PubMed database. The ligand information will be useful when considering catalytic mechanisms of the enzymes. Therefore, annotation of ligand molecules bound to the enzyme structures was performed manually for each PDB entry, and then tabulated for corresponding cofactors, substrates and products. In the search page those enzyme data with any PDB data, whose ligand molecules could be annotated as cofactors, substrates, products, intermediates or analogues of native ligand molecules, can be retrieved.

Meanwhile, entries without any PDB data, for which the complex structures with ligand molecules have been determined, can also be retrieved. This capability will be useful for structural biologists as they search for target enzymes. Information with reference to previous studies, especially those studies related to protein structures and catalytic mechanisms have been collected for each enzyme in this database. A link to the abstract page allows users to access that literature, which is maintained in the PubMed database.

Hierarchical Classification of Catalytic Mechanisms (RLCP)

A novel classification of enzyme catalytic mechanisms, which clusters catalytic mechanisms at four levels, has also been developed:

i. *Basic reaction (R)* "Basic Reaction" represents reaction types such as hydrolysis, phosphorolysis, and transfer, which are mostly related to the primary number of each EC number.

ii. *Ligand group involved in catalysis (L)* Reactive groups of substrates are classified at the second level (L). Whereas EC numbers have been based on whole chemical formulae of substrates and products, only the reactive parts of ligand molecules are considered at the "L" level.

iii. *Type of catalytic mechanism (C)* The catalytic mechanisms are classified systematically.

iv. *Residues/cofactors located on proteins (P)* Types of catalytic residues and cofactors with the ligating residues are classified.

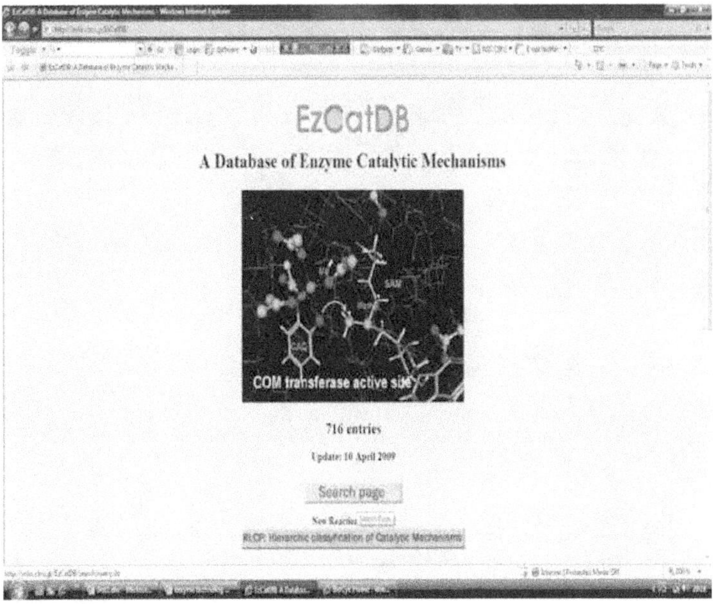

Enzyme data with tertiary structures deposited in the PDB, to which EC numbers and CATH domain classification (version 2.4) had been assigned, have been analysed and annotated in the EzCatDB. The EzCatDB groups enzyme data in the PDB and the Swiss-Prot database with identical domain compositions, EC numbers and catalytic mechanisms. Some enzyme sequences from different organisms can be homologous. Those PDB data and their corresponding Swiss-Prot codes from different organisms can be included in the same EzCatDB entries if they have the same EC numbers, catalytic mechanisms and domain compositions, and structures in terms of the CATH classification.

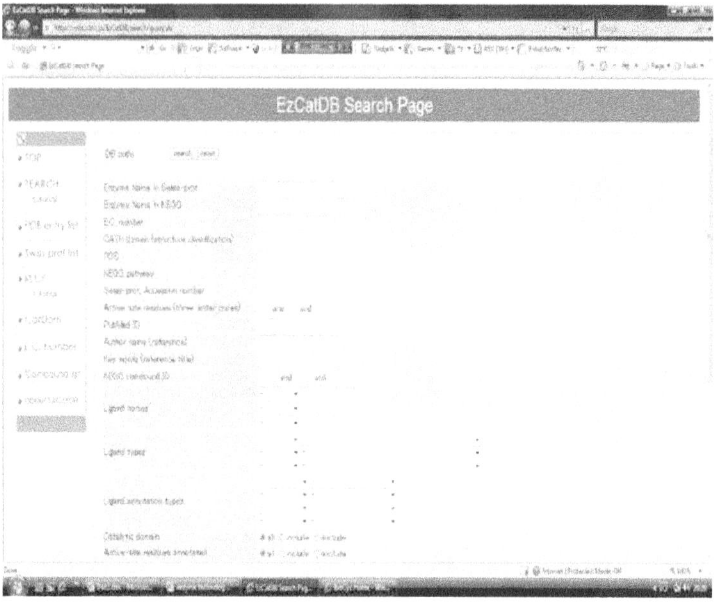

THE ENZYME DATA BANK

The ENZYME data bank is a repository of information relative to the nomenclature of enzymes. It is primarily based on the recommendations of the Nomenclature Committee of the International Union of Biochemistry and Molecular Biology (IUBMB) and it contains the following data for each type of characterized enzyme:

- EC number
- Recommended name
- Alternative names (if any)
- Catalytic activity
- Cofactors (if any)

- Pointers to the Swiss-Prot protein sequence entrie(s) that correspond to the enzyme (if any)
- Pointers to human disease(s) associated with a deficiency of the enzyme (if any)

The ENZYME data bank can be useful to anybody working with enzymes and it can be of help in the development of computer programs involved with the manipulation of metabolic pathways. The main source for the data in the ENZYME data bank comes from the recommendations of the IUBMB, but additional information has been extracted from the literature. The information concerning human diseases originates from the MIM database of Victor McKusick. The commission regularly updates additions to the nomenclature so that they can be integrated into the databank in a timely manner.

The entries in the database are structured so as to be usable by human readers as well as by computer programs. An entry in the database is composed of defined-line types, each with its own format; they are used to record the various types of data which make up the entry. The database is distributed with a user's manual (ENZUSER.TXT); a file describing the various classes, subclasses and sub-subclasses of enzymes (ENZCLASS.TXT); and a file that describes how the database can be obtained (ENZYME.GET). ENZYME is distributed on magnetic tape and on CD-ROM by the EMBL Data Library.

For all enquiries regarding the subscription and distribution of ENZYME one should contact:

EMBL Data Library
European Molecular Biology Laboratory
Postfach 10.22.09,
Meyerhofstrasse 1 D-69012 Heidelberg,
Germany

Telephone: (+49 6221) 387 258

Telefax: (+49 6221) 387 519 or 387 306

Electronic network address: datalib@EMBL-heidelberg.de

ENZYME can be obtained from the EMBL File Server ExPASy (Expert Protein Analysis System) server, University of Geneva, Switzerland Internet address: expasy.hcuge.ch (129.195.254.61), National Institute of Genetics (Japan) FTP server Internet address: ftp.nig.ac.jp (133.39.16.66)

A version of the database in the ASN.1 data exchange format compatible with the databases and software developed by the National Center for Biotechnology Information (NCBI) is also available on some of the above servers.

GLOSSARY

Abzyme Also called catmab (from catalytic monoclonal antibody), it is a monoclonal antibody with catalytic activity. Abzymes are usually artificial constructs, but are also found in normal humans (anti-vasoactive intestinal peptide autoantibodies) and in patients with autoimmune diseases such as systemic lupus erythematosus, where they can bind to and hydrolyse DNA. Abzymes are potential tools in biotechnology, e.g. to perform specific actions on DNA.

Acetoin 3-hydroxy 2-butanone, containing a ketone as well as a hydroxy group. It is an intermediate of the butanediol cycle in microorganisms. It is formed in the fermentation of glucose by certain bacteria (Enterobacteriaceae). In mammals it is oxidized to carbon dioxide.

Acetyl-CoA An important molecule in metabolism, used in many biochemical reactions. Its main use is to convey the carbon atoms within the acetyl group to the citric acid cycle to be oxidized for energy production.

Acid–base catalysis In acid catalysis and base catalysis a chemical reaction is catalysed by an acid or a base. The acid is often the and the base is often a hydroxyl ion. Typical reactions catalysed by proton transfer are esterfications and aldol reactions.

Activation energy The energy that must be overcome in order for a chemical reaction to occur. Activation energy may also be defined as the minimum energy required to start a chemical reaction. The activation energy of a reaction is usually denoted by E_a, and given in units of kilojoules per moles.

Activator A DNA-binding protein that regulates one or more genes by increasing the rate of transcription. The activator may increase transcription by virtue of a connected domain which assists in the formation of the RNA polymerase holoenzyme, or may operate through a co-activator.

Active site The active site of an enzyme contains the catalytic and binding sites. The structure and chemical properties of the active site allow the recognition and binding of the substrate. The active site is usually a big pocket or cleft surrounded by amino acid and other side chains at the surface of the enzyme that contains residues responsible for the substrate specificity (charge, hydrophobicity, steric hindrance) and catalytic residues which often act as proton donors or acceptors or are responsible for binding a cofactor such as PLP, TPP or NAD. The active site is also the site of inhibition of enzymes.

Acute myeloid leukaemia (AML) Also known as acute myelogenous leukaemia, or AML. It is a cancer of the myeloid line of blood cells, characterized by the rapid growth of abnormal white blood cells which accumulate in the bone marrow and interfere with the production of normal blood cells. AML is the most common acute leukaemia affecting adults, and its incidence increases with age.

Acyl-CoA (Formyl-CoA) A coenzyme involved in the metabolism of fatty acids. It is a temporary compound formed when coenzyme A (CoA) attaches to the end of a long-chain fatty acid, inside living cells. The CoA is then removed from the chain, carrying two carbons from the chain with it, forming acetyl-CoA. This is then used in the citric acid cycle to start a chain of reactions, eventually forming many adenosine triphosphates.

Adenosine triphosphate (ATP) A multifunctional nucleotide that plays an important role in cell biology as a coenzyme is the "molecular unit of currency" of intracellular energy transfer.

Adenoviruses Medium-sized (90–100 nm), non-enveloped (naked) icosahedral viruses composed of a nucleocapsid and a double-stranded linear DNA genome. There are over 52 different serotypes in humans, which are responsible for 5–10% of upper respiratory infections in children, and many infections in adults as well.

Adsorbent A material that sorbs another substance, i.e., that has the capacity or tendency to take it up by either absorption or adsorption.

Affinity The dissociation constant is commonly used to describe the affinity between a ligand (L) (such as a drug) and a protein (P), i.e., how tightly a ligand binds to a particular protein. Ligand–protein affinities are influenced by non-covalent intermolecular interactions between the two molecules such as hydrogen bonding, electrostatic interactions , hydrophobic and van der Waals forces.

Affinity chromatography A chromatographic method of separating biochemical mixtures, based on a highly specific biological interaction such as that between antigen and antibody, enzyme and substrate, or receptor and ligand.

Alkali A basic, ionic salt of an alkali metal or alkaline earth metal element. Alkalis are best known for being bases that dissolve in water. Bases are compounds with a pH greater than 7.

Allosteric enzymes Enzymes that change their conformation upon binding of an effector. An allosteric enzyme is an oligomer whose biological activity is affected by altering the conformation(s) of its quaternary structure. Allosteric enzymes tend to have several subunits.

Allosteric sites A site on a multi-subunit enzyme that is not the substrate-binding site but that when reversibly bound by an effector induces a conformational change in the enzyme, altering its catalytic properties.

Allozymes Variant forms of an enzyme that are coded for by different alleles at the same locus as opposed to isozymes which are enzymes that perform the same function, but which are coded for by genes located at different loci.

Alzheimer's disease The most common form of dementia. This incurable, degenerative, and terminal disease was first described by Alois Alzheimer. It is also called Alzheimer disease, senile dementia of the Alzheimer Type (SDAT) or simply Alzheimer's.

Amino acids The basic structural building units of proteins, which form short polymer chains called peptides or longer chains called either polypeptides or proteins.

Ampholytes Molecules that contain both acidic and basic groups (and are therefore amphoteric) and will exist mostly as zwitterions in a certain range of pH. The pH at which the average charge is zero is known as the molecule's isoelectric point.

Analogue A structural derivative of a parent compound that often differs from it by a single element.

Angiogenesis A physiological process involving the growth of new blood vessels from pre-existing vessels. Though there has been some debate over this, "vasculogenesis" is the term used for spontaneous blood-vessel formation, and "intussusception" is the term for new blood vessel formation by splitting off existing ones.

Anions Atoms or groups of atoms that have gained electrons. Having more negatively charged electrons than positively charged protons, they are negatively charged.

Antibiotics A substance or compound (also called chemotherapeutic agent) tha kills or inhibits the growth o bacteria. Antibiotics belong to the group of antimicrobia compounds used to treat infections caused by microorganisms, including fung and protozoa.

Antioxidant A molecule capable of slowing or preventing the oxidation of other molecules

Antisense RNA A single-stranded RNA that is complementary to a messenger RNA (mRNA) strand transcribed within a cell. Antisense RNA may be introduced into a cell to inhibit translation of a complementary mRNA by base-pairing to it and physically obstructing the translation machinery. This effect is therefore stoichiometric.

Antivitamins In simple terms "a substance that makes a vitamin ineffective." A vitamin antagonist is essentially the same thing as an antivitamin. It is a substance that lessens or negates the chemical action of a vitamin in the body.

Apoenzyme The protein part of an enzyme which requires a coenzyme for activity, and is therefore inactive if the coenzyme is absent.

Aptamers Oligonucleic acid or peptide molecules that bind to a specific target molecule. Aptamers are usually created by selecting them from a large random sequence pool, but natural aptamers also exist in riboswitches. Aptamers can be used for both basic research and clinical purposes as macromolecular drugs. Aptamers can be combined with ribozymes to self-cleave in the presence of their target molecule. These compound molecules have additional research, industrial and clinical applications.

More specifically, aptamers can be classified as:

- *DNA or RNA aptamers* They consist of (usually short) strands of oligonucleotides.
- *Peptide aptamers* They consist of a short variable peptide domain, attached at both ends to a protein scaffold.

Artificial enzymes A synthetic, organic molecule prepared to recreate the active site of an enzyme.

Ascorbate An ion of ascorbic acid required for a range of essential metabolic reactions in all animals and plants. It is made internally by almost all organisms, humans being a notable exception. Deficiency in this vitamin causes scurvy in humans. It is also widely used as a food additive.

Autolysis The destruction of a cell through the action of its own enzymes. It may also refer to the digestion of an enzyme by another molecule of the same enzyme.

Auxostat A continuous culture device which, while in operation, uses feedback from a measurement taken on the growth chamber to control the media flow rate, maintaining the measurement at a constant. *Auxo* was the Greek Goddess of spring growth, and as a prefix represents nutrients. However, the most typical auxostats are pH-auxostats, with feedback between the growth rate and a pH meter.

Bacteriophage Any one of a number of viruses that infect bacteria. The term is commonly used in its shortened form, "phage". Bacteriophages consist of an outer protein capsid enclosing genetic material. The genetic material can be ssRNA, dsRNA, ssDNA, or dsDNA ('ss-' or 'ds-' prefix denotes single-stranded or double-stranded) between 5,000 and 500,000 nucleotides long with either circular or linear arrangement. Bacteriophages are much smaller than the bacteria they destroy—usually between 20 and 200 nm in size.

Beta-oxidation The process by which fatty acids, in the form of Acyl-CoA molecules, are broken down in mitochondria and/or in peroxisomes to generate acetyl-CoA, the entry molecule for the Krebs cycle.

Bi-Bi reactions A reaction catalysed by a single enzyme in which two substrates and two products are involved; the ping-pong mechanism may be involved in such a reaction.

Bioluminescence The production and emission of light by a living organism. Its name is a hybrid word, originating from the Greek *bios* for "living" and the Latin *lumen* for "light". Bioluminescence is a naturally occurring form of chemi-luminescence where energy is released by a chemical reaction in the form of light emission. Adenosine triphosphate (ATP) is involved in most instances. The chemical reaction can occur either inside or outside the cell. In bacteria, the expression of genes related to bioluminescence is controlled by an operon called the Lux operon. Bioluminescence has appeared independently several times (up to 30 or more) during evolution.

Bioreactor A device or system that supports a biologically active environment. In one case, a bioreactor is a vessel in which is carried out a chemical process which involves organisms or biochemically active substances derived from such organisms. This process can either be aerobic or anaerobic. These bioreactors are commonly cylindrical, ranging in volume from litres to cubic metres, and are often made of stainless steel. A bioreactor may also refer to a device or system meant to grow cells or tissues in the context of cell culture. These devices are being developed for use in tissue engineering.

Biosensor A device for the detection of an analyte that combines a biological component with a physico-chemical detector component. It consists of 3 parts

—the sensitive biological element, the transducer and the detector element.

Biotin (Vitamin H/B$_{7)}$ A water-soluble B-complex vitamin which is composed of an ureido (tetrahydroimidizalone) ring fused with a tetrahydrothiophene ring. Biotin is a cofactor in the metabolism of fatty acids and leucine, and in gluconeogenesis.

BLAST In bioinformatics, **B**asic **L**ocal **A**lignment **S**earch **T**ool, or **BLAST**, is an algorithm for comparing primary biological sequence information, such as the amino-acid sequences of different proteins or the nucleotides of DNA sequences. A BLAST search enables a researcher to compare a query sequence with a library or database of sequences, and identify library sequences that resemble the query sequence above a certain threshold. For example, following the discovery of a previously unknown gene in the mouse, a scientist will typically perform a BLAST search of the human genome to see if humans carry a similar gene; BLAST will identify sequences in the human genome that resemble the mouse gene based on similarity of sequence. The BLAST program was designed by Eugene Myers, Stephen Altschul, Warren Gish, David J. Lipman and Webb Miller at the NIH and was published in *J. Mol. Biol.* in 1990.

Bronsted acid A molecular entity capable of donating a hydron (proton) to a base, (i.e., a hydron donor) or the corresponding chemical species.

Bronsted base A molecular entity capable of donating a hydron (proton) from an acid (i.e., a hydron acceptor) or the corresponding chemical species.

Buffer An aqueous solution consisting of a mixture of a weak acid and its conjugate base or a weak base and its conjugate acid. It has the property that the pH of the solution changes very little when a small amount of acid or base is added to it. Buffer solutions are used as a means of keeping pH at a nearly constant value in a wide variety of chemical applications.

Calvin cycle A series of biochemical reactions that take place in the stroma of chloroplasts in photosynthetic organisms. It is also known as Calvin–Benson–Bassham cycle, light-independent reaction, or carbon fixation.

Carboxypeptidase An enzyme (EC number 3.4.16–3.4.18) that hydrolyses the carboxy-terminal (C-terminal) end of a peptide bond. Humans, animals, and plants contain several types of carboxypeptidases with diverse functions ranging from catabolism to protein maturation. The first carboxypeptidases studied were those involved in the digestion of food (pancreatic carboxypeptidases A1, A2, and B). However, most of the known carboxypeptidases are not involved in catabolism; they help to mature proteins or regulate biological processes.

Catalase A common enzyme found in nearly all living organisms which are exposed to oxygen, where it functions to catalyse the decomposition of hydrogen peroxide to water and oxygen.

Catalysis Catalysis is the process in which the rate of a chemical reaction is either increased or decreased by means of a chemical substance known as a catalyst.

Catalyst Unlike other reagents that participate in the chemical reaction, a catalyst is not consumed by the reaction itself. The catalyst may participate in multiple chemical transformations. Catalysts that speed the reaction are called positive catalysts. Catalysts that slow down the reaction are called negative catalysts.

CATH Protein Structure Classification A semi-automatic, hierarchical classification of protein domains published in 1997 by Christine Orengo, Janet Thornton and their colleagues. CATH shares many broad features with its principal rival, SCOP, however there are also many areas in which the detailed classification differs greatly.

Cations Atoms that have lost an electron to become positively charged.

Cell division A process by which a cell, called the parent cell, divides into two or more cells, called daughter cells. Cell division is usually a small segment of a larger cell cycle. This type of cell division in eukaryotes is known as mitosis, and leaves the daughter cell capable of dividing again. The corresponding sort of cell division in prokaryotes is known as binary fission. In another type of cell division present only in eukaryotes, called meiosis, a cell is permanently transformed into a gamete and cannot divide again until fertilization.

Chelator Chelation is the binding or complexation of a bi- or multidentate ligand. These ligands, which are often organic compounds, are called chelants, chelators, chelating agents, or sequestering agents. Chelating agents form multiple bonds with a single metal ion.

Chemiluminescence/Chemoluminescence The emission of light with limited emission of heat (luminescence), as the result of a chemical reaction. Given reactants A and B, with an excited intermediate $\diamond$, $[A] + [B] \rightarrow [\diamond] \rightarrow$ [Products] + Light

Chemostat Word derived from "**Chemical** environment is **static**", and is a bioreactor to which fresh medium is continuously added, while culture liquid is continuously removed to keep the culture volume constant. By changing the rate with which medium is added to the bioreactor, the growth rate of the micro-organism can be easily controlled.

Chemotherapy In its most general sense, refers to treatment of disease by chemicals that kill cells, both good and bad, but specifically those of microorganisms or cancer. Chemotherapy acts by killing cells that divide rapidly, one of the main properties of cancer cells. This means that it also harms cells that divide rapidly under normal circumstances: cells in the bone marrow, and hair follicles; this results in the most common side-effects of chemotherapy–myelosuppression (decreased production of blood cells), mucositis (inflammation of the lining of the digestive tract) and alopecia (hair loss).

Chiral The term used to describe an object that is non-superposable on its mirror image.

Chromatogram The visual output of the chromatograph. In the case of an optimal separation, different peaks or patterns on the chromatogram correspond to different components of the separated mixture.

Chromatography The collective term for a set of laboratory techniques for the separation of mixtures. It involves passing a mixture dissolved in a "mobile phase" through a stationary phase, which separates the analyte to be measured from other molecules in the mixture and allows it to be isolated.

Chymotrypsin A digestive enzyme that can perform proteolysis. Chymotrypsin cleaves peptides at the carboxyl side of tyrosine, tryptophan, and phenylalanine because these three amino acids contain aromatic rings, which fit into a "hydrophobic pocket" in the enzyme. Over time, chymotrypsin also hydrolyses other amide bonds, particularly those with leucine-donated carboxyls.

Cis-trans **isomerism/geometric isomerism/configuration isomerism/E-Z isomerism** A form of stereoisomerism describing the orientation of functional groups within a molecule. In general, such isomers contain double bonds, which cannot rotate, but they can also arise from ring structures, wherein the rotation of bonds is greatly restricted.

Coenzyme A A coenzyme (CoA, CoASH, or HSCoA) notable for its role in the synthesis and oxidation of fatty acids, and the oxidation of pyruvate in the citric acid cycle. It is adapted from cystamine, pantothenate, and adenosine triphosphate.

Coenzyme B A coenzyme required for redox reactions in methanogens. The full chemical name of coenzyme B is 7-mercaptoheptanoylthreonine-phosphate. The molecule contains a thiol, which is its principal site of reaction.

Coenzyme F420 A coenzyme (8-hydroxy 5-deazaflavin) involved in redox reactions in methanogens. It is a flavin derivative. The coenzyme is a substrate for coenzyme F420 hydrogenase, 5,10-methylene-tetrahydromethanopterin reductase and methylene-tetrahydromethanopterin dehydrogenase.

Coenzyme M A coenzyme required for methyl-transfer reactions in the metabolism of methanogens. The coenzyme is an anion with the formula $HSCH_2CH_2SO_3^-$. It is named

2-mercaptoethanesulphonate and abbreviated HS-CoM.

Coenzyme Q Also known as ubiquinone or 2,3-dimethoxy 5-methyl 6-multiprenyl 1,4-benzoquinone. The reduced form (QH_2) is called ubiquinol and the partially reduced free-radical form ($\cdot Q^{\cdot}$) is called semiquinone. The quinone group allows Coenzyme Q to function as an electron-carrier, while the highly hydrophobic tail of isoprene units helps to confine coenzyme Q to lipid-rich areas of cells. CoQ is the only component of the electron transport chain that is lipid rather than protein, and CoQ is the only component that is not anchored to the inner mitochondrial membrane. CoQ is so hydrophobic, in fact, that it usually shuttles back and forth laterally in the middle of the phospholipid bilayer without getting close to the polar phosphate groups on the edges of the membrane. CoQ picks up reducing equivalents from protein complex I and protein complex II and shuttles these "electrons" (as $CoQH_2{=}QH_2$) to protein complex III and then returns (as CoQ=Q) to get more reducing equivalents (the "Q cycle").

Coenzymes A nonproteinaceous organic substance that usually contains a vitamin or mineral and combines with a specific protein, the apoenzyme, to form an active enzyme system.

Cofactor A non-protein chemical compound that is bound (either tightly or loosely) to a protein and is required for the protein's biological activity. These proteins are commonly enzymes and cofactors can be considered "helper molecules/ions" that assist in biochemical transformations.

Competitive inhibition A form of enzyme inhibition where binding of the inhibitor to the enzyme prevents binding of the substrate and vice versa. In competitive inhibition, the inhibitor binds to the same active site as the normal enzyme substrate, without undergoing a reaction. The substrate molecule cannot enter the active site while the inhibitor is there, and the inhibitor cannot enter the site when the substrate is there. In this case, the maximum speed of the reaction is unchanged, while the apparent affinity of the substrate to the binding site is decreased.

Conjugated proteins A protein that functions in interaction with other chemical groups attached by covalent bonds or by weak interactions. Glycoproteins are generally the largest and most abundant group of conjugated proteins. They range from glycoproteins in cell-surface membranes that constitute the glycocalyx, to important antibodies produced by leucocytes.

Continuous stirred-tank reactor (CSTR) Also known as vat- or backmix reactor, it is a common ideal reactor type in chemical engineering. A CSTR often refers to a model used to estimate the key unit operation variables when using a continuous agitated-tank reactor to reach a specified output. The mathematical model works for all fluids: liquids, gases, and slurries.

Cooperative binding A special case of allostery. Cooperative binding requires that the macromolecule have more than one binding site, since cooperativity results from the interactions between binding sites. If the binding of ligand at one site increases the affinity for ligand at another site, the macromolecule exhibits positive cooperativity. Conversely, if the binding of ligand at one site lowers the affinity for ligand at another site, the protein exhibits negative cooperativity. If the ligand binds at each site independently, the binding is non-cooperative.

Covalent bond A type of chemical bonding that is characterized by the sharing of pairs of electrons between atoms, or between atoms and other covalent bonds. In short, attraction-to-repulsion stability that forms between atoms when they share electrons is known as covalent bonding. Covalent bonding includes many kinds of interaction, including σ-bonding, π-bonding, metal to non-metal bonding, agostic interactions, and three-centre two-electron bonds.

Covalent catalysis A type of enzyme reaction with substrates to form very unstable, covalently joined enzyme–substrate complexes which undergo further reaction.

Crystallization The natural or artificial process of formation of solid crystals precipitating from a solution or melt, or more rarely deposited directly from a gas. Crystallization is also a chemical solid–liquid separation technique, in which mass transfer of a solute from the liquid solution to a pure solid crystalline phase occurs.

Cytochrome Membrane-bound haemoproteins that contain haem groups and carry out electron transport. They are found either as monomeric proteins (e.g. cytochrome *c*) or as subunits of bigger enzymatic complexes that catalyse redox reactions. They are found in the mitochondrial inner membrane and endoplasmic reticulum of eukaryotes, in the chloroplasts of plants, in photosynthetic microorganisms, and in bacteria.

Cytoplasm The part of a cell that is enclosed within the plasma membrane. In eukaryotic cells, the cytoplasm contains organelles, such as mitochondria, which are filled with liquid that is kept separate from the rest of the cytoplasm by biological membranes. The cytoplasm is the site where most cellular activities occur, such as many metabolic pathways, and processes such as cell division.

Cytoskeleton A cellular "scaffolding" or "skeleton" contained within the cytoplasm. The cytoskeleton is present in all cells; it was once thought this structure was unique to eukaryotes, but recent research has identified the prokaryotic cytoskeleton. It is a dynamic structure that maintains cell shape, protects the cell, enables cellular motion (using structures such as flagella, cilia and lamellipodia), and plays important roles in both intracellular transport (the movement of vesicles and organelles, for example) and cellular division.

de novo pathway A biochemical pathway where a complex biomolecule is synthesized anew from simple precursor molecules.

Deoxyribozymes/DNA enzymes/ catalytic DNA,/DNAzymes DNA molecules with catalytic action. In contrast to a RNA ribozyme that has many catalytic capabilities, DNA is only associated with gene replication and nothing else. The reasons are that DNA lacks specific functional groups and that DNA prefers the double coil conformation in which potential catalytic sites are shielded.

In comparison to proteins built up from 20 monomers, both RNA and DNA have a much more restricted set of monomers to choose from which limits the construction of interesting catalytic sites. For these reasons, DNAzymes exist only in the laboratory.

Derived proteins A derivative of protein effected by chemical change, e.g. hydrolysis.

Deuterium Also called heavy hydrogen, it is a stable isotope of hydrogen with a natural abundance in the oceans of approximately one atom in 6500 of hydrogen. The nucleus of deuterium, called a deuteron, contains one proton and one neutron, whereas the far more common hydrogen nucleus contains no neutrons.

Dialysis A process that works on the principle of the diffusion of solutes and ultrafiltration of fluid across a semi-permeable membrane. Blood flows by one side of a semi-permeable membrane, and a dialysate or fluid flows by the opposite side. Smaller solutes and fluid pass through the membrane. The blood flows in one direction and the dialysate flows in the opposite direction. The counter-current flow of the blood and dialysate maximizes the concentration gradient of solutes between the blood and dialysate, which helps to remove more urea and creatinine from the blood.

Diastase Any one of a group of enzymes which catalyses the breakdown of starch into maltose. It was the first type of enzyme discovered, in 1833, by Anselme Payen, who found it in malt solution. Today, diastase means any α-, β-, or γ- amylase (all of them hydrolases) that can break down carbohydrates.

DNA (Deoxyribonucleic acid) A nucleic acid that contains the genetic instructions used in the development and functioning of all known living organisms and some viruses. The main role of DNA molecules is the long-term storage of information. DNA is often compared to a set of blueprints or a recipe, or a code, since it contains the instructions needed to construct other components of cells, such as proteins and RNA molecules. The DNA segments that carry this genetic information are called genes, but other DNA sequences have structural purposes, or are involved in regulating the use of this genetic information.

DNA polymerase An enzyme that catalyses the polymerization of deoxyribonucleotides into a DNA strand. DNA polymerases are bestknown for their role in DNA replication, in which the polymerase "reads" an intact DNA strand as a template and uses it to synthesize the new strand. The newly polymerized molecule is complementary to the template strand and identical to the template's original partner strand. DNA polymerases use a magnesium ion for catalytic activity.

DNA polymerase II (DNA Pol II) A prokaryotic DNA polymerase most likely involved in DNA repair. The enzyme is 90 kDa in size and is coded by the polB gene. Strains lacking the gene show no defect in growth or replication. Synthesis of Pol II is induced during the stationary phase of cell growth. This is a phase in which little growth and DNA synthesis occurs.

DNA polymerase III DNA polymerase III holenzyme is the primary enzyme complex involved in prokaryotic DNA replication. The complex has high processivity (i.e., the number of nucleotides added per binding event) and, specifically referring to the replication of the *E.coli* genome, works in conjunction with four other DNA polymerases (Pol I, Pol II, Pol IV, and Pol V). Being the primary holoenzyme involved in replication activity, the DNA Pol III holoenzyme also has proofreading capabilities that correct replication mistakes by means of exonuclease activity working 3′→5′. DNA Pol III is a component of the replisome, which is located at the replication fork.

DNA repair A collection of processes by which a cell identifies and corrects damage to the DNA molecules that encode its genome. In human cells, both normal metabolic activities and environmental factors such as UV light and radiation can cause DNA damage, resulting in as many as 1 million individual molecular lesions per cell per day. Many of these lesions cause structural damage to the DNA molecule and can alter or eliminate the cell's ability to transcribe the gene that the affected DNA encodes. Other lesions induce potentially harmful mutations in the cell's genome, which affect the survival of its daughter cells after it undergoes mitosis. Consequently, the DNA repair process is constantly active as it responds to damage in the DNA structure.

Downstream The part of a bioprocess where the cell mass from the upstream are processed to meet purity and quality requirements. Downstream processing is usually divided into three main sections, a capture section, a purification section and a polishing section.

Drug A chemical substance used in the treatment, cure, prevention, or diagnosis of disease or used to otherwise enhance physical or mental well-being. They may be prescribed for a limited duration, or on a regular basis for chronic disorders.

EC numbers The Enzyme Commission number (EC number) is a numerical classification scheme for enzymes, based on the chemical reactions they catalyse. As a system of enzyme nomenclature, every EC number is associated with a recommended name for the respective enzyme. Strictly speaking, EC numbers do not specify enzymes, but enzyme catalysed reactions. If different enzymes (for instance from different organisms) catalyse the same reaction, then they receive the same EC number.

E.coli A gram-negative bacterium that is commonly found in the lower intestine of warm-blooded organisms (endotherms). Most *E. coli* strains are harmless, but some, such as serotype O157:H7, can cause serious food poisoning in humans, and are occasionally responsible for costly product recalls. The harmless strains are part of the normal flora of the gut, and are of benefit to their hosts since they can produce vitamin K_2 and prevent the establishment of pathogenic bacteria within the intestine.

Electron microscopy A type of microscope that uses a particle beam of electrons to illuminate a specimen and create a highly magnified image. They have much greater resolving power than light microscopes that use electromagnetic radiation and can obtain much higher magnifications of up to 2 million times, while the best light microscopes are limited to magnifications of 2000 times.

Electrons Subatomic particles that carry a negative electric charge. The electron has no known substructure and is believed to be a point particle. Electrons participate in gravitational, electromagnetic and weak interactions.

Electron transport chain An electron transport chain couples a chemical reaction between an electron donor (such as NADH) and an electron acceptor (such as O_2) to the transfer of H^+ ions across a membrane, through a set of mediating biochemical reactions. Electron transport chains are used for extracting energy from sunlight (photosynthesis) and from redox reactions such as the oxidation of sugars (respiration).

Electrophile A reagent attracted to electrons that participates in a chemical reaction by accepting an electron pair in order to bond to a nucleophile. Because electrophiles accept electrons, they are Lewis acids. Most electrophiles are positively charged, have an atom which carries a partial positive charge, or have an atom which does not have an octet of electrons.

Electrophoresis A technique used for the separation of deoxyribonucleic acid (DNA), ribonucleic acid (RNA), or protein molecules using an electric current applied to a gel matrix.

Electrostatic catalysis Stabilization of charged transition states can also be by residues in the active site forming ionic bonds (or partial ionic charge interactions) with the intermediate. These bonds can either come from acidic or basic side chains found on amino acids such as lysine, arginine, aspartic acid or glutamic acid or from metal cofactors such as zinc. Metal ions are particularly effective and can reduce the pKa of water enough to make it an effective nucleophile.

ELISA (Enzyme-linked immunosorbent assay) A biochemical technique used mainly in immunology to detect the presence of an antibody or an antigen in a sample. The ELISA has been used as a diagnostic tool in medicine and plant pathology, as well as a quality control check in various industries. In simple terms, in ELISA, an unknown amount of antigen is affixed to a surface, and then a specific antibody is washed over the surface so that it can bind to the antigen. This antibody is linked to an enzyme, and in the final step, a substance is added that the enzyme can convert to some detectable signal. Thus in the case of fluorescence ELISA, when light of the appropriate wavelength is shone upon the sample, any antigen/antibody complexes will fluoresce so that the amount of antigen in the sample can be inferred through the magnitude of the fluorescence.

EMBL The European Molecular Biology Laboratory (EMBL), is a molecular biology research institution supported by 20 European countries and Australia as associate member state. The EMBL was created in 1974 and is a non-profit organisation funded by public research money from its member states. Research at EMBL is conducted by approximately 85 independent groups covering the spectrum of molecular biology. The Laboratory operates from five sites: the main Laboratory in Heidelberg, and Outstations in Hinxton (the European Bioinformatics Institute (**EBI**)), Grenoble, Hamburg, and Monterotondo near Rome. Each of the sites has a research specific field. At EBI, the research is oriented towards computational biology and bioinformatics, at Grenoble and Hamburg the research is in the field of structural biology, at Monterotondo the research is using mainly mouse models for medical related problems and last but not least in Heidelberg, the headquarters, there are big departments in cell biology and gene expression as well as smaller complementing the aforementioned research fields.

Endocytosis The process by which cells absorb material (molecules such as proteins) from outside the cell by engulfing it with their cell membrane. It is used by all cells of the body because most substances important to them are large polar molecules that cannot pass through the hydrophobic plasma membrane or cell membrane. The process opposite to endocytosis is exocytosis.

Endoenzymes An enzyme that functions within the cell in which it was produced (also called intracellular enzyme).

Entropy The quantitative measure of disorder in a system. The concept comes out of thermodynamics, which deals with the transfer of heat energy within a system.

Enzyme Proteins that serve as catalysts and that speed up or slow down reactions, but remain unchanged.

Enzyme activity The catalytic effect exerted by an enzyme, expressed as units per milligram of enzyme or as molecules of substrate transformed per minute per molecule of enzyme.

Enzyme immunoassay (EIA) *See* ELISA.

Enzyme kinetics The study of the chemical reactions that are catalysed by enzymes, with a focus on their reaction rates. The study of an enzyme's kinetics reveals the catalytic mechanism of this enzyme, its role in metabolism, how its activity is controlled, and how a drug or a poison might inhibit the enzyme.

Erythrocytes Red blood cells, also known as red blood corpuscles, haematids or erythrocytes. They are the most common type of blood cells and the vertebrate body's principal means of delivering oxygen to the body tissues via the blood. They take up oxygen in the lungs or gills and release it while squeezing through the body's capillaries. The cells are filled with haemoglobin, a biomolecule that can bind to oxygen. The blood's red colour is due to the colour of haemoglobin.

Evaporation The slow vaporization of a liquid and the reverse of condensation.

Exocytosis The durable process by which a cell directs the contents of secretory vesicles out of the cell membrane. These membrane-bound vesicles contain soluble proteins to be secreted to the extracellular environment, as well as membrane proteins and lipids that are sent to become components of the cell membrane.

Exoenzymes Enzymes produced within the cell, then released outside of the cell to begin the process of extracellular digestion. It is usually used for breaking up large molecules that would not be able to enter the cell otherwise.

Exon A nucleic acid sequence that is represented in the mature form of an RNA molecule after i) portions of a precursor RNA, introns, have been removed by cis-splicing or ii) two or more precursor RNA molecules have been ligated by trans-splicing. The mature RNA molecule can be a messenger RNA or a functional form of a non-coding RNA such as rRNA or tRNA. Depending on the context, exon can refer to the sequence in the DNA or its RNA transcript.

Exonuclease activity Enzymes that work by cleaving nucleotides one at a time from the end of a polynucleotide chain. A hydrolysing reaction occurs that breaks phosphodiester bonds at either the 3′ or 5′ ends. Its close relative is the endonuclease, which cleaves phosphodiester bonds in the middle of a polynucleotide chain.

ExPASy (**Ex**pert **P**rotein **A**nalysis **Sy**stem) A proteomics server of the Swiss Institute of Bioinformatics (SIB) which analyses protein sequences and structures and two-dimensional gel electrophoresis (2D PAGE electrophoresis). The server functions in collaboration with the European Bioinformatics Institute. ExPASy also produces the protein sequence knowledgebase, UniProtKB/Swiss-Prot, and its computer annotated supplement, UniProtKB/Trembl. Between its installation on 1 August 1993 and 5 April 2007, ExPASy was consulted 1 billion times.

Factor X Also known by the eponym Stuart-Prower factor or as thrombokinase, it is an enzyme (EC 3.4.21.6) of the coagulation cascade. It is a serine endopeptidase. It is synthesized in the liver and requires vitamin K for its synthesis.

Fatty acids Carboxylic acids often with a long unbranched aliphatic tail (chain), which is either saturated or unsaturated. Fatty acids are produced by the hydrolysis of the ester linkages in a fat or biological oil (both of which are triglycerides), with the removal of glycerol.

Fermentation The process of energy production in a cell under anaerobic conditions (without oxygen). Energy is derived from the oxidation of organic compounds, such as carbohydrates, using an endogenous electron acceptor, which is usually an organic compound. This is in contrast to cellular respiration, where electrons are donated to an exogenous electron acceptor, such as oxygen, via an electron transport chain.

Ferredoxin Iron–sulphur proteins that mediate electron transfer in a range of metabolic reactions.

Flavin adenine dinucleotide (FAD) A redox cofactor involved in several important reactions in metabolism. FAD can exist in two different redox states and its biochemical role usually involves changing between these two states. Many oxidoreductases, called flavoenzymes or flavoproteins, require FAD as a prosthetic group which functions in electron transfers.

Flavin mononucleotide (FMN) Also known as riboflavin-5′-phosphate, this is a biomolecule produced from riboflavin (vitamin B_2) by the enzyme riboflavin kinase and functions as prosthetic group of various oxidoreductases including NADH dehydrogenase.

Flavonoids (Bioflavonoids) A class of plant secondary metabolites. They are most commonly known for their antioxidant activity. However, it is now known that the health benefits they provide against cancer and heart disease are the result of other mechanisms.

Flavoproteins Proteins that contain a nucleic acid derivative of riboflavin—the flavin adenine dinucleotide (FAD) or flavin mononucleotide (FMN). Flavoproteins are involved in a wide array of biological processes, including, but by no means limited to, bioluminescence, removal of radicals contributing to oxidative stress, photosynthesis, DNA repair, and apoptosis.

Fluorescence A luminescence that is mostly found as an optical phenomenon in cold bodies, in which the molecular absorption of a photon triggers the emission of a photon with a longer (less energetic) wavelength. The energy difference between the absorbed and emitted photons ends up as molecular rotations, vibrations or heat.

Fractionation A separation process in which a certain quantity of a mixture (solid, liquid, solute, suspension or isotope) is divided up in a number of smaller quantities (fractions) in which the composition changes according to a gradient. Fractions are collected based on differences in a specific property of the individual components.

Free energy The Gibbs free energy is a thermodynamic potential that measures the "useful" or process-initiating work obtainable from an isothermal, isobaric thermodynamic system. The Gibbs free energy is the maximum amount of non-expansion work that can be extracted from a closed system; this maximum can be attained only in a completely reversible process.

Freeze-drying Also known as lyophilization or cryodesiccation, it is a dehydration process typically used to preserve a perishable material or make the material more convenient for transport. Freeze-drying works by freezing the material and then reducing the surrounding pressure and adding enough heat to allow the frozen water in the material to sublime directly from the solid phase to gas.

Freezing The process in which a liquid turns into a solid when cold enough. The freezing point is the temperature at which this happens. Melting, the process of turning a solid to a liquid, is almost the exact opposite of freezing. All known liquids undergo freezing when the temperature is lowered enough with the sole exception of liquid helium, which remains liquid at absolute zero and can only be solidified under pressure.

Gel filtration Size exclusion chromatography (SEC) is a chromatographic method in which particles are separated based on their size, or in more technical terms, their hydrodynamic volume. It is usually applied to large molecules or macromolecular complexes such as proteins and industrial polymers. Typically, when an aqueous solution is used to transport the sample through the column, the technique is known as gel filtration chromatography, versus the name gel permeation chromatography which is used when an organic solvent is used as a mobile phase.

GenBank sequence database An open access, annotated collection of all publicly available nucleotide sequences and their protein translations. This database is produced at National Center for Biotechnology Information (NCBI) as part of the International Nucleotide Sequence Database Collaboration, or INSDC. GenBank and its collaborators receive sequences produced in laboratories throughout the world from more than 100,000 distinct organisms. GenBank continues to grow at an exponential rate, doubling every 18 months. Release 155, produced in August 2006, contained over 65 billion nucleotide bases in more than 61 million sequences. GenBank is built by direct submissions from individual laboratories, as well as from bulk submissions from large-scale sequencing centers.

Gene knockout (KO) A genetic technique in which an organism is engineered to carry genes that have been made inoperative (have been "knocked out" of the organism). Also known as knockout organisms or simply knockouts, they are used in learning about a gene that has been sequenced, but which has an unknown or incompletely known function.

Genes The basic unit of heredity in a living organism. All living things depend on genes. Genes hold the information to build and maintain their cells and pass genetic traits to their offspring.

Gene therapy The insertion of genes into an individual's cells and tissues to treat a disease, such as a hereditary disease in which a deleterious mutant allele is replaced with a functional one. Although the technology is still in its infancy, it has been used with some success. Antisense therapy is not strictly a form of gene therapy, but is a genetically mediated therapy and is often considered together with other methods.

Glutathione A tripeptide that contains an unusual peptide linkage between the amine group of cysteine and the carboxyl group of the glutamate side chain. Being, an antioxidant, it protects cells from toxins such as free radicals.

Gluten A composite of the proteins gliadin and glutenin. These exist, conjoined with starch, in the endosperms of some grass-related grains, notably wheat, rye, and barley. Gliadin and glutenin compose about 80% of the protein contained in wheat seed. Being insoluble in water, they can be purified by washing away the associated starch. Worldwide, gluten is an important source of nutritional protein, both in foods prepared directly from sources containing it, and as an additive to foods otherwise low in protein.

Glycolysis The metabolic pathway that converts glucose, $C_6H_{12}O_6$, into pyruvate, $C_3H_4O_3^-$. The free energy released in this process is used to form the high energy compounds, ATP (adenosine triphosphate) and NADH (reduced nicotinamide adenine dinucleotide).

Group I catalytic introns Large self-splicing ribozymes. They catalyse their own excision from mRNA, tRNA and rRNA precursors in a wide range of organisms. The core secondary structure consists of nine paired regions (P1-P9). These fold to essentially two domains — the P4–P6 domain (formed from the stacking of P5, P4, P6 and P6a helices) and the P3–P9 domain (formed from the P8, P3, P7 and P9 helices). The secondary structure mark-up for this family represents only this conserved core. Group I catalytic introns often have long open reading frames inserted in loop regions.

Group II intron A class of self-catalytic ribozymes and retroelements found in rRNA, tRNA, mRNA of organelles in fungi, plants, protists, and bacteria. Self-splicing occurs *din vitro* (for a few of the introns studied to date), but protein machinery is probably required *in vivo*. In contrast to group I introns, intron excision occurs in the absence of GTP and involves the formation of a lariat, with a branchpoint strongly resembling that found in lariats formed during splicing of nuclear pre-mRNA. It is thought that pre-mRNA splicing (see spliceosome) may have evolved from group II introns due to the similar catalytic mechanism as well the structural similarity of the Domain V substructure to the U6/U2 extended snRNA.

Haem A prosthetic group that consists of an iron atom contained in the centre of a large heterocyclic organic ring called a porphyrin. Not all porphyrins contain iron, but a substantial fraction of porphyrin-containing metalloproteins have haem as their prosthetic subunit; these are known as haemoproteins.

Haemoglobin An iron-containing oxygen-transport metalloprotein present in the red blood cells of vertebrates, and the tissues of some invertebrates.

Haemoprotein A metalloprotein containing a haem prosthetic group, either covalently or non-covalently bound to the protein itself. The iron in the haem is capable of undergoing oxidation and reduction.

Hairpin ribozyme A small section of RNA that can act as an enzyme known as a ribozyme. It was first identified in the minus strand of the tobacco ringspot virus (TRSV) satellite RNA where it catalyses a self-cleavage reaction to process the products of rolling circle virus replication to unit-length satellite RNA. Unlike other ribozymes that cleave RNA, the hairpin ribozyme does not require a metal ion for the reaction. The structure of the hairpin ribozyme has been solved by X-ray crystallography. The minimal hairpin ribozyme structure required for self-cleavage is composed of four base-paired helices, and 2 internal loops, A and B. The bond that is cleaved lies within loop A.

Hammerhead RNAs RNAs that self-cleave via a small conserved secondary structural motif termed a hammerhead because of its shape. Most hammerhead RNAs are subsets of two classes of plant pathogenic RNAs: the satellite RNAs of RNA viruses and the viroids. The self-cleavage reactions, first reported in 1986, are part of a rolling circle replication mechanism. The hammerhead sequence is sufficient for self-cleavage and acts by forming a conserved three-dimensional tertiary structure.

Hepatitis delta virus (HDV) ribozyme A non-coding RNA that is necessary for viral replication and is thought to be the only catalytic RNA known to be required for viability of a human pathogen. The ribozyme acts to process the RNA transcripts to unit lengths in a self-cleavage reaction. The ribozyme is found to be active *in vivo* in the absence of any protein factors and is the fastest known naturally occurring self-cleaving RNA. The crystal structure of this ribozyme has been solved using X-ray crystallography and shows five helical segments connected by a double pseudoknot.

Heterotropic It is characterized by an enzyme activity in which the substrate binds to the enzyme at only one site and a different molecule modifies the reaction by binding to an allosteric site.

Holoenzyme The complete enzyme including all subunits often used in reference to RNA and DNA polymerases.

Homotropic It is characterized by an enzyme activity in which the substrate binds to the enzyme at two different sites of which one is the normal reactive site and the other is an allosteric site.

Hybridization A process of combining different varieties or species of organisms to create a hybrid, the result of interbreeding between two animals or plants of different taxa.

Hydrolases An enzyme that catalyses the hydrolysis of a chemical bond.

Hydrolysis A chemical reaction during which one or more water molecules are split into hydrogen and hydroxide ions which may go on to participate in further reactions. It is the type of reaction that is used to break down certain polymers, especially those made by step-growth polymerization. Such polymer degradation is usually catalysed by either acid, e.g. concentrated sulphuric acid [H_2SO_4] or alkali, e.g. sodium hydroxide [NaOH] attack, often increasing with their strength or pH.

Hygroscopy The ability of a substance to attract water molecules from the surrounding environment through either absorption or adsorption. Hygroscopic substances include sugar, honey, glycerol, ethanol, methanol, sulphuric acid, methamphetamine, iodine, many chloride and hydroxide salts, and a variety of other substances.

Idiotype A shared characteristic between a group of immunoglobulin or T-cell receptor (TCR) molecules based upon the antigen-binding specificity and therefore structure of their variable region. The variable region of antigen receptors of T cells (TCRs) and B cells (immunoglobulins) contains a complementarity determining region (CDR) with a unique amino acid structure that determines the antigen specificity of the receptor. The structure formed by the CDR is known as the idiotope.

Imidazole An organic compound with the formula $C_3H_4N_2$. This aromatic heterocyclic is classified as an alkaloid. Imidazole refers to the parent compound whereas imidazoles are a class of heterocyclics with similar ring structure but varying substituents. This ring system is present in important biological building blocks such as histidine, and the related hormone histamine. Imidazole can serve as a base and as a weak acid.

Immobilization The technique used for the physical or chemical fixation of cells, organelles, enzymes or other proteins (e.g. monoclonal antibodies) onto a solid support, into a solid matrix or retained by a membrane, in order to increase their stability and make possible their repeated or continued use. Enzyme immobilization is the attachment of an enzyme to an inert, insoluble material such as calcium alginate (produced by reacting a mixture of sodium alginate solution and enzyme solution with calcium chloride). This can provide increased resistance to changes in conditions such as pH or temperature. It also allows enzymes to be held in place throughout the reaction, following which they are easily separated from the products and may be used again—a far more efficient process and so is widely used in industry for enzyme-catalysed reactions.

Induced fit hypothesis Hypothesis which proposes that an enzyme can be induced to change the shape of its active site slightly if the substrate does not fit the active site exactly.

Inhibitor Molecules that bind to enzymes and decrease their activity. Since blocking an enzyme's activity can kill a pathogen or correct a metabolic imbalance, many drugs are enzyme inhibitors. They are also used as herbicides and pesticides. Not all molecules that bind to enzymes are inhibitors.

Insertional mutagenesis Mutagenesis of DNA by the insertion of one or more bases.

Insulin A hormone that has extensive effects on metabolism and other body functions, such as vascular compliance. Insulin causes cells in the liver, muscle, and fat tissue to take up glucose from the blood, storing it as glycogen in the liver and muscle, and stopping use of fat as an energy source. When insulin is absent (or low), glucose is not taken up by body cells, and the body begins to use fat as an energy source, for example, by transfer of lipids from adipose tissue to the liver for mobilization as an energy source.

Intron A DNA region within a gene that is not translated into protein. These non-coding sections are transcribed to precursor mRNA (pre-mRNA) and some other RNAs (such as long noncoding RNAs), and subsequently removed by a process called splicing during the processing to mature RNA. After intron splicing (i.e., removal), the mRNA consists only of exon-derived sequences, which are translated into a protein. The word intron is derived from the term intragenic region and also called intervening sequence (IVS).

Invertase A sucrase enzym that catalyses the hydrolysi (breakdown) of sucrose (tabl sugar) to fructose and glucose usually in the form of invertе sugar syrup. For industrial use invertase is usually derived fron yeast.

Ion exchange An exchange o ions between two electrolytes o between an electrolyte solutio and a complex. In most cases th term is used to denote th processes of purification separation, and decontaminatio of aqueous and other ion containing solutions with solic polymeric or mineralic "ior exchangers".

Ion-exchange chromatography A process that allows th separation of ions and pola molecules based on the charge properties of the molecules. It can be used for almost any kind of charged molecule including large proteins, small nucleotides and amino acids.

Irreversible inhibition Amino acids with key catalytic functions in the active site can sometimes be identified by determining which amino acid is covalently linked to an inhibitor after the enzyme is inactivated.

Irreversible inhibitors are those that combine with or destroy a functional group on an enzyme that is essential for the enzyme's activity or that form a particularly stable non-covalent association.

Isoelectric focusing Isoelectric focusing (IEF), also known as electrofocusing, is a technique for separating different molecules by their electric charge differences. It is a type of zone electrophoresis, usually performed in a gel, that takes advantage of the fact that a molecule's charge changes with the pH of its surroundings.

Isoelectric point The pH at which a particular molecule or surface carries no net electrical charge. The pI value can affect the solubility of a molecule at a given pH. Such molecules have minimum solubility in water or salt solutions at the pH which corresponds to their pI and often precipitate out of solution.

Isoenzymes Enzymes that differ in amino acid sequence but catalyse the same chemical reaction. They usually display different kinetic parameters (e.g. different values), or different regulatory properties. Isozymes are usually the result of gene duplication, but can also arise from polyploidization or nucleic acid hybridization.

Isomerases An isomerase is an enzyme that catalyses the structural rearrangement of isomers. Isomerases thus catalyse reactions of the form A → B where B is an isomer of A.

Isomerization A process by which one molecule is transformed into another molecule which has exactly the same atoms, but the atoms are rearranged, e.g. A — B — C → B — A — C (these related molecules are known as isomers). Under some conditions, isomerization occurs spontaneously. Many isomers are equal or roughly equal in bond energy, and so exist in roughly equal amounts, provided that they can interconvert relatively freely, that is the energy barrier between the two isomers is not too high.

Isoprenoids Class of organic compounds composed of two or more units of hydrocarbons, with each unit consisting of five carbon atoms arranged in a specific pattern. Isoprenoids play widely varying roles in the physiological processes of plants and animals. They also have a number of commercial uses. Isoprenoids are probably the most diverse and largest family of natural products and include many important drugs (e.g. taxol, artemisinin), valuable flavour and fragrance compounds, as well as pigments, antioxidants and steroids.

Isotachophoresis A technique in analytical chemistry used to separate charged particles. It is a further development of electrophoresis. It is a powerful separation technique using a discontinuous electrical field to create sharp boundaries between the sample constituents.

Isotope exchange A chemical reaction in which the reactant and product chemical species are chemically identical but have different isotopic composition. In such a reaction the isotope distribution tends towards equilibrium (as expressed by fractionation factors) as a result of transfers of isotopically different atoms or groups.

Isotopic exchange A process in which two atoms belonging to different isotopes of the same element exchange valency states or locations in the same molecule or different molecules.

IUBMB (International Union of Biochemistry and Molecular Biology) An international non-governmental organization concerned with biochemistry and molecular biology. Formed in 1955 as the **International Union of Biochemistry**, the union has presently 77 member countries (as of 2008).

IUPAC (International Union of Pure and Applied Chemistry) An international scientific organization, not affiliated with any government. It strives to advance chemistry, in part by setting global standards for names, symbols, and units. Nearly 1200 chemists are involved in IUPAC projects.

Katal (symbol: kat) The SI unit of catalytic activity. It is a derived SI unit for expressing quantity values of catalytic activity of enzymes and other catalysts. Its use is recommended by the General Conference on Weights and Measures and other international organizations. It replaces the non-SI enzyme unit. Enzyme units are, however, still more commonly used than the katal in practice at present, especially in biochemistry. The katal is not used to express the rate of a reaction, that is expressed in moles per second.

Kornberg enzyme The enzyme DNA polymerase, isolated from *Escherichia coli* in 1958 by A. Kornberg and his colleagues, which functions in repair synthesis of damaged DNA.

Lagging strand The strand that is synthesized during DNA replication, apparently in the 3′ to 5′ direction, but actually in the 5′ to 3′ direction by ligating short fragments synthesized individually. Strand of DNA being replicated discontinuously.

Leadzyme A small ribozyme that was artificially made using *in vitro* selection techniques. Leadzyme is able to cleave RNA in the presence of lead. The structure of leadzyme has been determined by X-ray crystallography. It has been proposed that a naturally occurring leadzyme occurs in the 5S rRNA and further that this may be an important mechanism in lead toxicity.

Lewis acid An electron pair acceptor.

Ligand An atom, ion, or molecule to a central metal to produce a coordination complex. The bonding between the metal and ligand generally involves formal donation of one or more of the ligand's electrons. The metal–ligand bonding ranges from covalent to more ionic.

Ligase An enzyme that can catalyse the joining of two large molecules by forming a new chemical bond, usually with accompanying hydrolysis of a small chemical group pendant to one of the larger molecules.

Light reaction The first stage of photosynthesis. In this process light energy is converted into chemical energy in the form of the energy-carriers ATP and NADPH. In the light-independent reactions, the formed NADPH and ATP drive the reduction of CO_2 to more useful organic compounds, such as glucose.

Lipopolysaccharide Also known as lipoglycans, these are large molecules consisting of a lipid and a polysaccharide joined by a covalent bond; they are found in the outer membrane of gram-negative bacteria, act as endotoxins and elicit strong immune responses in animals.

Lipoprotein A biochemical assembly that contains both

proteins and lipids. The lipids or their derivatives may be covalently or non-covalently bound to the proteins. Many enzymes, transporters, structural proteins, antigens, adhesins and toxins are lipoproteins.

Liposome A tiny bubble (vesicle), made out of the same material as a cell membrane. Liposomes can be composed of naturally derived phospholipids with mixed lipid chains (like egg phosphatidylethanolamine), or of pure surfactant components like DOPE (dioleoylphosphatidylethanolamine). Liposomes, usually but not by definition, contain a core of aqueous solution; lipid spheres that contain no aqueous material are called micelles, however, reverse micelles can be made to encompass an aqueous environment. Liposomes can be filled with drugs, and used to deliver drugs for cancer and other diseases.

Lock-and-key hypothesis The specific action of an enzyme with a single substrate can be explained using a Lock and Key analogy first postulated in 1894 by Emil Fischer. In this analogy, the lock is the enzyme and the key is the substrate. Only the correctly sized key (substrate) fits into the key hole (active site) of the lock (enzyme).

Luciferase A generic term for the class of oxidative enzymes used in bioluminescence and is distinct from a photoprotein. One famous example is the firefly luciferase (EC 1.13.12.7) from the firefly *Photinus pyralis*.

Lyases An enzyme that catalyses the breaking of various chemical bonds by means other than hydrolysis and oxidation, often forming a new double bond or a new ring structure.

Lysozyme An enzyme found in egg white, tears, and other secretions. It is responsible for breaking down the polysaccharide walls of many kinds of bacteria and thus it provides some protection against infection.

Maillard reaction A chemical reaction between an amino acid and a reducing sugar, usually requiring heat. It is vitally important in the preparation or presentation of many types of food, and, like caramelization, it is a form of non-enzymatic browning. The reaction is named after the chemist Louis-Camille Maillard who discovered it in the 1910s while attempting to reproduce biological protein synthesis.

Malonyl-CoA A coenzyme A derivative. It plays a key role in chain elongation in fatty acid biosynthesis and polyketide biosynthesis. Malonyl-CoA is also used in transporting alpha-ketoglutarate across the mitochondrial membrane into the mitochondrial matrix.

Metabolism The set of chemical reactions that occur in living organisms in order to maintain life. These processes allow organisms to grow and reproduce, maintain their structures, and respond to their environments. Metabolism is usually divided into two categories. Catabolism breaks down organic matter, for example to harvest energy in cellular respiration. Anabolism, on the other hand, uses energy to construct components of cells such as proteins and nucleic acids.

Metal ion catalysis Mechanism used by enzymes to catalyse reactions, which involves the use of metal ions to activate bound water through the formation of a nucleophilic hydroxide ion.

Metalloenzymes Proteins which function as an enzyme and contain metals that are tightly bound and always isolated with the protein. In proteins such as haemoglobins and cytochromes, the metal is Fe^{2+} or Fe^{3+} and it is part of the haem prosthetic group. In other metalloenzymes the metal is built into the structure of the enzyme molecule. The metal ion cannot be removed without destroying the structure of the enzyme. Metals built into the molecule include Mg^{2+} in most phosphotransferases; Zn^{2+} in alcohol dehydrogenase; Mn^{2+} in arginase; Fe^{2+} in ferredoxin; and Cu^{2+} in cytochrome oxidase.

Methanogens Archaea that produce methane as a metabolic by-product in anoxic conditions. They are common in wetlands, where they are responsible for marsh gas, and in the guts of animals such as ruminants and humans, where they are responsible for the methane content of flatulence.

Mismatch repair A system for recognizing and repairing erroneous insertion, deletion and mis-incorporation of bases that can arise during DNA replication and recombination, as well as repairing some forms of DNA damage. Mismatch repair is strand-specific.

Mitochondria A membrane-enclosed organelles found in most eukaryotic cells. They are sometimes described as "cellular power plants" because they generate most of the cell's supply of adenosine triphosphate (ATP), used as a source of chemical energy.

Molecular markers Any kind of molecule indicating the existence of a chemical or physical process. They are used in molecular biology and biotechnology experiments where they are used to identify a particular sequence of DNA.

Molecular weight The mass of one molecule of that substance, relative to the unified atomic mass unit.

Monocytes A type of white blood cell, part of the human body's immune system. Monocytes are produced by the bone marrow from haematopoietic stem cell precursors called monoblasts. Monocytes circulate in the bloodstream for about one to three days and then typically move into tissues throughout the body.

Mutarotation The term given to the change in the specific rotation of a cyclic monosaccharide as it reaches an equilibrium between its α and β anomeric forms.

Mutations Changes to the nucleotide sequence of the genetic material of an organism. Mutations can be caused by copying errors in the genetic material during cell division, by exposure to ultraviolet or ionizing radiation, chemical mutagens, or viruses, or can be induced by the organism itself, by cellular processes such as hyper-mutation.

NAD Nicotinamide adenine dinucleotide, abbreviated NAD^+, is a coenzyme found in all living cells. The compound is a dinucleotide, since it consists of two nucleotides joined through their phosphate groups: with one nucleotide containing an adenine base, and the other containing nicotinamide.

NADP Nicotinamide adenine dinucleotide phosphate ($NADP^+$, in older notation triphospho-pyridine nucleotide, TPN) is used in anabolic reactions, such as lipid and nucleic acid synthesis, which require NADPH as a reducing agent. NADPH is the reduced form of $NADP^+$, and $NADP^+$ is the oxidized form of NADPH. $NADP^+$ differs from NAD^+ by the presence in $NADP^+$ of an additional phosphate group on the 2′ position of the ribose ring that carries the adenine moiety.

NCBI The National Center for Biotechnology Information is part of the United States National Library of Medicine (NLM), a branch of the National Institutes of Health. The NCBI is located in Bethesda, Maryland, and was founded in 1988 through legislation sponsored by Senator Claude Pepper. The NCBI houses genome sequencing data in GenBank and an index of biomedical research articles in PubMed Central and PubMed, as well as other information relevant to biotechnology. All these databases are available online through the Entrez search engine.

Neutrophils The most common type of white blood cells, comprising about 50–70% of all white blood cells. They are phagocytic, meaning that they can ingest other cells, though they do not survive the act. Neutrophils are the first immune cells to arrive at a site of infection, through a process known as chemotaxis.

Nitrogenous base An organic compound that owes its property as a base to the lone pair of electrons of a nitrogen atom. Notable nitrogenous bases include purine bases. Pyrimidine and purine bases include the nucleobases (building blocks of DNA and RNA)—principally adenine, guanine, thymine, cytosine, and uracil. Adenine and thymine/guanine and cytosine are complementary to each other.

Nomenclature A system of names or terms, or the rules used for forming the names, as used by an individual or community, especially those used in a particular science (scientific nomenclature).

Non-competitive inhibition A type of enzyme inhibition that reduces the maximum rate of a chemical reaction (V_{max}) without changing the apparent binding of the catalyst for the substrate.

Non-histone proteins In chromatin, those proteins which

remain after the histones have been removed, are classified as non-histone proteins. Scaffold proteins, DNA polymerase, heterochromatin Protein 1 and Polycomb are common non-histone proteins. This classification group also includes numerous other structural, regulatory, and motor proteins.

Nucleophile A reagent that forms a chemical bond to its reaction partner (the electrophile) by donating both bonding electrons. Because nucleophiles donate electrons, they are by definition Lewis bases. All molecules or ions with a free pair of electrons can act as nucleophiles. Neutral nucleophilic reactions with solvents such as alcohols and water are named solvolysis.

Nucleoprotein Any protein which is structurally associated with nucleic acid (either DNA or RNA). The prototypical example is any of the histone class of proteins, which are identifiable on strands of chromatin. Telomerase, an RNP (RNA/protein complex), and protamines are also nucleoproteins.

Nucleoside Glycosylamines consisting of a nucleobase bound to a ribose or deoxyribose sugar.

Nucleotide Molecules that when joined together, make up the structural units of RNA and DNA. Additionally, nucleotides play central roles in metabolism. In that capacity, they participate in cellular signalling (cyclic guanosine monophosphate and cyclic adenosine monophosphate), and are incorporated into important cofactors of enzymatic reactions (coenzyme A, flavin adenine dinucleotide, flavin mono-nucleotide, and nicotinamide adenine dinucleotide phosphate).

Okazaki fragments A relatively short fragment of DNA (with an RNA primer at the 5′terminus) created on the lagging strand during DNA replication.

Opsonin Any molecule that acts as a binding enhancer for the process of phagocytosis, for example, by coating the negatively-charged molecules on the membrane.

Optical density Optical density is the measure of the transmission of an optical medium for a given wavelength.

Higher OD, lower transmittence and vice versa, e.g. optical density of 1 means 90% of incident light is absorbed.

Optical isomer Two molecules are optical isomers of one another if they are mirror images of one another, and non-superimposible on one another.

Optically active Optical rotation is the rotation of linearly polarized light as it travels through certain materials. It occurs in solutions of chiral molecules such as sucrose (sugar), solids with rotated crystal planes such as quartz, and spin-polarized gases of atoms or molecules. It is capable of rotating the plane of vibration of polarized light to the right or left.

Osmotic pressure The hydrostatic pressure produced by a difference in concentration between solutions on the two sides of a surface such as a semipermeable membrane. Jacobus Henricus van 't Hoff first proposed a formula for calculating the osmotic pressure, but this was later improved upon by Harmon Northrop Morse.

Oxidation The interaction between oxygen molecules and all the different substances they may contact, from metal to living tissue.

Oxidative damage An oxidative stress caused by an imbalance between the production of reactive oxygen and a biological system's ability to readily detoxify the reactive intermediates or easily repair the resulting damage.

Oxidizing agent A chemical compound that readily transfers oxygen atoms, or a substance that gains electrons in a redox chemical reaction, the oxidizing agent being reduced in the process in both cases. It is also called an oxidant, oxidizer or oxidiser.

Oxidoreductases An enzyme that catalyses the transfer of electrons from one molecule (the reductant, also called the hydrogen or electron donor) to another (the oxidant, also called the hydrogen or electron acceptor).

Pantothenate (Pantothenic acid, Vitamin B_5) A water-soluble vitamin required to sustain life (essential nutrient). Pantothenic acid is needed to form coenzyme-A (CoA), and is critical in the metabolism and synthesis of carbohydrates, proteins, and fats. In chemical structure, it is the amide between D-pantoate and beta-alanine.

Parkinson's disease A degenerative disease of the brain (central nervous system) that often impairs motor skills, speech, and other functions.

Pepsin An enzyme that is released by the chief cells in the stomach and that degrades food proteins into peptides. Pepsin was discovered in 1836 by Theodor Schwann who also coined this enzyme's name from the Greek word *pepsis*, meaning digestion (*peptein*: to digest). It was the first animal enzyme to be discovered, and, in 1929, it became one of the first enzymes to be crystallized, by John H. Northrop. Pepsin is a digestive protease.

Peptide Short polymers formed from the linking, in a defined order, of α-amino acids. The link between one amino acid residue and the next is known as an amide bond or a peptide bond.

Peptidoglycans Also known as murein, it is a polymer consisting of sugars and amino acids that form a meshlike layer outside the plasma membrane of bacteria, forming the cell wall. The sugar component consists of alternating residues of ($\beta 1 \rightarrow 4$)-linked *N*-acetylglucosamine and *N*-acetylmuramic acid residues.

Phage display A method for the study of protein–protein, protein–peptide, and protein–DNA interactions that uses bacteriophages to connect proteins with the genetic information that encodes them. This connection between genotype and phenotype enables large libraries of proteins to be screened and amplified in a process called *in vitro* selection, which is analogous to natural selection. The most common bacteriophages used in phage display are M13 and fd filamentous phage, though T4, T7, and λ phage have also been used.

Phagocytosis Phagocytosis is the cellular process of phagocytes and protists of engulfing solid particles by the cell membrane to form an internal phagosome, which is a food vacuole, or pteroid. It is a specific form of endocytosis involving the vesicular internalization of solid particles, such as bacteria, and is therefore distinct from other forms of endocytosis such as pinocytosis, the vesicular internalization of various liquids.

Photo-oxidation Oxidation is a process in which something (an atom or molecule or substance) loses an electron to something else. Photo-oxidation is therefore the process of oxidation which is caused by shining light on it. Often, light can be used to cause reactions to happen, such as oxidation. The term "photo" comes from "photon" which is light.

Photosynthesis A process that converts carbon dioxide into organic compounds, especially sugars, using the energy from sunlight. Photosynthesis occurs in plants, algae, and many species of bacteria.

pI The isoelectric point (pI), sometimes abbreviated to IEP, is the pH at which a particular molecule or surface carries no net electrical charge. The pI value can affect the solubility of a molecule at a given pH. Such molecules have minimum solubility in water or salt solutions at the pH which corresponds to their pI and often precipitate out of solution.

Ping-Pong reactions A special multisubstrate reaction in which, for a two-substrate, two-product (i.e., bi-bi) system, an enzyme reacts with one substrate to form a product and a modified enzyme, the latter then reacting with a second substrate to form a second, final product, and regenerating the original enzyme.

Plasmids An extra-chromosomal DNA molecule separate from the chromosomal DNA, which is capable of replicating independently of the chromosomal DNA. In many cases, it is circular and double-stranded. Plasmids usually occur naturally in bacteria, but are sometimes found in eukaryotic organisms.

Polyacrylamide A polymer ($-CH_2CHCONH_2-$) formed from acrylamide subunits that can also be readily cross-linked. Polyacrylamide is not toxic, but unpolymerized acrylamide can be present in the polymerized acrylamide. Therefore it is recommended to handle it with caution. In the cross-linked form, it is highly water-absorbent, forming a soft gel used in such applications as polyacrylamide gel electrophoresis and in manufacturing soft contact lenses. In the straight-chain form, it is also used as a thickener and suspending agent.

Polypeptide Chains of amino acids linked covalently by peptide bonds. Proteins are made up of one or more polypeptide molecules.

Polysaccharide Polymeric carbohydrate structures, formed of repeating units (either mono- or disaccharides) joined together by glycosidic bonds. Polysaccharides are often quite heterogeneous, containing slight modifications of the repeating unit. Depending on the structure, these macromolecules can have distinct properties from their monosaccharide building blocks.

Population genetics The study of the allele frequency distribution and change under the influence of the four evolutionary processes: natural selection, genetic drift, mutation and gene flow. It also takes account of population subdivision and population structure in space. As such, it attempts to explain such phenomena as adaptation and speciation.

Precipitation A widely use phenomenon in downstrear processing of biological products such as proteins. This uni operation serves to concentrat and fractionate the targe product from variou contaminants.

Primase DNA primase is a RNAP enzyme involved in th replication of DNA. Primas synthesizes a short RNA segmen (called a primer) complementar to an ssDNA template. Primas is of key importance in DNA replication because no know DNA polymerases can initiate th synthesis of a DNA stran without an initial RNA or DNA primer (for temporary DNA elongation).

Primer A strand of nucleic acid that serves as a starting point for DNA replication. They are required because the enzymes that catalyse replication, DNA polymerases, can only add new nucleotides to an existing strand of DNA. The polymerase starts replication at the 3´-end of the primer, and copies the opposite strand.

Prodrug A pharmacological substance (drug) that is administered in an inactive (or significantly less active) form. Once administered, the prodrug is metabolized *in vivo* into an active metabolite. The rationale behind the use of a prodrug is generally for absorption, distribution, metabolism, and excretion (ADME) optimization. Prodrugs are usually designed to improve oral bioavailability, with poor absorption from the gastrointestinal tract usually being the limiting factor. Additionally, the use of a prodrug strategy increases the selectivity of the drug for its intended target.

Proenzymes Inactive form of an enzyme which can then be converted to the active form, usually by excision of a polypeptide, e.g. trypsinogen is the zymogen of trypsin.

Proof reading The error-correcting processes involved in DNA replication. In bacteria, all three DNA polymerases (I, II, and III) have the ability to proofread, using $3' \rightarrow 5'$ exonuclease activity. In eukaryotes only the polymerases that deal with the elongation (γ, δ, and ε) have proofreading ability ($3' \rightarrow 5'$ exonuclease activity).

Prosthetic groups A characteristic non-amino acid substance that is strongly bound to a protein and necessary for the protein portion of an enzyme to function; often used to describe the function, as in haemprotein for haemoglobin.

Protein Data Bank (PDB) A repository for the 3D structural data of large biological molecules, such as proteins and nucleic acids. The data typically obtained by X-ray crystallography or NMR spectroscopy and submitted by biologists and biochemists from around the world, can be accessed at no charge on the internet. The PDB is overseen by an organization called the Worldwide Protein Data Bank, wwPDB. The PDB is a key resource in areas of structural biology, such as structural genomics.

Protein engineering The process of developing useful or valuable proteins. It is a young discipline, with much research currently taking place into the understanding of protein folding and protein recognition for protein design principles.

Protein folding The physical process by which a polypeptide folds into its characteristic and functional three-dimensional structure from random coil. Each protein exists as an unfolded polypeptide or random coil when translated from a sequence of mRNA to a linear chain of amino acids.

Proteins They are organic compounds made of amino acids arranged in a linear chain polymer and joined together by peptide bonds between the carboxyl and amino groups of adjacent amino acid residues also known as polypeptides.

Proteolytic enzymes A group of enzymes that break the long chainlike molecules of proteins into shorter fragments (peptides) and eventually into their components, amino acids. Proteolytic enzymes are present in bacteria and plants but are most abundant in animals. In the stomach, protein materials are attacked initially by the gastric enzyme pepsin. When the protein material is passed to the small intestine, proteins, which are only partially digested in the stomach, are further attacked by proteolytic enzymes secreted by the pancreas.

Proteome The entire complement of proteins expressed by a genome, cell, tissue or organism. The term has been applied to several different types of biological systems. A cellular proteome is the collection of proteins found in a particular cell type under a particular set of environmental conditions such as exposure to hormone stimulation. It can also be useful to consider an organism's complete proteome, which can be conceptualized as the complete set of proteins from all of the various cellular proteomes. This is very roughly the protein equivalent of the genome. The term "proteome" has also been used to refer to the collection of proteins in certain subcellular biological systems. For example, all of the proteins in a virus can be called a viral proteome. The proteome is larger than the genome, especially in eukaryotes, in the sense that there are more proteins than genes. This is due to alternative splicing of genes and post-translational modifications like glycosylation or phosphorylation.

Protons The proton is a subatomic particle with an electric charge of +1 elementary charge. It is found in the nucleus of each atom but is also stable by itself and has a second identity as the hydrogen ion, $^{1}H^{+}$.

Pseudogene They are defunct relatives of known genes that have lost their protein-coding ability or are otherwise no longer expressed in the cell. Although some do not have introns or promoters (these pseudogenes are copied from mRNA and incorporated into the chromosome and are called processed pseudogenes), most have some gene-like features (such as promoters, CpG islands, and splice sites), they are nonetheless considered nonfunctional, due to their lack of protein-coding ability resulting from various genetic disablements (stop codons, frameshifts, or a lack of transcription) or their inability to encode RNA (such as with rRNA pseudogenes).

Purine A heterocyclic aromatic organic compound, consisting of a pyrimidine ring fused to an imidazole ring.

Pyrimidine A heterocyclic aromatic organic compound similar to benzene and pyridine, containing two nitrogen atoms at positions 1 and 3 of the six-member ring. It is isomeric with two other forms of diazine.

Pyrophosphothiamine (Thiamine pyrophosphate (TPP)) A coenzyme vital to tissue respiration. It is required for the oxidative decarboxylation of pyruvate to form acetyl-coenzyme A, providing entry of oxidizable substrate into the Krebs cycle for the generation of energy. It is also a coenzyme for transketolase, which functions in the pentose-phosphate pathway, an alternate pathway for glucose oxidation.

Pyruvate Pyruvic acid ($CH_3COCOOH$) is an organic acid. It is also a ketone, as well as being the simplest alpha-keto acid. The carboxylate (COOH) ion (anion) of pyruvic acid, CH_3COCOO^-, is known as pyruvate, and is a key inter-section in several metabolic pathways. It can be made from glucose through glycolysis, supplies energy to living cells in the citric acid cycle, and can also be converted to carbohydrates via gluconeogenesis, to fatty acids or energy through acetyl-CoA, to the amino acid alanine and to ethanol.

Quinone Compounds having a fully conjugated cyclic dione structure, such as that of benzoquinones, derived from aromatic compounds by conversion of an even number of –CH= groups into –C(=O)– groups with any necessary rearrangement of double bonds (polycyclic and heterocyclic analogues are included).

Redox reactions All chemical reactions in which atoms have their oxidation number (oxidation state) changed. This can be either a simple redox process such as the oxidation of carbon to yield carbon dioxide or the reduction of carbon by hydrogen to yield methane (CH_4), or it can be a complex process such as the oxidation of sugar in the human body through a series of very complex electron transfer processes.

Rennet A natural complex of enzymes produced in any mammalian stomach to digest the mother's milk, and is often used in the production of cheese.

Replication The basis for biological inheritance. It is a fundamental process occurring in all living organisms to copy their DNA. This process is "semi-conservative" in that each strand of the original double-stranded DNA molecule serves as template for the reproduction of the complementary strand. Hence, following DNA replication, two identical DNA molecules have been produced from a single double-stranded DNA molecule.

Reporter gene A gene that researchers attach to another gene of interest in cell culture, animals or plants. Certain genes are chosen as reporters because the characteristics they confer on organisms expressing them are easily identified and measured, or because they are selectable markers. Reporter genes are generally used to determine whether the gene of interest has been taken up by or expressed in the cell or organism population.

Resin An ion-exchange resin or ion-exchange polymer is an insoluble matrix (or support structure) normally in the form of small (1–2 mm diameter) beads, usually white or yellowish, fabricated from an organic polymer substrate. The material has highly developed structure of pores on the surface of which are sites with easily trapped and released ions. The trapping of ions takes place only with simultaneous releasing of other ions; thus the process is called ion-exchange.

Ribonuclease (RNase) A type of nuclease that catalyses the degradation of RNA into smaller components. Ribonucleases can be divided into endo-ribonucleases and exo-ribonucleases, and comprise several subclasses within the EC 2.7 (for the phosphorolytic enzymes) and 3.1 (for the hydrolytic enzymes) classes of enzymes.

Ribonuclease P (RNase P) A type of ribonuclease which cleaves RNA. RNase P is unique from other RNases in that it is a ribozyme, a ribonucleic acid that acts as a catalyst in the same way that a protein-based enzyme would. Its function is to cleave off an extra, or precursor, sequence of RNA on tRNA molecules. Further RNase P is one of two known multiple turnover ribozymes in nature (the other being the ribosome), the discovery of which earned Professor Sidney Altman the Nobel Prize in Chemistry in 1989.

Ribozyme (Derived from **ribo**nucleic acid en**zyme**, also called RNA enzyme or catalytic RNA) is an RNA molecule that catalyses a chemical reaction. Many natural ribozymes catalyse either the hydrolysis of one of their own phosphodiester bonds, or the hydrolysis of bonds in other RNAs, but they have also been found to catalyse the aminotransferase activity of the ribosome.

RNA (Ribonucleic acid) A type of molecule that consists of a long chain of nucleotide units. Each nucleotide consists of a nitrogenous base, a ribose sugar, and a phosphate. RNA is very similar to DNA, but differs in a few important structural details: in the cell, RNA is usually single-stranded, while DNA is usually double-stranded; RNA nucleotides contain ribose while DNA contains deoxyribose (a type of ribose that lacks one oxygen atom); and RNA has the base uracil rather than thymine that is present in DNA.

Salvage pathway A recycling biochemical pathway where a salvaged intermediate from the degradative pathway is used to be converted back into being a complex biomolecule.

Schiff's base Named after Hugo Schiff, it is a functional group that contains a carbon–nitrogen double bond with the nitrogen atom connected to an aryl or alkyl group, but not hydrogen. Schiff bases are of the general formula $R_1R_2C{=}N - R_3$, where R_3 is an aryl or alkyl group that makes the Schiff base a stable imine.

Single-nucleotide polymorphism (**SNP**, pronounced *snip*) A DNA sequence variation occurring when a single nucleotide — A, T, C, or G — in the genome (or other shared sequence) that differs between members of a species (or between paired chromosomes in an individual). For example, two sequenced DNA fragments from different individuals, AAGC**C**TA to AAGC**T**TA, contain a difference in a single nucleotide. In this case we say that there are two alleles : C and T. Almost all common SNPs have only two alleles.

Site-directed mutagenesis A molecular biology technique in which a mutation is created at a defined site in a DNA molecule, usually a circular molecule known as a plasmid. In general, site-directed mutagenesis requires that the wild-type gene sequence be known. This technique is also known as site-specific mutagenesis or oligonucleotide-directed mutagenesis.

Specific activity The number of enzyme units per ml divided by the concentration of protein in mg/ml. Specific activity values are therefore quoted as units/mg. Specific activity is of no relevance as far as setting up assays is concerned, though it is an important measure of enzyme purity and quality.

Stereoisomers Isomeric molecules that have the same molecular formula and sequence of bonded atoms (constitution) but which differ only in the three dimensional orientations of their atoms in space.

Sterols An important class of organic molecules. They occur naturally in plants, animals and fungi, with the most familiar type of animal sterol being cholesterol which has been shown to contribute to high blood pressure and heart disease. Within the past decade, interest in plant sterols as a dietary supplement has increased, due to studies showing that they can contribute to lower cholesterol levels.

Streptavidin A 52,800 dalton tetrameric protein purified from the bacterium *Streptomyces avidinii*. It finds wide use in molecular biology through its extraordinarily strong affinity for biotin (also known as vitamin H); the dissociation constant (K_d) of the biotin–streptavidin complex is on the order of ~10^{-15} mol/L, ranking among one of the strongest non-covalent interactions known in nature. The crystal structure of streptavidin with biotin bound was first solved in 1989 by Hendrickson *et al.* and as of May 2009, there are 134 structures deposited on the RCSB Protein Data Bank. The N and C termini of the 159 residue full-length protein are processed to give a shorter 'core' streptavidin, usually composed of residues 13–139; heterogeneous products can result when the termini are processed differently.

Substrate A molecule upon which an enzyme acts. Enzymes catalyse involving the substrate(s). In the case of a single substrate, the substrate binds with the enzyme active site, and an enzyme–substrate complex is formed. The substrate is transformed into one or more products, which are then released from the active site.

Superoxide dismutase A class of enzymes that catalyse the dismutation of superoxide into oxygen and hydrogen peroxide. As such, they are an important antioxidant defence in nearly all cells exposed to oxygen.

Swiss-Prot A manually curated biological database of protein sequences. Swiss-Prot was created in 1986 by Amos Bairoch during his PhD and developed by the Swiss Institute of Bioinformatics and the European Bioinformatics Institute. Swiss-Prot strives to provide reliable protein sequences associated with a high level of annotation (such as the description of the function of a protein, its domains structure, post-translational modifications, variants, etc.), a minimal level of redundancy and high level of integration with other databases.

Tautomerism Isomers of organic compounds that readily interconvert by a chemical reaction called tautomerization.

Tautomerization The isomerization by which tautomers are interconverted. It is a heterolytic molecular re-arrangement and is frequently very rapid.

Telomerase An enzyme that adds specific DNA sequence

repeats ("TTAGGG" in all vertebrates) to the 3´ end of DNA strands in the telomere regions, which are found at the ends of eukaryotic chromosomes. The telomeres contain condensed DNA material, giving stability to the chromosomes. The enzyme is a reverse transcriptase that carries its own RNA molecule, which is used as a template when it elongates telomeres, which are shortened after each replication cycle. Telomerase was discovered by Carol W. Greider and Elizabeth Blackburn in 1985 in the ciliate *Tetrahymena*. There are some indicators that telomerase is of retroviral origin.

Temperature coefficient The relative change of a physical property when the temperature is changed by 1 K.

Thiamine pyrophosphate (TPP) Also known as thiamine diphosphate (ThDP), is a thiamine (vitamin B_1) derivative which is produced by the enzyme thiamine pyrophosphatase. It is a coenzyme that is present in all living systems, in which it catalyses several biochemical reactions.

Thrombin (Activated Factor II [IIa]) Commonly called pro-thrombin, it is a coagulation protein in the bloodstream, which has many effects in the coagulation cascade. It is a serine protease (EC 3.4.21.5) that converts soluble fibrinogen into insoluble strands of fibrin, as well as catalyses many other coagulation-related reactions.

TPCK (Tosyl phenylalanyl chloromethyl ketone) A protease inhibitor. Its structural formula is 1-chloro-3-tosylamido-4-phenyl-2-butanone. TPCK is the irreversible inhibitor of chymotrypsin. It also inhibits some cysteine proteases such as papain, bromelain or ficin. It does not inhibit trypsin or zymogens. TPCK is chosen for the chemical labelling of active histidine in enzyme analysis. The phenylalanine moiety is bound to the enzyme because of specificity for aromatic amino acid residues at the active site.

Transamination The reaction between an amino acid and an alpha-keto acid. The amino group is transferred from the former to the latter; this results in the amino acid being converted to the corresponding α-keto acid, while the reactant α-keto acid is converted to the corresponding amino acid.

Transcription The synthesis of RNA under the direction of DNA. RNA synthesis, or transcription, is the process of transcribing DNA nucleotide sequence information into RNA sequence information. Both nucleic acid sequences use complementary language, and the information is simply transcribed, or copied, from one molecule to the other. DNA sequence is enzymatically copied by RNA polymerase to produce a complementary nucleotide RNA strand, called messenger RNA (mRNA), because it carries a genetic message from the DNA to the protein-synthesizing machinery of the cell. One significant difference between RNA and DNA sequence is the presence of U, or uracil in RNA instead of the T, or thymine of DNA. In the case of protein-encoding DNA, transcription is the first step that usually leads to the expression of the genes, by the production of the mRNA intermediate, which is a faithful transcript of the gene's protein-building instruction. The stretch of DNA that is transcribed into an RNA molecule is called a transcription unit. A DNA transcription unit that is translated into protein contains sequences that direct and regulate protein synthesis in addition to coding the sequence that is translated into protein. The regulatory sequence that is before (upstream (–), towards the 5´ DNA end) the coding sequence is called 5´untranslated region (5´UTR), and sequence found following (downstream (+), towards the 3´ DNA end) the coding sequence is called 3´ untranslated region (3´UTR). Transcription has some proofreading mechanisms, but they are fewer and less effective than the controls for copying DNA; therefore, transcription has a lower copying fidelity than DNA replication.

Transferases An enzyme that catalyses the transfer of a functional group (e.g. a methyl or phosphate group) from one molecule (called the donor) to another (called the acceptor).

Transition state A particular configuration along the reaction coordinate and the state corresponding to the highest energy along this reaction coordinate. At this point, assuming a perfectly irreversible reaction, colliding reactant molecules will always go on to form products.

Transketolase An enzyme of both the pentose phosphate pathway in animals and the Calvin cycle of photosynthesis, which catalyses two important reactions, which operate in opposite directions in these two pathways.

Translation The first stage of protein biosynthesis (part of the overall process of gene expression). Translation is the production of proteins by decoding mRNA produced in transcription. Translation occurs in the cytoplasm where the ribosomes are located. Translation proceeds in four phases: activation, initiation, elongation and termination (all describing the growth of the amino acid chain, or polypeptide that is the product of translation). Amino acids are brought to ribosomes and assembled into proteins.

Trypsin A serine protease Trypsin (EC 3.4.21.4) found in the digestive system of many vertebrates, where it hydrolyses proteins. Trypsin is produced in the pancreas as the inactive proenzyme trypsinogen. Trypsin predominantly cleaves peptide chains at the carboxyl side of the amino acids lysine and arginine, except when either is followed by proline.

Trypsinogen The precursor form of the pancreatic enzyme trypsin or a zymogen which is found in pancreatic juice, along with amylase, lipase, and chymotrypsinogen. It is activated by enteropeptidase, which is found in the intestinal mucosa, to form trypsin. Once activated, the trypsin can activate more trypsinogen into trypsin. Trypsin cleaves peptide bond on carboxyl side of basic amino acids.

Turbidostat A continuous culture device, similar to a chemostat or an auxostat, which has feedback between the turbidity of the culture vessel and the dilution rate. The theoretical relationship between growth in a chemostat and growth in a turbidostat is somewhat complex, in part because it is similar. A chemostat technically has a fixed volume and flow rate, thus a fixed dilution rate. When the cells are uniform and at equilibrium, operation of a chemostat and turbidostat should be identical. It is only when classical chemostat assumptions are violated (for instance, out of equilibrium; or the cells are mutating) that a turbidostat is functionally different. One case may be while cells are growing at their maximum growth rate, in which case it is difficult to set a chemostat to the appropriate constant dilution rate.

Turnover number K_{cat} The maximum number of molecules of substrate that an enzyme can convert to product per catalytic site per unit of time and can be calculated as follows: $k_{cat} = V_{max}/[E]_T$.

Ultracentrifugation The ultracentrifuge is a centrifuge optimized for spinning a rotor at very high speeds, capable of generating acceleration as high as 1,000,000 g (9,800 km/s^2).

Ultrafiltration A variety of membrane filtration in which hydrostatic pressure forces a liquid against a semipermeable membrane. Suspended solids and solutes of high molecular weight are retained, while water and low molecular weight solutes pass through the membrane. This separation process is used in industry and research for purifying and concentrating macromolecular (10^3–10^6 Da) solutions, especially protein solutions.

Ultraviolet Electromagnetic radiation with a wavelength shorter than that of visible light, but longer than X-rays, in the range 10 nm to 400 nm, and energies from 3 eV to 124 eV. It is so named because the spectrum consists of electromagnetic waves with frequencies higher than those that humans identify as the colour violet.

Uncompetitive inhibition In uncompetitive inhibition the inhibitor cannot bind to the free enzyme, but only to the ES-complex. The EIS-complex thus formed is enzymatically inactive. This type of inhibition is rare, but may occur in multimeric enzymes.

Urease An enzyme that catalyses the hydrolysis of urea into carbon dioxide and ammonia, which is found in bacteria, yeast and several higher plants.

Valency A measure of the number of chemical bonds formed by the atoms of a given element, and is also known as valence or valency number.

van der Waals force The attractive or repulsive force between molecules (or between parts of the same molecule) other than those due to covalent bonds or to the electrostatic interaction of ions with one another or with neutral molecules, and named after Johannes Diderik van der Waals and also known as van der Waals interaction.

Vascular endothelial growth factor (VEGF) A sub-family of growth factors, more specifically of platelet-derived growth factor family of cystine-knot growth factors. They are important signalling proteins involved in both vasculogenesis (the de novo formation of the embryonic circulatory system) and angiogenesis (the growth of blood vessels from pre-existing vasculature).

Vitamins An organic compound required as a nutrient in tiny amounts by an organism. A compound is called a vitamin when it cannot be synthesized in sufficient quantities by an organism, and must be obtained from the diet. Thus, the term is conditional both on the circumstances and the particular organism.

Xenobiotic A chemical which is found in an organism but which is not normally produce or expected to be present in it. I can also cover substances whic are present in much highe concentrations than are usual Specifically, drugs such a antibiotics are xenobiotics in humans because the human bod does not produce them itself, no are they part of a normal diet.

X-ray crystallography An extremely precise, but also difficult and expensive means of imaging the exact structure of a given molecule or macromolecule in a crystal lattice.

X-ray diffraction The atomic planes of a crystal cause an incident beam of X-rays (if wavelength is approximately the magnitude of the interatomic distance) to interfere with one another as they leave the crystal This phenomonen is called X-ray diffraction.

Zone electrophoresis A type of electrophoresis used by physical chemists, in which the components of a mixture are separated into distinct zones by moving the solution through a porous medium such as filter paper.

Zwitterions A chemical compound that carries a total net charge of 0, thus being electrically neutral but carries formal positive and negative charges on different atoms.

Zymogen An inactive enzyme precursor that requires a biochemical change (such as a hydrolysis reaction revealing the active site, or a change in the configuration to reveal the active site) to become an active enzyme. The biochemical change usually occurs in a lysosome where a specific part of the precursor enzyme is cleaved in order to activate it. The amino acid chain that is released upon activation is called the activation peptide.

REFERENCES

Agarwal, P.K. (2005). "Role of protein dynamics in reaction rate enhancement by enzymes." *J. Am. Chem. Soc.* 127 (43): 15248.

Anfinsen, C.B. (1973). "Principles that govern the folding of protein chains." *Science.* 181: 223–30.

Bairoch, A. (2000). "The ENZYME database in 2000." *Nucleic Acids Res.* 28: 304–5.

Baynes, W. John. (2005). *Medical Biochemistry*, 2nd (edn.). Elsevier-Mosby.

Bergmeyer, H.U. (1974). *Methods of Enzymatic Analysis*, Vol. 4. Academic Press, New York. pp.2066–2072.

Blackbum, G.M., Datta, A. and Partridge, L.J. (1996). "The medical potential of catalytic antibodies." *Pure & Appl. Chem.*, Vol. 68(11), pp. 2009–2016.

Boyer Rodney (2002). *Concepts in Biochemistry.* 2nd (edn.). John Wiley & Sons. New York, Chichester, Weinheim, Brisbane, Singapore, Toronto. Inc. pp. 137–8.

Briggs, G.E. and Haldane, J.B.S. (1925). "A note on the kinetics of enzyme action." *Biochem. J.* 19: 339–339.

Bugg, T. (2004). *Introduction to Enzyme and Coenzyme Chemistry*, 2nd (edn.). Blackwell Publishing Limited.

Cambou, B. and Klibanov, A.M. (1984). "Unusal catalytic properties of usual enzymes." In: *Enzyme Engineering*, Vol 5, (eds.). Laskin,A.I., Tsao,G.T. and Wingard,L.B. Jr. New York Academy of Sciences,New York. pp. 219–33.

Carl Woese. (1967). *The Genetic Code.* Harper and Row, New York.

Cech, T. (2000). "Structural biology. The ribosome is a ribozyme." *Science.* 289 (5481): 878–9.

Churchwella, M., Twaddlea, N., Meekerb, L. and Doergea, D. (2005). "Improving sensitivity in liquid chromatography-mass spectrometry." *Journal of Chromatography* B Vol. 825: Issue 2, 25. pp. 134–143.

Cleland, W.W. (1963). "The kinetics of enzyme-catalyzed reactions with two or more substrates or products 2. {I}nhibition: nomenclature and theory." *Biochim. Biophys. Acta.* 67: 173–87.

Cornish-Bowden, Athel.(2004). *Fundamentals of Enzyme Kinetics.* 3rd (edn.) Portland Press.

Cowan, D.A. (1997). "Thermophilic proteins: stability and function in aqueous and organic solvents." *Comp. Biochem. Physiol. A Physiol.* 118 (3): 429–38.

Dixon, M.and Webb, E.C. (Eds.). (1979). *Enzymes.* Longman, London.

Dulieu, C., Moll, M., Boudrant, J.and Poncelet, D. (2000). "Improved performances and control of beer fermentation using encapsulated alpha-acetolactate decarboxylase and modeling." *Biotechnology Progress.* 16 (6): 958–65.

Dunaway-Mariano, D. (2008). "Enzyme function discovery." *Structure.* 16 (11): 1599–600.

Eisenmesser, E.Z., Bosco, D.A., Akke, M. and Kern, D. (2002). "Enzyme dynamics during catalysis." *Science.* 295 (5559): 1520–3.

Enzyme Nomenclature, Recommendations for enzyme names from the Nomenclature Committee of the International Union of Biochemistry and Molecular Biology.

Eric, E., Conn Paul, K. and Stumpf, G.B. *Outlines of Biochemistry,* 5th (edn.). Wiley, New York.

Fersht, Alan. (1985). *Enzyme Structure and Mechanism*.W.H. Freeman and Co., San Francisco. pp. 50–2.

Fersht, Alan (1999). *Structure and Mechanism in Protein Science: A Guide to Enzyme Catalysis and Protein Folding.* W.H. Freeman and Co., San Francisco.

Fischer E. (1894). "Einfluss der Configuration auf die Wirkung der Enzyme." *Ber. Dt. Chem. Ges.* 27: 2985–9.

Fisher, Z., Hernandez Prada, J.A., Tu, C., Duda, D., Yoshioka, C., An, H., Govindasamy, L., Silverman D.N. and McKenna, R. (2005). "Structural and kinetic characterization of active-site histidine as a proton shuttle in catalysis by human carbonic anhydrase II." *Biochemistry.* 44 (4): 1097–115.

Grisham, Charles M. and Reginald, H. Garrett. (1999). *Biochemistry.* Saunders College Pub.Philadelphia. pp. 426–7.

Groves, J.T. (1997). "Artificial enzymes. The importance of being selective." *Nature.* 389 (6649): 329–30.

Guzmán-Maldonado, H. and Paredes-López, O. (1995). "Amylolytic enzymes and products derived from starch: A review." *Critical Reviews in Food Science and Nutrition.* 35 (5): 373–403.

Hean, J. and Weinberg, M.S. (2008). "The hammerhead ribozyme revisited: new biological insights for the development of therapeutic agents and for reverse genomics applications." *RNA and the Regulation of Gene Expression: A Hidden Layer of Complexity.* Caister Academic Press, U.K.

Hilvert, D. (2001). "Enzyme engineering."*Chimia.* 55, 867–869.

Janda, K.D. (1991). "Catalytic antibodies and enzyme inhibitors." *Pure & Appl. Chem.* Vol. 66, No. 4, pp. 703–708.

Jencks, W.P. (1987)."*Catalysis in Chemistry and Enzymology.* Dover, New York.

Jin Tang and Ronald, R. Breaker (1997). "Structural diversity of self-cleaving ribozymes." Proceedings of the National Academy of Sciences. 97 (11): 5784–5789.

Johnston, W., Unrau, P., Lawrence, M., Glasner, M. and Bartel, D. (2001). "RNA-catalyzed RNA polymerization: accurate and general RNA-templated primer extension." (PDF). *Science.* 292 (5520): 1319–25.

Koshland, D.E. (1958). "Application of a theory of enzyme specificity to protein synthesis." *Proc. Natl. Acad. Sci.* 44 (2): 98–104.

Koshland, D.(1959). *The Enzymes*. Academic Press, New York.

Lerner, R.A. and Tramontano, A.(1987). "Antibodies as enzymes." *Trends in Biochemical Sciences*. 12, 427–30.

Lilley, D. (2005). "Structure, folding and mechanisms of ribozymes." *Curr. Opin. Struct. Biol.* 15(3): 313–23.

Lincoln,A.Tracey. and Joyce,F.Gerald. (2009). "Self-sustained replication of an RNA enzyme." *Science.* 323: 1229.

Lowe, C.R.(1981). "Immobilized coenzymes." In: *Topics in Enzyme and Fermentation Biotechnology,* vol 5, (ed.). Wiseman, A. pp. 13–146. Ellis Horwood Ltd., Chichester, U.K.

Lubert Stryer. (1995). *Biochemistry,* 4th (edn.). W.H. Freeman & Company, San Francisco.

Minton, A.P. (2001). "The influence of macromolecular crowding and macromolecular confinement on biochemical reactions in physiological media." *J. Biol. Chem.* 276 (14): 10577–80.

Murray, R.K., Granner, D.K., Peter, M.A. and Rodwell, V.W. (eds.). (2005). *Harper's Biochemistry.* McGraw-Hill.

Neet, K.E. (1980). " Cooperativity in enzyme function: Equilibrium and kinetic aspects." *Methods Enzymol.* 64, 139–192.

Nielsen, H., Westhof, E. and Johansen, S. (2005). "An mRNA is capped by a 2´, 5´ lariat catalyzed by a group I-like ribozyme." *Science.* 309 (5740): 1584–7.

Nomenclature Committee of the International Union of Biochemistry (NC-IUB) (1979). "Units of Enzyme Activity." *Eur. J. Biochem.* 97: 319–20.

Page, M.I. and Williams, A. (1987). (eds.). *Enzyme Mechanisms.* Royal Society of Chemistry.

Palmer, T. (1985). *Understanding Enzymes.* Wiley,New York. pp. 257–274.

Passonneau, J.V. and Lowry, O.H.(1993). *Enzymatic Analysis. A Practical Guide.* Humana Press, Totowa, NJ. pp.85–110.

Perutz, M. (1990). *Mechanisms of Cooperativity and Allosteric Regulation in Proteins.* Cambridge University Press, New York.

Pike, V.W. (1987). "Synthetic enzymes." In: *Biotechnology*, Vol.7a *Enzyme Technology*, Kennedy, J.F. (ed.). VCH Verlagsgesellschaft mbH.Weinheim. pp. 465–85.

Radzicka, A. and Wolfenden, R. (1995). "A proficient enzyme." *Science.* 6 (267): 90–931.

Schnell, S., Chappell, M.J., Evans, N.D. and Roussel, M.R.(2006). The mechanism distinguishability problem in biochemical kinetics: The single-enzyme, single-substrate reaction as a case study." *Comptes Rendus Biologies.* 329, 51–61.

Segel, I.H. (1975). *Enzyme Kinetics*. Wiley, New York. pp. 346–464.

Segel Irwin, H. (1993)*Enzyme Kinetics: Behavior and Analysis of Rapid Equilibrium and Steady-State Enzyme Systems.* Wiley-Interscience, New York.

Shaw, W.V. (1987). "Protein engineering: the design, synthesis and characterisation of factitious proteins." *Biochemical Journal.* 246, 1–17.

Smith, S. (1994). "The animal fatty acid synthase: one gene, one polypeptide, seven enzymes." *Faseb. J.* 8 (15): 1248–59.

Todd, M.J. and Gomez, J.(2001). "Enzyme kinetics determined using calorimetry: a general assay for enzyme activity?" *Anal Biochem.* 296(2): 179–87.

Villa, J., Strajbl, M., Glennon, T.M., Sham, Y.Y., Chu, Z.T.and Warshel,A. (2000). "How important are entropic contributions to enzyme catalysis?." *Proc. Natl. Acad. Sci., U.S.A.* 97 (22): 11899–904.

Wagner, L.Arthur. (1975). *Vitamins and Coenzymes.* Krieger Pub. Co., New York.

Walsh, C.(1979). *Enzymatic Reaction Mechanisms.*W.H. Freeman and Co., San Francisco.

Warshel, A., Sharma P.K., Kato, M., Xiang, Y., Liu, H.and Olsson M.H. (2006). "Electrostatic basis for enzyme catalysis." *Chem. Rev.* 106 (8): 3210–35.

Warshel, A.(1991). *Computer Modeling of Chemical Reactions in Enzymes and Solutions.* John Wiley & Sons Inc., New York.

Winter, G. and Fersht, A.R. (1984). "Engineering enzymes." *Trends in Biotechnology.* 2, 115–19.

Zaher, H.S., Unrau, P. (2007). "Selection of an improved RNA polymerase ribozyme with superior extension and fidelity." *RNA.* 13 (7):1017–26.

Website

http://www.ncbi.nlm.nih.gov/sites/entrez?db=PubMed&cmd=Retrieve&list_uids=17586759.

http://adsabs.harvard.edu/abs/1984OrLi...14..291V.

http://www.blackwell-synergy.com/doi/pdf/10.1111/j.1432-1033.1979.tb13116.x.

http://www.biochemj.org/bj/019/0338/bj0190338_browse.htm.

http://www.brendaenzymes.org

http://www.brendaenzymes.org/soap

http://www.jbc.org/cgi/content/full/276/14/10577.

http://www.sciencemag.org/cgi/content/abstract/1167856.

http://web.wi.mit.edu/bartel/pub/publication_reprints/Johnston_Science01.pdf.

http://www.pnas.org/cgi/content/full/97/11/5784.

http://gallica.bnf.fr/ark:/12148/bpt6k90736r/f364.chemindefer.

http://www.expasy.ch

http://www.expasy.ch/enzyme/

http://kinasedb.ontology.ims.u-tokyo.ac.jp/.

http://pkr.sdsc.edu/html/index.shtml

http://www.enzome.com.

http://ergo.integratedgenomics.com/ERGO/.

http://www-mitchell.ch.cam.ac.uk/macie/

http://bioinformatics.ai.sri.com/ptools/.

http://mbs.cbrc.jp/ezCatDB

INDEX

A
Abacavir sulphate 479
abl gene 363
Absolute specificity 66
Absorption 454, 458, 461
Abzyme 385, 386, 388, 389, 390, 391, 397, 398
Abzyme prodrug therapy (ADAPT) 395
Abzymes 397
Acceptors 123
Accession numbers 516
Acetoacetate 168
Acetylation 118, 199
Aacetylcholine 84
Acetyl transacylase 101
Acetyl-CoA 98, 100, 101, 117, 118
Acetyltransferases 118
Acid 465
Acid azide derivative 299
Acid catalysis 161
Acid catalyst 367
Acid–base catalysis 161
Acoustic systems 421
Acoustic waves 439
Actinomycetes 403
Activation energy 62, 64, 144, 159
Activator 79, 494
Active site 62, 68, 70, 83, 143, 144, 145, 146, 159, 248, 400, 245, 397, 399, 483
Activity tests 273
Acyl-CoA 119, 120, 125
Acyl groups 115
Adenosine 5´triphosphate 123
Adenosine monophosphate 123
Adenosine triphosphate 119
Adenoviral vectors 370
Adenylyl transferase 117
ADEPT 395
ADP 433
ADP-ribose 113
Adrian Brown 209
Adsorbent 278
Adsorption 278, 279, 411, 440
Affinity-based biosensors 421
Affinity chromatography 281
Affinity sensors 418
Ag/AgCl electrode 428
Agarose 281, 299
Agitator 333
Agitator shaft 335
AIDS drug 479
Airlift reactor systems 331
Albinism 478

Alcohol 288, 404
Alcohol dehydrogenase (NADP) 21
Alcohol oxidase 429
Alcohol 115, 459
Aldehydes 473
Aldolase 90
Aldol-reaction 391
Aldose-ketose isomerases 26
Aldrich 393
Alexandar Flemming 182
Alginate gel method 299 309
Alkaline pH 99
Alkaline phosphatase 366
Alkaline phosphatase antibody 418
Alkylation method 299
Alleles 88
Alliinase 129
Allosteric 102, 202
Allosteric activation 249
Allosteric activator 253
Allosteric competitive inhibition 225
Allosteric effectors 244
Allosteric enzyme 203, 244, 245, 362
Allosteric inactivator 253
Allosteric inhibition 93, 250
Allosteric ligand 254
Allosteric modification 202
Allosteric modulation 254
Allosteric modulator 255
Allosteric regulation 248
Allosteric ribozymes 362
Allosteric site(s) 243, 251, 254
Allozymes 97
Alzheimer's 113
α-Amylase 458, 475
AMENDA 527, 528
Amidase 389
Amides 388
Amine oxidases 14
Amino acid 87, 130, 145, 317, 343, 406, 479, 480, 481, 519, 524
Amino acid sequence 401
Aminopenicillanic acid 473
AMP 198
Amperometric biosensors 416, 427, 428
Ampholytes 285
Amylase 3, 463, 464, 456, 461, 463, 465
Anabaena ribozyme 350
Anaemia 478
Anaerobic fermentations 110
Analytes 417, 421, 447, 448
Andrew Elington 361
Angiozyme 374
Animal feed 404
Anions 84
Annotation 532
Anode 89
Anti-anxiety drugs 254
Anti-cancer drugs 373
Anti-HIV 349
Anti-idiotypic 386
Anti-pernicious factor (AP factor) 121
Anti-redepositors 404
Anti-virus therapy 398
Antibiotics 470, 472
Antibodies 306, 370, 382, 385, 386, 389, 392, 397, 392, 395, 396, 411, 414, 418

Antibody–antigen interactions 443
Antibody-directed abzyme prodrug therapy 394
Antibody–hapten complex 388
Antigen 421
Antigen-directed abzyme prodrug therapy (adapt) 393
Antigen-tracer 421
Antineoplastic agents 372
Antitopoisomerase 390
Antivitamins 141
Apoenzyme 63, 112, 186
Apoptosis 525
Aptamer 362, 378, 379, 380, 382
Aptamer database 377
Arabinogalactan 458
Arabinose 472
Arabinoxylan 458
Archaea 187, 403, 512, 536
Aromatic compound degradation 537
Arthur Harden 112
Arthur Kornberg 189, 190
Artificial enzymes 407
Artificial ribozymes 360, 379
Artizymes 407
Arylsulphatase 80
Ascorbate 132, 139, 140
Ascorbate oxidase 131
Ascorbate reductase 133
Ascorbic acid 132, 430
ASN.1 546
ASP Server 525
Asparaginase 477
Aspartate 530
Aspartate carbamoyl transferase 202
Aspartate transcarbamoylase 250
Aspartic acid protease 455
Aspartic proteinase 525
Aspergillus niger 458, 475
Aspergillus oryzae 3, 464
AtCase 251
Ativan 254
Atoms 533
ATP 64, 93, 192, 251, 252, 430, 436, 442, 470
ATP/ADP 321
ATPase 145
Autoimmune diseases 386
Auxostat 329
Avidin 418
Axen 302
Azomethine 128

B

Bacillus 464
Bacillus subtilis 4, 467
Bacteria 5, 182, 436, 455, 460, 512, 536
Bacterial cell walls 181
Bacterial genomes 512
Bacterial protease 4
Bacteriophage 182, 350
Baddiley 128
Baffles 333
Baking 458
Bangham 409
Barker 122
Base catalysis 162, 164
Basic reaction 542
Batch 326, 329
Batch bioreactors 327
Bating 4

Beans 455
Beer 461, 462
Beer-making 462
Benzodiazepine receptor 254
Benzoquinone 129, 297
Bernard Green 386
Bernfeld 308
Beta-oxidation 124, 125
Beverage industry 404
Bi-Bi reactions 236, 237
Bile 454
Binding site 68, 159
Bioaffinity-based biosensors 420
Biocatalysis 418
Biocatalysis-based biosensors 418
Biocatalyst 294
Biochemical networks 522
Biocomposting 404
BioCyc 494, 537, 538
Bio-finishing 404
Biologic Grignard reagent 122
Bioluminescence 442
Bio-polishing 404, 465
Biopulping 468
Bioreaction 325
Bioreactor 325, 326, 330, 337
Bioremediation 408
Biosensors 294, 319, 411, 420, 450
Biospecific binding method 299, 306
Bio-stoning 466
Biosynthesis 537
Biotechnology 386
Biotin 13, 100, 320
Bisubstrate reactions 235
Biuret method 271
BIAcore 439
BIAcoreIM 446
BLAST 380, 382, 540
BLAST hits 540
Bleach boosting 469
Bleaching 467, 468
Bleaching agents 465
Blood clots 478
Blood glucose 95
Blood pressure 130
Blood proteins 275
BOD 421
Boidin 4, 467
Brassinosteroids 119
Braunschweig 526
Braunschweig enzyme database 526
BRCA 1 373
Bread making 459
BRENDA 487, 523, 526, 527
Briggs G.E. 211
Briggs and Haldane 218
Britton Chance 7
Bronsted acid 161, 368
Bronsted base 162
Buchner 2
Buffer 94, 266, 434

C

Calf stomach 460
Calorimetric biosensor 415, 430, 431
Calvin cycle 115
Camptothecin 396
Cancer 130, 368, 373, 502
Candida 470

Carbamoyl-transferases 40
Carbodimides 298
Carbohydrases 459, 465
Carbohydrate-active enzymes 517
Carbohydrate-binding modules 517
Carbohydrates 461
Carbon dioxide 463
Carbonic anhydrase 171
Carbon source 326
Carboxylase 19, 155
Carboxymethyl cellulose 280
Carboxypeptidase 63, 68, 293
Carnitine 120
κ-Carrageenan 310
κ-Carrageenan gel method 299
Carrier 300
Carrier cross-linking method 299, 304
CAS 535
Cascade systems 200
Cat ELISA 397
Catalase 5, 21, 137, 466, 467
Catalysis 348, 355, 399, 542
Catalyst 61, 386, 397
Catalytic activity 261, 262, 293, 382, 544
Catalytic antibodies 393, 395
Catalytic efficiency 76, 224
Catalytic oligonucleotides 369
Catalytic RNA 344, 346, 349, 365, 366, 376
Catalytic site 68, 232, 243, 244
Catalytic site atlas 483, 531
Catalytic trimers 250
Catalytic turnover 244
CATH 541, 543
CATH classification 531
CATH codes 532
Cation 84, 424
Cationic exchangers 280
Catmab 386
Cazy 517
CD-ROM 545
$CD4^+$ lymphocytes 349
Celite 280
Cell 343
Cell cycle 370, 504, 525
Cell differentiation 504
Cell division 123
Cell-mediated immunity 397
Cell membranes 454
Cell mutation 327, 328
Cell proliferation 504
Cell receptors 414
Cellular receptors 370
Cellular respiration 118, 123
Cellulase 455, 456, 464, 465, 466, 468, 480
Cellulose 455, 458
Changuex 245
Chaotropic agents 273
Cheese making 460, 479
Chemical database 533
Chemical modification 146, 368, 403
Chemical transformation 385
Chemiluminescence 440, 442
Chemostat 329
Chemotherapy 130
Chibata 293
Chitodextrin 181
Chloroacetazolamide 153

Chloroform 288
Chloroketone 153
Chloroplast tRNA 350
Cholecystokinin-pancreazymin 192
Cholesterol 430
Cholesterol oxidase 111, 429
Cholinesterases 14
Christian 107
Christian Hansen 1
Chromatograms 268
Chromatographic methods 268
Chromatography 89, 268, 278, 279, 280
Chromogen 435
Chromophore 153
Chromosome 525
Chymosin 460
Chymotrypsin 5, 63, 68, 148, 151, 153, 154, 184, 238
Chymotrypsinogen 63, 184, 194, 195
Cinacalcet 254
cis–trans isomerism 136
cis-acting hammerhead ribozyme 347
Citrate lyase 118
Citric acid cycle 119, 124, 125
Clark oxygen electrode 428
Class 534
Classification 523, 526, 527, 531
Cleavage 346
Cleland's shorthand notation 236
Clement Markert 87
Clinical trials 398
Clostridium sticklandii 179
Clotting factor 197
CMLReact 531, 532
Co-factor 350
CoA 119
CoASH 115, 117
Cobalt (III) 349
Cocaine 397
Codase 12
Codehydrogenase II 114
Coding sequences 529
Coenzyme 39, 63, 73, 76, 79, 91, 105, 106, 111, 131, 244, 257, 259, 267, 306, 417, 449
Coenzyme I 107, 109, 110, 112, 131, 139
Coenzyme II 105, 107, 110, 113, 131, 139
Coenzyme III 139
Coenzyme 94, 115
Coenzyme A 107, 117, 119, 120, 140, 269
Coenzyme B 120, 121
Coenzyme B_{12} 121, 122
Coenzyme F_{420} 123
Coenzyme M 120
Coenzyme M reductase 121
Coenzyme Q10 129, 130
Cofactor 39, 62, 110, 121, 135, 258, 260,, 321, 322,430, 449, 494, 507, 530, 533, 541, 544
Collagen 465, 467
Colorimetric methods 270
Colowick 108
Column chromatography 288

COMMENT 20, 534
Committees on biochemical Nomenclature 516
Common names 15
Competitive inhibition 224, 228
Competitive inhibitor 225, 227, 229
Complement activation 200
Compound section 533, 535
Concanavalin A 307
Concerted feedback inhibition 205
Condensing reagent 299
Condensing reagent method 303
Conductimetric biosensors 434
Congruence analysis 540
Contigs 513
Continuous bioreaction 328
Continuous bioreactors 333
Continuous flow reactors 338
Continuous stirred tank reactor 331
Cooperative binding 247
Cori 200
Covalent binding 301
Covalent bonds 144
Covalent catalysis 161, 165, 166, 391
Cracking 462
Cross-linked enzyme crystals 319
Cross-linking 298
Cross-linking method 307
Crotonase 18
Crystallization 4, 251, 286
CSTR 333, 340
CTP 201, 202, 251, 252
Cumulative feedback inhibition 206
Cyanocobalamine 121, 122
Cyanogen bromide method 302
Cysteine 139
Cystein protease 455
Cysteine proteinase 525
Cytochrome 137, 450 115
Cytochrome *c* 110, 131, 139
Cytochrome *c* peroxidase 138
Cytochrome oxidase 12, 131, 138
Cytochrome reductases 110
Cytochrome system 132
Cytochromes b_3 and c_1 139
Cytoplasmic localization 370
Cytoskeleton 124
Cytotoxic compounds 394

D

10-23 DNA enzyme 362
D- and L-amino acid oxidases 429
Darwinian 361
Data extraction 502
Database 381, 402, 524, 525, 526, 534
David Chilton Phillips 5
David Greenberg 6
DBGET/LinkDB 535, 536
DBGET/LinkDB system 534
DDBJ 377
De novo pathway 112
De-pilling 404
DEAE–Cellulose 323
DEAE-Sephadex 293, 304
Deazaflavin 123
Decarboxylation 118, 168, 388
Degradation/utilization/assimilation 537
Dehairing 467
Dehydrase 19
Dehydroascorbate 131

Dehydroascorbic acid 132
Density gradient technique 285
Deoxy-haemoglobin 247
Deoxyribozyme 362, 363
Dephosphocoenzyme A kinase 117
Depolymerization 143
Desizing 404, 467
Detector element 414
Detergents 464
Detergents industry 404
Deuterium 108, 109
Dextran 286, 299, 439, 474
Dextrins 462
Diabetes 502
Dialysis 63, 276, 286
Diaphorase 111
Diastase 2, 293, 462
Diazo coupling method 299, 303
Diazosulphonic acid 153
Dielectric constant 174
Diels–Alder addition 388
Diethylaminoethyl cellulose 280
Diffusion 331, 428
Digestion 454
Digoxin 446
Diisopropylfluorophosphate 84
Diisopropylphosphofluoridate 148
Dimerizing 19
Dimers 250
3-dimethylaminopropionitrile 308
Dimethyloxaloacetate 168
Dinitrofluorobenzene 157
Dipoles 175
Direct optical sensing 440
Disaccharide 181
Disinfectants 470
Dissociation constant 147, 225
Dissolved oxygen 329
Disulphide bridges 181
Dithionite 110, 139
Dixon 10
DNA 187, 343, 386
DNA mismatch repair 186
DNA polymerase 186, 187
DNA polymerase activity 189
DNA polymerase I 189
DNA polymerase II 190
DNA polymerase III 190
DNA polymerase III holoenzyme 190
DNA polymerases 187
DNA repair 190
DNA replication 186, 189
DNA ribozymes 369
DNA sequence 411
DNA sequencing 97
DNAzyme 358, 363, 364
Domain 508, 541
Donors 161
Double-displacement 238
Downstream processing 326
Doxorubicin 396
Draining 469
Drug 115, 478

E

E. coli 98, 101, 187, 200, 201, 251, 403, 406
Eadie–Hofstee plot 221, 222
EC hierarchy 530
EC number 11, 491, 507, 508, 528, 529, 532, 531, 534, 541, 542, 543, 544
EcoCyc 540

EC system 523
EDTA 84, 349
Edward Rickes 121
Edwin Webb 10
Effectors 243, 306, 533, 534
Efficiency 417
Effluent 328
Effront 4
Elastin 465
Electrochemical 411, 414
Electrochemical biosensors 448
Electrode 412, 448
Electrodecantation 285
Electrode method 266, 267
Electrolytes 283
Electrons 412
Electron carriers 537
Electron transport chain 126, 129
Electron transport reactions 537
Electron transport system 321
Electron waves 447
Electrophiles 170
Electrophilic catalysis 165
Electrophoresis 87, 89, 93, 283, 289, 308
Electrophoretic mobility 91
Electrophoretic separation 289
Electrophoretogram 94
Electro-optical 411
Electrostatic 65
Electrostatic catalysis 161, 174, 175
Electrostatic interactions 174
Elimination 25
ELISA 430
Ellipsometry 443
EMBL 377, 382, 545, 546
EMBL data library 545
Emil Fisher 69
Emulsifiers 318
Enantiomers 392, 479
Encapsulation 298
END 516
Endocytosis 124, 369
Endoenzymes 62
Endogenous enzymes 393
Endopeptidases 46
Endotoxin LPS 397
Enteropeptidase 62, 194
Entrapment 311, 312, 320
Entrapment method 307
Entropy 405
ENTRY 534
ENZCLASS.TXT 545
ENZUSER.TXT 545
Enzymatic activity 530
Enzymatic catalysis 159
Enzyme 136, 411, 414, 453, 532, 533
Enzyme assays 209, 263
Enzyme catalysis 339
Enzyme catalytic-mechanism database, 541
Enzyme Commission Classification 532
Enzyme concentration 75, 79
Enzyme Data Bank 544
Enzyme database 497, 523, 530
Enzyme engineering 402, 403, 409
Enzyme entries 534

Enzyme immobilization 293, 294, 296, 331, 404
Enzyme inhibitors 81, 508
Enzyme isolation 259
Enzyme kinetics 81, 209, 212
Enzyme Nomenclature 10
Enzyme Nomenclature Database 516
Enzyme purification 288
Enzyme replacement therapy 478
Enzyme section 533
Enzyme specificity 65
Enzyme structure 526, 527
Enzyme technology 405
Enzyme unit 261
Enzyme–substrate 81
Enzyme-linked field effect transistors 427
ENZYME.GET 545
Enzymic catalysis 407
Epimerase 18, 26, 55
Epimerization 472
Epithelial cells 194
ERGO 511, 512, 513, 514
Erlanger 155
Error correction 187
Erythematosus 390
Erythrocyte 173, 288, 438
Escherichia coli 323, 405, 430
ESR 171
Essentiality data 513
Esters 388, 459
Esterase 84, 389
Estradiol 446
Ethanol 459, 463, 480
Eubacteria 350
Eukarya 512
Eukaryotes 499
Eukaryotic 188
Eukaryotic polymerases 188
Euryarchaeota 187
Evanescent wave 441, 445, 448
Exergonic 120
Exocytosis 124
Exoenzymes 62
Exon 351, 353
3´–5´ exonuclease 187
5´–3´ exonuclease activity 190
Exonuclease activity 189
Exothermic reactions 338
ExPASY 497, 546
Experiment 378
Expression data 513
External transesterification 367
EzCatDB 541, 543

F

Fabrics 466
Fabric softening 404
Factor 136
FAD 99, 107, 412
$FADH_2$ 125, 412
Family A polymerases 187
Family B polymerases 187
Family C polymerases 187
Family D polymerases 187
Family RT 188
Family X 187
Family Y polymerases 187
Faraday 426
Fat digestion 454
Fats 454
Fatty acid synthase 12

Fed-batch bioreactions 326
Fed-batch system 330
Feedback control 329
Feedback inhibition 102, 201, 203, 251
Fermentation 1, 2, 3, 325, 450, 480
Fermenters 330, 453
Ferredoxin 115
Ferricyanide 131
FET 427
Ficin 453
Filmer 245
Firefly 442
Firefly luciferase 264
First-generation biosensor 412
Fiske 269
Fixed volume fed-batch 330
Flat file 535
Flavin 13, 111, 140
Flavin adenine dinucleotide 125
Flavin mononucleotide 126
Flavoenzymes 125
Flavonoids 119
Flavoprotein 12, 125
Fluidized bed reactor 332, 338, 340
Fluorescence 443
Fluorescence resonance energy transfer 358
Fluorimetric methods 263
Fluorophore 441
FMN 126
FMNH 126
$FMNH_2$ 126
Folic acid 140
Folin–Ciocalteau 270
Food enzymes 453
Food industry 404
Food processing 323
Food proteins 525
Fractional adsorption 278
Fractional denaturation 275
Fractional precipitation 274, 275, 276
Fractionation 274
Fractionation methods 273, 279
Fred L Crane 129
Free-boundary electrophoresis 283, 284
Free energy 120, 161, 175
Free radical 173
Freeze-drying 285
Freezing 272
FRENDA 527
Fructose syrup 315
Fumarase 268
Functional domains 517
Fungal mitochondria 350
Fungi 455, 460, 536

G

Galactoflavin 141
Galactose 480
Galactose oxidase 429
Galactosidase 423
Gastric lipase 454
Gaucher's disease type 478
Gel electrophoresis 94, 285
Gel filtration 289
Gel filtration chromatography 282
GEMS 423
Genbank 377, 382
Gene 88, 529, 532

Gene duplication 88
Gene expression 89
Generations of biosensors 412
Genes 535
Gene therapy 371
Genetic engineering 404, 405
Genome 540
Genomic annotation 517
Genomic data 513
Genomic RNA 357
Geometrical isomers 66
Geometric specificity 67
George Minot 121
George William Murphy 121
GIF 535
GIF image 535
Glass electrode 266, 424
Gliadin 459
GLMS ribozyme 344
β-Glucanase 460
Glucanases 461, 463
Glucans 458
Glucoamylase 4, 469, 475
Glucokinase 95
Gluconic acid 470
Glucose 430, 480
Glucose isomerase 315, 405, 455m 472
Glucose oxidase 9, 111, 411, 429, 433, 435, 458, 459, 469, 479
Glucose 6-phosphate 95
Glucose syrups 323
Glutathione 111, 131, 136
Glutathione dehydrogenase 132
Glutathione peroxidase 400
Glutathione reductase 131
Gluten 459
Glycation end-products 182
Glycogen synthase 199
Glycolysis 93, 124
Glycoproteins 307
Glycosidases 474
Glycosidic bonds 517
Glycosyltransferases 474
Glyoxylase 136
GOD 429
Gornall 270
GPCR 255
Gram-negative bacteria 182
Gram-positive 182
Granulocytes 438
Griffin 293
Group I introns 344, 350, 353, 354, 365, 367
Group II introns 344, 350, 353, 354, 365, 367
Group III introns 350
Group ribozyme 351, 352
Group specificity 66
Grubhofer 293
GTN 408
Guanosine co-factor 365
Guinea pig 140
Gustav Lienhard 177

H

Haem 247
Helix I 348
Helix III 348
Haemo-abzymes 390
Haemochromogen 137
Haemoenzymes 390

Haemoglobin 90, 137, 247, 249, 250
Haemophilia 390
Haemoprotein 12, 137
Hairpin 344, 365
Hairpin motif 348
Hairpin ribozyme 344, 348, 349, 356, 364, 372
Haldane 219
Haldane relationship 220
Hammerhead 344, 348, 349, 356, 360, 364, 365
Hammerhead ribozyme 344, 345, 346, 347, 349, 371
Hanes plot 221
Hapten 387, 391, 396, 446
Harris 109
Hashimoto thyroiditis 390
hCG 442, 446
HDV ribozyme 344, 356, 357
Heart attacks 478
Heart LDH 92
Heat exchanger 335
Heat-sensing 430
Helix I 347
Helix III 347
Helmholtz centre 526
Hemicelluloses 458
Hepatitis 438
Hepatitis B 356, 446
Hepatitis delta ribozymes 365
Hepatitis delta virus 356
Hepatitis delta virus ribozyme 344
Hepatocytes 95
Her-2 374
Herzyme 374
Heterotropic 244
Hexokinase 95, 433
HFCS 474
Hierarchical 519
Hierarchical tree 526
High-fructose corn syrup 474
High-fructose syrup 323, 479
Hill coefficient 245
Hind III 406
H4 isoenzyme 93
HIV 438
HIV infection 371
HIV treatment 397
HIV virus 397
HIV-1 349
HMG-CoA/mevalonic acid pathway 118
hnrNA 371
Hoffmann-Ostenhoff 10
Holoenzyme 63, 111, 166
Homocysteine 122
Homologous 507
Homologous factor VII 390
Homology 540
Homotropic 244
Hopkins 131, 136
Horecker 145
Horseradish peroxidase 435
HPLC 448
HSCoA 115, 117
HS-CoM 121
Human herpes 438
Human serum 389
Humicola 464
Humicola fungi 464
Humicola lanuginose 464
Hunter R.l. 87
Hydrazine 232, 303
Hydrochloric acid 454

Hydrogen-bonding 65, 184
Hydrogen bonds 144, 508
Hydrogen peroxide 467, 470
Hydrolase 10, 18, 27, 44, 532
Hydrolysing factor VIII 390
Hydrophobic 65
Hydrophobic interactions 144
Hydroxyapatite 280
Hydroxymethyltransferase 130
Hyperbola 80
I
IAsys 439
Idiotypes 388
IgG 442
Imidazole 166, 176
Imidazole group 150, 172, 301
Imidazole ring 154
Immobilization 295, 300
Immobilization of enzymes 301
Immobilized enzymes 293, 294, 296, 297, 313, 320, 331, 332, 337, 338, 479, 340
Immobilized fumarase 323
Immobilized glucose isomerase 323
Immobilized glucose oxidase 322, 324
Immobilized invertase 315
Immobilized lipases 318
Immobilized raffinase 324
Immunization 386, 396
Immunoassays 418
Immunoglobulin 388
Immunosensors 435, 436, 442, 446
in cis 346, 349
in situ 419
in trans 346, 347, 349
in vitro 321, 361, 371
in vitro substrates 524
in vivo 367, 371, 440
Induced fit hypothesis 65, 70
Induced-fit 248
Influenza virus 442
Infrared 443
Inhibition kinetics 224
Inhibitor constant 227, 233
Inhibitors 306, 494, 524, 533, 534
Insertion elements 513
Institute of Bioinformatics 487, 526
Insulin 477
Integrated genomics 512
IntEnz 487
Internal guide sequence 351
Internal reflection spectroscopy 443
Internal transesterification 366
International Union of Biochemistry 10, 11, 261
International Union of Biochemistry and Molecular Biology 533, 544
Intestinal peptide 390
Introns 343, 350
Invertase 315, 475
Iodoacetamide 152
Ion exchange 279, 280
Ion exchangers 280
Ion-exchange cellulose 304
Ion-exchange chromatography 280, 323
Ion-selective field effect transistors 427

Ionic binding method 299, 304
Ionization 73
Irreversible 81
Irreversible inhibition 83
Irreversible inhibitor 149
ISIS/Base 532
Isocitrate dehydrogenase 97
Isoelectric focusing 285, 289
Isoelectric points 285
Isoenzymes 87, 89, 95, 101
Isoforms 527
Isofunctional enzymes 203
Isomer 479
Isomerases 18, 26, 28, 55, 532
Isomerization 25, 26, 123, 323, 388, 455, 472
Isoprene repeats 129
Isotachophoresis 284
Isotope exchange 238
Isotopic 13
Isotopic exchange 238
Isozymes 87, 88, 90, 540
IUBMB 487, 523, 526, 533, 534, 545
IUBMB/IUPAC 516
IUPAC 15

J

Jaag 4
James B.S. Haldane 211
James B.Sumner 5
Japanese Genome Net 536
Japanese Genome Net Database 535
Java 532
John Alfred Valentine Butler 6
Joinase 12
Jones proteins 390
Jowett 136

K

Karl Lohmann 123
Katal 261, 262, 294
Kcat 364
KEGG 530, 533, 534, 535
KEGG map 530
Keilin 137
Kelly Kruger 343
Keto acids 317
Ketones 392, 473
Keywords 525
Kilning 462
Kinase 19
Kinase pathway database 499
Kinetic efficiency 363
Kinetic studies 209
Kjeldahl method 269, 270, 271
KNF 248
Kohler 385
koji 3
Koshland 245
Koshland–Nemethy–Filmer model 248
Kretschmann 444
Kretschmann configuration 443, 444
Kuhnau 136
Kuhne 2

L

Laccase 466, 467
Lactate 93
Lactate dehydrogenase 87, 90, 91, 109, 433

Lactate oxidase 433
Lactic dehydrogenase 66
Lactone ring 132
Lactones 132
Lactose 315, 460, 479
Lagging strand 189
Lamella 456
Lariat structure 353
Laser light 447
Lattice 307
Lattice type 307, 308
Laundry detergent 466
LC/MS 448
LD1 95
LD2 95
LDH 91
Leading electrolyte 284
Leadzyme 344, 361
Leather industry 404
Leathers 466
Lectin 306, 418
Legume 172
Leibig 2
Leonor Michaelis 211
Lester Smith E. 121
Leuconostoc 474
Leukaemia 95, 477
Lewis acid 168, 366
Ligand 201, 246, 247, 250, 254, 279, 281, 370, 507, 508, 524, 526, 528, 533, 536, 541, 542
Ligand database 533
Ligase 18, 26, 28, 190, 400, 477, 489, 532
Ligation 349
Lignin 468
Linear competitive inhibition 228
Lineweaver–Burk 233
Lineweaver–Burk equatio 227, 231, 234
Lineweaver–Burk plot 231
Lineweaver–Burk plot equatio 217
Lingual lipase 454
Linus Pauling 177, 385
Linux 529
Lipase 318, 454, 459, 464 465, 467, 473
Lipids 369, 422, 478
Lipmann 115, 269
Lipoamide cofactor 100
Lipoate 126, 127
Lipolytic enzymes 467
Lipopolysaccharide 182
Lipoprotein 196
Liposomes 307, 409
Lipoxygenase 453
Liquefaction 475
Local hot spots 339
Lock-and-key hypothesis 65
Lock-and-key theory 69
Loewus 109
Lohmann 134, 136
Luciferase 442
Luciferin 264
Luciferin–luciferase 442
Luminescence method 264
Lyases 10, 18, 24, 28, 52, 532
Lysine 41 186
Lysosomal storage disease 409
Lysozyme 5, 180, 181, 182, 272, 369, 409

M

M4 94
Macie 531, 532, 533
Magnetic-based biosensors 449
Magrate Otto Rohm 4
Maillard reaction 317, 324
Malcolm Dixon 10
Malcolm Gefter 190
Malonyl-CoA 101, 119
Maltose 2, 459, 462, 463
Mammalian CPEB3 ribozyme 344
Manometric method 264
Maraviroc 254
Marcel Florkin 10
Margan 136
Mashing 462
Mastication 454
Mathas 128
Maturation 461
Maude Menten 211
Maximal catalytic efficiency 216
Maximal velocity 212
MDL file 532
Media 327
Menten 7, 80
Metabolic diseases 478
Metabolic pathways 321, 514, 538, 539
Metabolites 538
MetaCyc 536, 537, 540
Metal-activated enzymes 167
Metal cofactors 353
Metal ion catalysis 161, 167, 175
Metal ions 84, 345, 366
Metallo- and haemo-abzymes 386
Metalloenzymes 167, 170, 172, 390
Metalloporphyrins 391
Metalloproteinase 525
Metalloproteins 391
Metastasis 373
Metastatic melanoma cells 373
Metazoan 536
Methanogenesis 120, 121, 537
Methanogens 120, 121, 123
Methosulphate 94
Methotrexate 130
Methyl coenzyme M 121
Methylene blue 147
Methylenetetrahydrofolate 130
Methylglyoxal 136
Methyltransferases 40
Mevalonate pathway 129
Micelle 307, 319
Michaelis 80
Michaelis constants 296
Michaelis theory 80
Michaelis–Menten 218, 345
Michaelis–Menten catalysis 399
Michaelis–Menten constant 221
Michaelis–Menten equation 212, 213, 215, 216, 217, 219, 227, 230, 233,234
Michaelis–Menten kinetics 216, 245
Michael Robertson 361
Microarray 448
Microbial cells 337
Microbial growth 325
Microbicides 398
Microcapsule 307
Micrococcus lysodeikticus 272
Microdroplets 313

Microfibrils 464
Micromoles 262
Microorganisms 411, 414, 453
Microsatellites 97
Microsoft Access 525
Milstein 385
MIM database 545
Missing enzymes 409, 540
Mitochondria 118, 119, 120, 129
Mitochondrial DNA 188
Mitochondrial DNA polymerase 187
Mitochondrial enzymes 271, 272
Mitosis 525
Mixed inhibition 232
Modellers 522
Modulators 243
Molecular markers 97
Molecular oxygen 137
Molecular scissors 375
Molecular sieves 283
Molecular switches 362
Molecular unit of currency 123
MOL files 535
Molybdate reagent 268
Monoclonal antibodies 387, 390, 477
Monoclonal antibody 385
Monocytes 181
Monod 245
Motif 349, 356, 535
Motility 123
Mozilla 530
Mozilla firefox 530
mRNA 350, 371
Mucus 182
Multienzyme complex 99, 102
Multienzyme complex system 97
Multiple myeloma 390
Murein 181
Muscle LDH 91
Mutagenesis 356, 397, 404
Mutagenic PCR 361
Mutant DNA 360
Mutants 526
Mutarotation 163, 164
Mutase 18
Mutation 89, 330, 354
MWC model 245, 247
Mycotoxins 451
Myoglobin 137
MySQL 378

N

N-acetylglucosamine 182, 183
N-acetylmuramate 182
N-acetylmuramic acid 183
NAD 22, 91, 111, 321
NAD(P)H 442, 470
NAD^+ 93, 99, 112, 113, 263
NADH 99, 112, 263, 321, 433, 442
NADP 22, 101
$NADP^+$ 112, 114
Nanokatals 262
Nanomoles 262
National Center for Biotechnology information 546
National institute of genetics 546
Natural language processing 499
NC-IUBMB 529, 530
NCBI-Pubmed 378
Negative cooperativity 247
Nelson 293
Nemethy 245

Neurodegenerative conditions 502
Neurospora 358
Neurotransmitter gamma-aminobutyric acid 254
Neutrophil 181
Nickel 407
Nicotinamide cofactors 516
Nicotinamide 105, 139
Nictotinamide adenine dinucleotide 417
Niels Jerne 388
Nitro blue tetrazolium salt 94
Nitrogen 537
Nitrogenase 172
NMR 171
N,N,N´N´-tetramethylethylenediamine 308
Nomenclature 89, 523, 526, 527
Noncompetitive inhibition 83, 232, 233, 234
Noncompetitive inhibitor 233
Non-histone proteins 118
Non-labelled affinity sensors 418
Non-metabolic pathways 514
Non-radiative plasmons 444
Non-redundant 536
Northrop 5
Nucleic acid 130, 201, 288, 343, 414
Nucleic acid hybridization 88
Nucleic acid polymerases 361
Nucleophile 168, 170, 351, 354, 365, 366, 367
Nucleophilic attack 184, 351, 367
Nucleophilic catalysis 165
Nucleophilic reaction 165
Nucleoprotein 274
Nucleoside 350
Nucleotides 13, 106, 108, 253, 350

O

Obesity 502
Occurrence 523, 526
Okazaki fragments 189
Old enzyme unit 262
Old yellow enzyme 408
Oligosaccharides 474
Omics 539
OMIM database 535
Oncogenic DNA viruses 372
Ontologies 514
Open reading frames 512
Opsonin 182
Optical 411, 414
Optical biosensors 416, 435, 447
Optical isomers 66, 127, 267
Optical sensing 440
Optical transducers 440, 441
Optimum pH 71, 73, 326
Optimum temperature 73
Oracle 511
Ordered mechanism 236
ORENZA 528, 529, 530
ORFs 513
Organelles 414
Organic fertilizers 404
Orientation effects 175
Orphan 529, 530
Orphan enzyme activity 529
Orphan enzymes 529
Ortholog 499, 513
Orthosteric ligand 255
Otto Hofmann-Ostenhof 9
Ovarian cancer 373
Oxidase 19, 21

Oxidation 388, 391
Oxidation–reduction 296
Oxidation–reduction reactions 167, 235
Oxidation–reduction systems 114
Oxidative damage 400
Oxidative phosphorylation 124
Oxidizing agents 465
Oxidoreductases 10, 18, 27 123, 532
Oxyanion 368
Oxygen 326
Oxygenase 18, 22, 31
Oxygen electrode 416

P

Packed bed 337
Packed bed reactor 332, 337, 338
Palindromic 491
Palladium 407
Pancreatic amylase 454
Pancreatic trypsins 467
Pantetheine 115
Pantothenate 117
Pantothenic acid 140
Papain 455
Paper 404, 468
Paralog 513
Parkinson's disease 113
Partition chromatography 283
Pasteur 2
Pathologic 538
Pathologic software 537
Pathway 534, 535
Pathway data 513
Pathway holes 540
Pathway/genome database 536, 537, 539
Payen 2
PBR322 406
P-chymotrypsin 195
PC/Linux 538
PC/Windows 538
PDB 483, 541, 543
PDB code 531, 532
Pdbsum database 491
Pectate lyases 458
Pectin 458
Pectin lyases 458
Pectinase 456
Pectinesterases 458
Penicillin 473
Penicillin G or V 316
Penicillin g-amidase 405, 406
Penicillium 473
PEP 174
Pepsin 5, 155, 293, 453, 454, 460
Pepsinogen 196, 454
Peptidases 45, 519
Peptide bonds 463, 524
Permease 12
Peroxidase 19, 21, 137, 322, 391
Peroxidases 468
Persoz 2
Pesticides 436
Peter Schultz 386, 393
PETN 408
PGDB 537, 538, 540
pH 266, 296, 297
pH electrode 416
Phage display 397
Phagocytic cells 181
Phagocytosis 182
Pharmaceutical agents 81

Pharmaceutical industry 404
Pharmacokinetics 374
Phenylalanine 184, 478
Phenylketonuria (pku) 478
Philips mechanism 183
Phosphatases 269, 502
Phosphobase 502
Phosphodiester bond 361
Phosphoesterase 461
Phosphofructokinase 263
Phosphoglucoisomerase 95
Phospholipase 366
Phosphomutase coenzyme 135
Phosphomutases 55
4´-phospho-n-pantothenoyl cysteine (ppc) 117
Phosphopyridoxal 128, 129, 135
Phosphorylase 191, 198, 199
Phosphorylase a 198
Phosphorylase b 198
Phosphorylase kinase 199
Phosphorylases 269
Phosphotransferases 55
Photo-crosslinkable resin prepolymer method 299, 312
Photo-oxidation 147, 154
Photodiode systems 436
Photomultiplier tubes 436
Photosensitizer 154
Photosynthesis 115, 123, 127
PHP 378, 529
pH probe 424
Physical adsorption 299, 305
Physical adsorption method 299
Phytase 461
Phytochemicals 119
Picokatals 262
Picomoles 262
Piezoelectric 414
Piezoelectric biosensors 416
Piezoelectric crystal 416, 437
Piezoelectric immunosensors 438
Piezoelectric sensors 449
Pigments 470
Ping 237
Ping-pong 238, 239
Ping-pong mechanism 238, 239
Ping-pong reactions 237
PIR 540
Pituitary hormones 436
PKA 72, 346
Plant tissues 456
Plant viruses 348
Plasmids 480
Plasmin 478
Plasmons—radiative 444
P1 nuclease 366
Pol 187, 188
Pol I 188
Pol II 188
Pol III 188
Pol IV 188
Pol V 188
Polarimetric method 267
Pollution 404, 450, 455
Polyacrylamide gel 94, 281
Polyacrylamide gel method 281, 308
Polyanionic 365
Polyethylene glycol 322
Polygalacturonase 458
Polyketides 119

Polymerase 187, 361
Polymerase II 370
Polymerase III promoters 370
Polymers 422
Polypeptide 89, 180
Polypeptide chain 90, 144, 194, 525
Polyphenols 139
Polyploidization 88
Polysaccharides 181, 456
Population genetics 97
Porphyrin ring 247
Positive cooperativity 247
Positive modulator 251
Post-translational modification 118
Potentiate 254
Potentiometric biosensors 415, 423, 425, 428, 448
PPC 117
PPi 120
P1 ribozyme 357
Pre-biotic RNA world 361
Pre-steady-state kinetics 223
pre-tRNAs 343
Precipitation 274
Precursors 192
Predicted and consensus interaction sites in enzymes 507
Primary cell wall 456
Primary literature 378
Primase 187, 188
Primer 186
Procarboxypeptidase 63
Prodrugs 394, 395, 396
Products 524, 533, 534, 541
Proenzyme 62
Proofreading 186, 188
Prosite 530, 535
Prosthetic group 13, 63, 12 137, 140, 408
Protease 192, 398, 454, 45 463, 465, 524, 525
Protein 422, 479, 513
Proteinase 78, 458, 461, 464, 46
Protein data ban 491, 507, 508, 531, 535
Protein degradation 504
Protein domain structures 531
Protein engineering 390, 406
Protein enzymes 366
Protein kinases 499
Protein localization 504
Protein scaffolds 399
Proteolysis 524
Proteolytic degradation 258
Proteolytic enzyme 97, 295, 453, 524, 525
Prothrombin 196
Proton relaxation rate 171
Protozoans 455
Pseudo first-order rate constan 177
Pseudogene 89
Pseudoknot 356
Pseudomonas putida 472
PSI-BLAST 483, 486
PSI-BLAST hit 483
Ptyalin 454
PubMed 525, 527, 541, 542
PubMed id 378
Pullman 108
Pullulanase 475
Purification 89, 258

Purine 107, 201
Purine–pyrimidine 363
Pyridoxal phosphate 102, 111, 166
Pyrimidine 107, 201, 253
Pyrophosphothiamine 127, 134, 140
Pyruvate 92, 93, 109
Pyruvate dehydrogenase 97
Pyruvate kinase 173, 433

Q

Q10 129
Quastel 136
Quinone 129
Quinonoid 108

R

Racemase 10, 18, 55, 179
Racemic mixtures 479
Rack mechanism 177
Raffinase 324
Random mechanism 237
Ratcliff 6
Reaction 20
Reactor 327, 330, 334, 341, 453
Rebase 489
Receptor 254
Recombinant antibodies 390
Recrystallization 288
Recycle reactor 332, 340
Redox 112, 125, 400, 417, 449, 537
Redox potential 267, 537
Redox reactions 113, 121, 123, 161
Reduced glutathione 131, 136
Reducing agent 115
Reductase 19
Reduction 388
Reference database 537
Reference electrode 428
Reference pathways 539
Reflectance 443
Regulatory data 513
Regulator molecules 192
Relational database 522, 523, 529
Rennet 460
Rennin 318, 453, 455
Re-oxidation 265
Reparase 12
Replicase 12
Resins 280
Resistance 341
Resonance 440
Restriction enzyme database 489
Restriction enzymes 489
Retro-aldol reaction 392
Retro-aldol-retro-michael reactions 396
Retrohoming 354
Retroviruses 188
Reverse transcription 354
Reversible inhibition 82
Reversible inhibitor 77
Rheumatism 130
Rheumatoid arthritis 390
Rhodium 407
RiboaptDB 377, 382
Riboflavin 125, 126, 141
Ribonuclease 143, 150, 154, 293
Ribonuclease P 344
Ribonuclease P ribozyme 344

Ribonucleic acids 343
Riboswitch 382
Ribozyme 343, 378, 379
Ribozyme libraries 372
Ribozymes 343, 354, 367, 368, 370, 374
Richard Gerber 6
Richard Lerner 386, 393
RLCP 542
RNA 187, 343, 353, 363, 513
RNA ligase ribozymes 361
RNA polymers 361
RNA primer 189
RNA ribozymes 364, 369
RNAse H 366
RNAse P 344, 354, 355, 356, 365, 367
RNAse P complex 343
Rochelle salt 437
Rohm 4
Rose bengal 147, 154
Rossmann fold 124
Roughton 7
Rpi 374
rRNA 350, 355
Ruminants 461
Rye 459

S

S. uvarum 463
SABIO-reaction kinetics database 522
SABIO-RK 522
Saccharification 405, 462, 475, 480
Saccharomyces 463
Saccharomyces amurcea 301
Saccharomyces carlsbergensis 301, 408
Saccharomyces cerevisiae 301, 463
Salivary amylase 454
Salts 326
Salvage pathways 112
SanPietro 108
Satellite DNA 345
Satellite RNA 348, 349, 358
Satellite viruses 348
SBML 522
Scalar 537
Schiff's base 128, 135, 157, 165, 304
Schleith 293
Schoellmann 152
Schuster 134
Schütz J. 78
Schütz law 78
Schwann 2
Scientific literature 486
Scopec 508
Second-generation biosensor 412
Secretin 192
Selective binding unit 418
Selectivity 417
Selenocysteine 400
Selenoenzyme 400
Selenosubtilisin 400
Selex 376, 379
Self-splicing 358
Self-splicing introns 350
Semi-continuous 326
Sephadex 286
Sequence retrieval 540
Sequence specificity 363

Sequential feedback control 204
Sequential model 245, 248
Sequential reactions 236
Serine enzymes 148
Serine protease 454
Serine proteinase 525
Serratia marcescens 301
Serum proteins 438
Sesquiterpenes 119
Shaw 152
Sidney Altman 343, 355
Signal transducer 421
Signal transduction 504
Signalling pathways 499
Single-displacement 238
Single-nucleotide polymorphisms 97
Single-stranded DNA 418
Single-stranded RNA 358
Sirna therapeutics 375
Site-directed 404
Site-directed mutagenesis 157, 400, 406
Sizing 466
Sludge 419
Small nuclear RNAs 370
Small nucleolar RNAs 370
Smooth surfacing 404
Solid-phase immunoassay 436
Solid-state electrodes 425
Solubility test 290
Sonic oscillations 272
Soya bean milk 324
Spacers 284
Sparging 462
Specific activity 260, 261
Specific affinity 279
Specificity 63, 143, 258, 347, 368 417, 523, 526, 527, 532
Spectrofluorimetric method 263
Spectrophotometer 266
Spectrophotometric measurements 270
Spectrophotometric method 263, 268
Spectroscopy 440
Spectrum 443
Spermidine 345
Spermine 345
Sponges 459
SPR 441, 444, 445
SPR fluoro-immunoassay 446
SQL 529
SRI 538
SRP 355
SRP RNA 355
5S rRNA 361
Stachyose 324
Stain removing 404
Stanley 5
Starch 422, 454, 458, 464
Steady state 211
Steady-state kinetics 223, 225
Stereo-isomers 479
Stereochemical 65
Stereoisomers 10, 109
Stereospecific 388
Stereospecificity 66, 67
Sterilization 328
Steroids 115, 436
Sterols 119, 243
Stirred tank 338

Stirred tank reactors 331, 333
Stoichiometry 533
Stone washing 404
Stopped-flow method 7
Streptavidin 320, 361
Streptokinase 478
Subbarow 269
Substrate analogue 147
Substrate concentration 80
Substrate elution 280
Substrates 533, 534, 541
Subtilisin 400
Succinate dehydrogenase 12
Sucrase 267
Sulphation 199
Sulphur 537
Sumner 3, 258
Super acids 168
Superoxide dismutase 172
Surface plasmon resonance 421, 444, 447
Surface plasmons 444
Sweet wort 462
Sweetener 323
Swiss-Prot 483, 508, 530, 540, 543, 545
Synthases 10
Synthetases 10, 24
Synthetic resin prepolymer method 299, 311
Synthetic ribozymes 360
Synzymes 386
SYSNAME 534
Systematic name 13, 15
Systemic evolution of ligands by exponential 376
Systems biology 526

T

T7 DNA polymerase 187
Takadiastase 3
Takamine 3, 4
Tanning 4, 466
Target identification 382
Tautomerase 19
Tautomerism 128
Tautomerization 162
Taxonomy tree explorer 526
TCA 93
TCA cycle 93
Telomerase 373
Temperature coefficient q10 74
Terminating electrolyte 284
Tertiary structures 543
Test method 75
Tetrahydrofolic acid 130
Tetrahymena 351
Tetrahymena group I intron 344
Tetrahymena ribozyme 351
Tetrahymena thermophila 350
Text file 381
Thawing 272
ThDP 133
Thermodynamics of enzyme-catalyzed reactions database 509
The thunberg method 265
Therapeutic RNAs 371
Thermistor 430
Thermistor biosensor 434
Thermolysin 474
Thermometric 449
Thermostability 405
THF 130

Thiamine (vitamin B_1) 140
Thiamine pyrophosphate 111, 133, 166
Thiazolinium ring 107
Thiazolium ring 133
Thioesters 117, 118
Thiol group 136
Thiophosgene 298
Thiophosphate 84
Third-generation biosensor 412
Thomas Kornberg 190
Thomas R. Cech 343
Threonine dehydrases 129
Thrombin 196
Thunberg tubes 265, 266
Thyroglobulin 390
Tiselius apparatus 283
Tissue 414
T lymphocytes 397, 438
TLCK 153
Tobacco ring spot 348
Toluidine 322
Tosyl 149
Total internal reflection 444
TPCK 149, 153, 154
Trans-acting ribozymes 347
Transacylases 115
Trans-cleaving ribozymes 373
Transaminases 97, 238
Transamination 111
Transcarbamyolase 250
Transcription 343, 360
Transducer 320, 411, 412, 414, 415, 430, 437
Transesterification 319, 350, 356, 353, 365
Transferase 10, 18, 27, 39, 532
Transhydrogenase 21
Transient state 333
Transition state 64, 159, 161, 165, 168, 175, 177, 196, 366, 367
Transition state analogs 179, 180
Transition state analogues 507
Transition state complex 161
Transition-state 386
Transition-state analogue 388
Transketolase 111
Transketolase 133
Translation 343, 360, 513
Translesion synthesis 187
Translocase 12
TREMBL 530
Tricarboxylic acid cycle 93
Trichoderma 464
Triglycerides 430
Trigonopsis variabilis 317
Trimers 250
2,4,6–trinitrotoluene 409
Triphosphopyridine nucleotide 114
Trivial 91
Trivial names 9
TRNA 350, 355, 370
Trypsin 4, 5, 62, 63, 78, 153, 193, 196, 235, 454
Trypsinogen 62, 193, 194
Tryptophan 184
Tsuchihashi method 288
Tumour cells 373, 394
Turbidostat 329
Turnover 346, 347, 355, 368
Turnover number 224
Type 1 ligases 361

Type 2 ligases 361
Tyrosinase 478
Tyrosine 184
Tyrosinolation 199
Tyrptophan 101

U

U6 snRNA 355
Ubiquinone 129
UDP 251
Ultracentrifugation 289
Ultrafiltration 286, 479
Ultraviolet light 127
UMP 251
Uncompetitive inhibition 229
Uncompetitive inhibitor 231
UniProtKB/Swiss-Prot accession codes 527
Units of enzyme activity 260
Urate oxidase 429
Urease 3, 9, 258
Urethane prepolymer 311, 312
Urethrane prepolymer method 299
Uridine phosphate 135
UTP 251

V

Vacancies 398
van der Waals 65, 144
van der Waals force 305
van der Waals interactions 143
Variable volume fed-batch 330
Varkud satellite ribozyme 358
Vectoral etrs 537
VeGF receptor 374
Vennesland 109
Victor mckusick 545
Viridiplantae 536
Viroids 345, 358
Viruses 438, 512
Vitamin B_6 140
Vitamin B_{12} 121, 140, 172
Vitamins 105, 139, 326, 470
VS ribozyme 344

W

Wan 308
Warburg 107
Waveguides 440, 442
Webb 10
Whey 460, 479
William Kuhne 2
William Youndin 112
Wobble base pair 357
Wobble pair 357
Wyman 245

X

X-ray crystal 392
X-ray crystallography 5, 124
X-ray diffraction 184
Xanax 254
Xenobiotic compounds 408
Xylanase 456, 460, 461, 466, 468

Y

Yamazoye 136
Yeast 3, 459

Z

Zeroth-order 210
Zone electrophoresis 89, 284, 289
Zwitterions 280
Zymogens 62, 63, 192, 193, 194, 198

www.ingramcontent.com/pod-product-compliance
Lightning Source LLC
LaVergne TN
LVHW020858300926
849194LV00024B/210

* 9 7 9 8 2 2 4 3 4 8 2 6 8 *